Design, Manufacturing and Applications of Composites

PROCEEDINGS OF THE
FOURTH JOINT CANADA–JAPAN
WORKSHOP ON COMPOSITES

VANCOUVER, BRITISH COLUMBIA
SEPTEMBER 2002

E D I T E D   B Y

J. Lo
H. Hamada
A. Poursartip
A. Nakai
S.V. Hoa

CRC Press
Taylor & Francis Group
Boca Raton London New York

CRC Press is an imprint of the
Taylor & Francis Group, an **informa** business

CRC Press
Taylor & Francis Group
6000 Broken Sound Parkway NW, Suite 300
Boca Raton, FL 33487-2742

First issued in hardback 2018

© 2003 by Taylor & Francis Group, LLC
CRC Press is an imprint of Taylor & Francis Group, an Informa business

No claim to original U.S. Government works

ISBN-13: 978-1-138-41287-3 (hbk)
ISBN-13: 978-0-8493-1534-3 (pbk)

**Visit the Taylor & Francis Web site at
http://www.taylorandfrancis.com**

**and the CRC Press Web site at
http://www.crcpress.com**

Library of Congress Card Number 2002031095

Library of Congress Cataloging-in-Publication Data

Joint Canada-Japan Workshop on Composites (4th : 2002 : Vancouver, B.C.)
    Design, manufacturing and applications of composites : proceedings of the Fourth Joint
Canada-Japan Workshop on Composites, Vancouver, British Columbia, September 2002
/ edited by J. Lo ... [et al.].
      p. cm.
    Includes bibliographical references and index.
    ISBN 0-8493-1534-4
    1. Composite materials--Congresses. 2. Manufacturing processes--Congresses. I. Lo, J.
II. Title.

TA418.9.C6 J573 2002
620.1'18--dc21

2002031095

Proceedings of the Fourth Joint Canada–Japan Workshop on Composites held in Vancouver, British Columbia, Canada, September 19-21, 2002, organized by CANMET, the Kyoto Institute of Technology and the University of British Columbia

ORGANIZING COMMITTEE

Co-Chairmen
J. Lo, CANMET, Canada

H. Hamada, Kyoto Institute of Technology, Japan

A. Poursartip, University of British Columbia, Canada

EXECUTIVE COMMITTEE

S.V. Hoa,
Concordia University

M.S Cheung,
Public Works & Government Services Canada

C. Taggart,
National Research Council Canada

N. Amiji,
Toshiba Corporation

I. Narisawa,
Yamagata University

T. Nishino,
Kobe University

M. Zako,
Osaka University

G. Fernlund,
University of British Columbia

A. Johnston,
National Research Council Canada

J. Jackman,
CANMET

C. Poon,
National Research Council Canada

H. Seumasu,
Sophia University

T. Ishikawa,
National Aerospace Laboratory

M. Ochi,
Kansai University

J. Beddoes,
Carleton University

R. Vaziri,
University of British Columbia

J. Denault,
Industrial Materials Institute, NRC

D. Nikanpour,
Canadian Space Agency

V. D'Arienzo,
Bell Helicopter Textron Canada

H. Mostaghaci,
Canadian Department of Foreign Affairs and International Trade

Y. Fujii,
Seikow Chemical Eng. & Machinery Ltd.

N. Takeda,
University of Tokyo

H. Hatta,
Institute of Space and Astronautical Science

A. Yokoyama,
Kyoto Institute of Technology

WORKSHOP SECRETARIES

A. Nakai, *Kyoto Institute of Technology*

S. Dionne, *CANMET*

Preface

It is without a doubt that the frontier of science can be pushed faster and further with international collaboration and technology transfer. It is based on this notion that the Canada–Japan Workshop on Composites was first held in Japan (1996). With the growing interest among Canadian and Japanese researchers in presenting their discoveries and sharing their findings, two more workshops were held in Canada (1998) and Japan (2000). With the rapid pace of scientific development in the conventional composite and new innovative composites, the fourth Canada–Japan Workshop on Composites was held to stimulate new and closer interactions among researchers from the two countries. The fourth Canada–Japan Workshop on Composites was co-chaired by Dr. Jason Lo (CANMET; Natural Resources Canada), Dr. Hiroyuki Hamada (Kyoto Institute of Technology) and Dr. Anoush Poursartip (University of British Columbia).

Compared to the previous workshops, there was an expanding coverage in research topics. Besides the conventional topics on metal, ceramic, polymer composite and composites in civil construction, development in the emerging fields of wood and nanocomposites were the new additions. Further, a new session for review papers on special topics was also organized.

The editorial committee would like to express their appreciation to many organizations and individuals who had contributed to the successful completion of the workshop. Among them are CANMET (Natural Resources Canada), Concordia Center for Composites, Kyoto Institute of Technology and University of British Columbia. Thanks are also extended to Powertech Labs for making arrangements for the participants to visit; and Sophie Mérineau and Margo Armstrong for their administrative and technical support.

Co-Editors

J. Lo
Materials Technology Laboratory
CANMET
Natural Resources Canada
Ottawa, Ontario, Canada

S.V. Hoa
Concordia Center for Composites
Department of Mechanical and
Industrial Engineering
Concordia University
Montreal, Quebec, Canada

H. Hamada
Advance Fibro-Science
Kyoto Institute of Technology
Kyoto, Japan

A. Poursartip
Department of Metals & Materials
Engineering
University of British Columbia
Vancouver, British Columbia, Canada

A. Nakai
Advance Fibro-Science
Kyoto Institute of Technology
Kyoto, Japan

Table of Contents

1. REVIEW PAPERS

2. MANUFACTURING METHODS

3. MATERIALS CHARACTERIZATION

5. CODE DEVELOPMENT

6. FATIGUE, FRACTURE, IMPACT

7. SENSORS AND CONTROL

8. NANOCOMPOSITES, NATURAL FIBER COMPOSITES

9. APPLICATIONS

REVIEW PAPERS

SOME VIEWS ON THE FUTURE OF COMPOSITE MATERIALS

J.D. EMBURY
McMaster University Hamilton CANADA.

A wide variety of composite materials based on metals, polymers and ceramics are now an essential party of the materials economy. The basic concepts by which these materials function are clearly identified, and much effort has been spent on optimizing the fabrication and design of these materials.

When we look to future developments a variety of possibilities emerge. These include the further integration of the fabrication and function processes, aspects of the scale of components which improve the competitive position of composites relative to conventional materials and the exploitation of new types of composite such as nanocomposites which exploit a variety of new length scales to achieve their functionality. This also gives rise to multifunctional composites which have attributes other than structural properties. In this talk these aspects of the future of composites will be explored and illustrated.

Recent FRP research in support of the Canadian Forces

Maj. R.G. Wight, Prof. M.A. Erki, Maj. P.J. Heffernan
Department of Civil Engineering
Royal Military College of Canada
PO Box 17000, Station Forces
Kingston, Ontario, Canada K7K 7B4
wight-g@rmc.ca, erki-m@rmc.ca, heffernan-p@rmc.ca

ABSTRACT

The Canadian Forces (CF) is targeting the use of FRP materials for the benefit of construction engineers sent overseas on peacekeeping missions. Recent work at the Royal Military College of Canada in support of the CF has been: (i) external strengthening using non-prestressed and prestressed carbon FRP (CFRP) sheets bonded to severely damaged concrete beams and slabs and (ii) standard pultruded glass FRP members for development of an FRP vehicle bridge. The CFRP strengthening of severely damaged 8 m continuous reinforced and prestressed concrete beams showed that continuous beam behaviour can be restored in single spans after continuity had been cut. In field testing of single span 17 and 18 m prestressed double-tee beams severely damaged and strengthened with non-prestressed and prestressed CFRP sheets, the loss of up to $1/3^{rd}$ of the prestressing strands could be compensated with bonded CFRP sheets. The behaviour of impact damaged reinforced concrete slabs (3 m) also improved with CFRP strengthening. For the FRP vehicle bridge, the testing of the fatigue-prone connections has yielded a negative inverse slope of approximately 10 for a derived pseudo S-N plot. Much more severe than steel connections, small changes in loading range for pultruded FRP connections could lead to large changes in fatigue life.

INTRODUCTION

The Canadian Forces (CF) is targeting the use of FRP materials for structures that have been severely damaged (1). Securing damaged bridges has been the primary concern of construction engineers sent overseas on peacekeeping missions, with the goal of re-establishing freedom of movement. The recent work at the Royal Military College of Canada (RMC) in support of the CF is: (i) external strengthening using non-prestressed and prestressed carbon FRP (CFRP) sheets bonded to severely damaged concrete beams and slabs and (ii) standard pultruded glass FRP members for development of an FRP vehicle bridge. The RMC work consists of laboratory and field testing, concurrent with the development of analytical models for design.

SECURING BRIDGES

During the period 1996 to 1999, in support of the NATO-lead Stabilization Force in (BiH), the CF contributed a specialist engineering team to re-establish freedom of movement for military and civilian traffic throughout Bosnia-Herzegovina (BiH). Owing to years of infrastructure neglect in BiH, during and after the hostilities, and the effects of battle damage, there was widespread and often severe infrastructure degradation. In the course of their work, the specialist engineering team uncovered a problem arising from the remediaton strategy being used for partially damaged continuous bridges.

An expedient method of repair for partially damaged bridges is the use of a pre-fabricated modular steel bridge to bridge-over, or overbridge, the whole of or a portion of a damaged span (Figure 1). Overbridging allows movement over the damaged span. However, in the case of continuous bridges, with the loss of continuity from the damage of a span and the associated collateral damage, the spans adjacent to the damaged span must carry increased internal shear and bending moments, for which they were not designed. Even with traffic restrictions on overbridged structures, the continued loading of these spans results in their steady deterioration and sometimes failure. Strengthening of the adjacent spans with CFRP sheets can permit them to carry the redistribution of loads. Also, strengthening with CFRP sheets can an alternative or complementary repair to overbridging a damaged span, depending on the type and extent of the damage.

CFRP Sheets for Strengthening Severely Damaged Concrete Structures

<u>Laboratory Testing of Damaged Continuous Beams</u>

For a continuous span bridge with a severely damaged span that requires overbridging, the adjacent spans can be stabilised with (i) external post-tensioning with steel tendons and bars; and (ii) external reinforcement using CFRP sheets. Both methods have been investigated at RMC. In the laboratory, 8m reinforced concrete and prestressed concrete beams were constructed as two-span continuous members, to be supported over two equal spans. The continuity between the two spans was discontinued using explosive cutting charges, manually cutting, or impact loading using an hydraulic jack. The remaining single spans were tested without and with strengthening for flexure and shear with bonded CFRP sheets. In addition, the CFRP sheets were either non-prestressed or prestressed using a prestressing method developed at RMC (2). As expected, comparing the behaviour of the continuous control beams with that of the unstrengthened spans of the damaged beams, the loss of continuity naturally resulted in lower yield and lower ultimate strengths and higher deflections. However, the CFRP strengthening successfully re-established structural behaviours in the damaged beams that were similar to, and sometimes improved upon, those observed for the undamaged continuous beams (3,4,5,6). Figure 2 shows the test results for 2-span reinforced concrete Tee-beams with stengthening of the remaining spans using different techniques.

Field Testing of Full-scale Prestressed Double-Tee Beams

On the grounds of the CF Base Kingston, two full-scale model bridges were constructed. These bridges consist of two precast, pretensioned concrete double-tee girders, with a cast-in-place concrete topping (Figure 3a). The two bridges were each single-lane, with spans of 17 m and 18 m, respectively (7). The bridges were load tested by passing a two-axle dump truck loaded with aggregate over the structure. While the load exceeded the normal service load for the structure, it was kept below the ultimate load capacity. One web in each structure was then intentionally damaged at mid-span, and two of the six steel prestressing strands were cut. The bridges were tested in this damaged state. As expected, a portion of the total load migrated from the damaged web to the undamaged web as the test vehicle passed over the mid-span. The damaged web was then repaired by using a cement-based patching material and CFRP sheets. For one bridge, the damaged web was strengthened with non-prestressed sheets, and for the other bridge, the damaged web was strengthened with sheets prestressed to a total of 40 kN. The bonded CFRP sheets restored stiffness to the damaged web and load sharing between the two webs of the double-tee girders of the two bridges (Figure 3b).

Laboratory Testing of Reinforced Concrete Slabs

Since concrete bridge decks are also vulnerable to damage in times of conflict, one-way reinforced concrete slabs 90 x 1000 x 3000 mm were constructed, severely damaged and repaired using carbon fibre reinforced polymer (CFRP) sheets (8,9,10). Damage was induced by severe reverse cyclic loading. Undamaged and damaged slabs, without CFRP strengthening, and undamaged and damaged slabs strengthened with non-prestressed and prestressed CFRP sheets were tested in flexure. The testing program included monotonic loading to failure and fatigue loading to validate the effectiveness of strengthening severely damaged slabs with CFRP sheets. The maximum load capacities of all the slabs that were pre-damaged by impact loading were lower than for the corresponding slabs tested without pre-damage (Figure 4). Also, the ductility of the pre-damaged slabs was lower than for the slabs without pre-damage. However, CFRP strengthening of these slabs improved their behaviour, and by prestressing the CFRP sheets fatigue life of the slabs was significantly improved.

DEVELOPMENT OF AN FRP VEHICLE BRIDGE

The CF has need of a lightweight, inexpensive, short span, gap-crossing aids. FRP materials may be used to advantage for small military vehicle bridges. Researchers at RMC have proposed for this need commercially available pultruded products that could be assembled using modular principles. These structures would be mounted on vehicles and erected with minimal support. The project has developed an expertise at RMC on the design of short span (up to 12 m) FRP gap-crossing structures (11,12,13).

Fatigue Behaviour of Connections for FRP Pultruded Members

While FRP pultruded structural members are well suited for bridge construction in remote or inaccessible locations where heavy construction equipment is neither practical nor desirable, the repetitive loading on bridges may cause premature fatigue failure, especially of the FRP connections. In structural steel connections, prior to the fatigue limit, small changes in fatigue loading amplitudes have little effect on their fatigue lives, even though steel is the most fatigue sensitive of the traditional building materials. However, the same small changes in fatigue loading can lead to shorter fatigue lives of equivalent non-yielding FRP connections. Moreover, owing to the lightweight of FRP structures, the live load to dead load ratio may be larger than that of an equivalent structure built using traditional building materials. Hence, there is a greater chance that the FRP structure will undergo higher loading ranges and possible stress reversals due to moving live loads, thereby exacerbating the fatigue sensitivity of its connections.

Monotonic and fatigue tests of double lap connections for glass FRP pultruded members have been conducted at RMC (14,15). Three types of connections were tested, namely bolted, bonded, and combined bolted/bonded connections. For the bolted and combined bolted/bonded connections, two mild steel bolts joined the three members. For the bonded connections, the bonding agent was a polyurethane structural adhesive. For the fatigue tests, loading ranges varied between 40 and 70% of the quasi-static connection strength, in tension and compression. Fatigue tests for the connections of FRP pultruded members were conducted in cyclic tension, cyclic compression, and reversed tension-compression, and the results were used to derive a pseudo S-N plot through which a linear log-log plot could be established. This log-log relationship has a negative inverse slope of approximately 10, whereas for the fatigue prone details of steel structures the negative inverse slope would be approximately 3. Therefore, small changes in loading range for pultruded FRP structures could lead to larger changes in fatigue life than expected from experience with steel structures. Also, with an increasing number of cycles, there was observed increased slip in all connection types tested, although this was greatest for the bolted connections. As a safety measure, owing to the sudden failure of bonded connections, supplementary bolting is probably needed for bonded connections, provided that load redistribution can occur prior to the bond failure.

CONCLUSIONS

This paper has summarised the work at RMC in support of the CF in the areas of (i) external strengthening using non-prestressed and prestressed carbon FRP (CFRP) sheets bonded to severely damaged concrete beams and slabs and (ii) standard pultruded glass FRP members for development of an FRP vehicle bridge. The CFRP strengthening of severely damaged 8 m continuous reinforced and prestressed concrete beams showed that continuous beam behaviour can be restored in single spans after continuity had been cut. In field testing of single span 17 and 18 m prestressed double-tee beams severely damaged and strengthened with non-prestressed and prestressed CFRP sheets, the loss of up to $1/3^{rd}$ of the prestressing strands could be compensated with bonded CFRP sheets. The behaviour of impact damaged reinforced concrete slabs (3 m) also improved with CFRP strengthening. For the FRP vehicle bridge, the testing of the fatigue-prone connections has yielded a negative inverse slope of approximately 10 for a derived pseudo S-N plot. Much more severe than steel connections, small changes in loading range for pultruded FRP connections could lead to larger changes in fatigue life.

REFERENCES

1. Wight, R.G., Erki, M.A., and Heffernan, P.J. 2002. Canadian Federal Interest in FRP for Structures. Structural Engineering International, Vol 12 No 2, 99-101.

2. El-Hacha, R., Wight, R.G., and Green, M.F., 2001. "Prestressed Fibre-Reinforced Polymer Laminates for Strengthening Structures." Progress in Structural Engineering and Materials, Vol. 3, Issue: 2, Apr/Jun 2001, pp. 111-121.

3. Porteous, J., 2001. External Shear Strengthening the Remaining Span of Severely Damaged Continuous Concrete Beams, MEng Thesis, Royal Military College of Canada, Kingston, Ontario, 166 pp.

4. El-Hacha, R., Wight, R.G., and Green, M.F., 2000, "Retrofitting of Severely Damaged Concrete Beams Using Prestressed CFRP Sheets." The 3rd International Conference on Advanced Composite Materials in Bridges and Structures (ACMBS-III), August 15–18, 2000, Ottawa, Ontario, Canada, pp 529-536.

5. Wight, R.G., Erki, M.A. and El-Hacha, R., 2001 "CFRP Sheet Strengthening Damaged Continuous Reinforced Concrete Beams." FRP Composites in Civil Engineering (CICE 2001), 12-14 Dec 2001, Hong Kong, pp 433-440.

6. Wight, R.G., Bouffard, P., Erki, M.A. and El-Hacha, R. 2002. Repair of Battle Damaged Over-bridged Structures. Structures under Shock and Impact VII. WIT

Press. 26-29 May 2002, pp 509-518.

7. Honorio, U., Wight, R.G., and Erki, M.A. 2002. CFRP Sheets for Strengthening Full-Scale Severely Damaged Concrete Structures. 4th Structural Specialty Conference (30th Annual Conference of the CSCE), Montreal, 5-8 Jun, CD ROM.

8. Wight, R.G., and Erki, M.A. 2001. CFRP Strengthening of Severely Damaged Reinforced Concrete Slabs. The 3rd International Conference on Concrete Under Severe Conditions Environment and Loading (CONSEC'01), Jun. 18–20, Vancouver, Canada, pp. 2191-2198.

9. Wight, R.G. and Erki, M.A., 2001. "Prestressed CFRP Sheets for Strengthening Concrete Slabs in Fatigue." FRP Composites in Civil Engineering (CICE 2001), 12-14 Dec 2001, Hong Kong. pp 1093-1100.

10. Heffernan, P.J., Wight, R.G., and Erki, M.A., 2002, Fatigue Behaviour of Concrete Slabs Strengthened with Prestressed CFRP Sheets. Proceedings of the 2002 Durability of Fibre Reinforced Polymer (FRP) Composites for Construction, Montreal, Quebec, 29-31 May 02. pp. 465 - 474.

11. Tanovic, R., Erki, M.A., Penstone, S. 1998. Fatigue Behaviour of Pultruded FRP Components for Short and Medium Span Bridges, 5th International Conference on Short and Medium Span Bridges, SMSB-V. Calgary, Canada. (CD).

12. Erki, M.A., Tanovic, R., Penstone, S.R., Johansen, G.E., and Wilson, R. 1997. Fatigue Evaluation of FRP Roadway Bridges, Recent Advances in Bridge Engineering, ed. U. Meier and R. Betti, Zurich, Switzerland, pp. 193-201.

13. Tanovic, R., Erki, M.A., Johansen, G.E., and Wilson, R. 1997. Extreme and Fatigue Traffic Loading of a Reinforced Plastic Vehicle Bridge, Proceedings of the Annual Conference of the CSCE, Sherbrooke, Canada, Vol. 6, pp. 61-70.

14. Erki, M.A., Shyu, C., and Tanovic, R. 2000. Fatigue Behaviour of Bolted and Bonded Connections for FRP Pultruded Members. Proceedings of the 3rd International Conference on Advanced Composite Materials in Bridge and Structures, Aug 00, Ottawa, ON, pp. 743-750.

15. Erki, M.A., Shyu, C., and Wight, R.G., 2002, Fatigue Behaviour of Connections for FRP Pultruded Members. Proceedings of the 2002 Durability of Fibre Reinforced Polymer (FRP) Composites for Construction, Montreal, Quebec, 29-31 May 02. pp. 549 - 561.

Figure 1 Over-bridged structure at Komar, Bosnia (courtesy of 1 Construction
Engineering Unit)

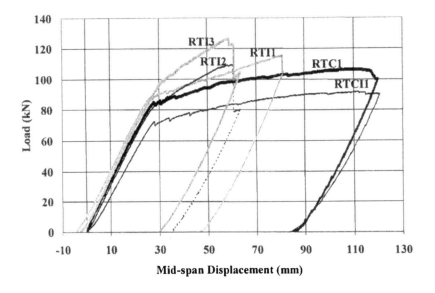

Figure 2. Load-displacement curves for reinforced concrete T-beams (RT): RTC1 – 2-
span (4 + 4 m) control (C) undamaged/no strengthening; remaining span (4 m)
with 2nd span damaged by impact (I): RTCI1 – no strengthening; with
stengthening RTI1 – prestressing wire; RTI2 –non-prestressed CFRPsheets;
RTI3 –prestressed CFRP sheets.

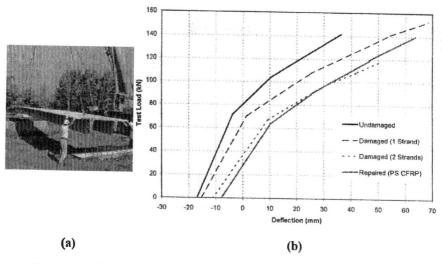

(a)   **(b)**

Figure 3 Placement of prestressed double-tee beams and load versus midspan deflection for 18 m bridge.

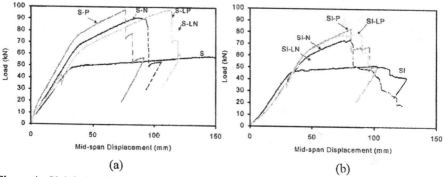

(a)   (b)

Figure 4. Slab behaviour without and with impact pre-damage: -N with non-prestressed CFRP sheets; - with prestressed CFRP sheets; -L with preloading; I with impact pre- damage.

The State of Art for the Research of Degradation for GFRP in Environmental Condition in Japan

Yoshimichi Fujii
 Sikow chemical eng. & machinery,
 Shioe 3-1-16, Amagasaki-city, Hyogo 661-0976, Japan

ABSTRACT

GFRP is a good material for anti corrosive environmental usage for civil engineering for example protection for concrete bridge and seashore structure and sewer construction. I have been studied for 30yeares in the field of corrosion GFRP. In the anti corrosive usage, the life time of GFRP may be only 5~10 years. But, in the civil engineering, the lifetime of GFRP may be 100years. I have been many experience for corrosion problems. These problems may be same category, so our anti corrosive data applicable to civil engineering for long time period. The degradation of GFRP is same category but only difference for the timetable. The degradation of GFRP have to each research for 3 materials; matrix resin, reinforced fiber and interface. In many case the competition of degradation in these 3 materials, so we must practice for many case study. I will show you several case studies and the revue research in Japan.

DEGRADATION OF RESIN

3 types degradation pattern (Figure I)

The type of reaction on the surface

The type of formation of corroded layer

The diffusion type

DEGRADATION OF GLASS FIBER

Stress corrosion of E and C and ECR in acid and alkali conditions (Figure II)

DEGRADATION OF INTERFACE (INTERPHASE)

DEGRADATION OF COMPOSITES

Which is faster degradation resin, fiber or interface? (Figure III)

WHAT IS LIFE TIME?

Effect of loading

Some proposal in sewer pipe acceleration test

WHAT IS NEED FOR BIG MARKET IN CIVIL ENGINEERING?

Estimation of lifetime

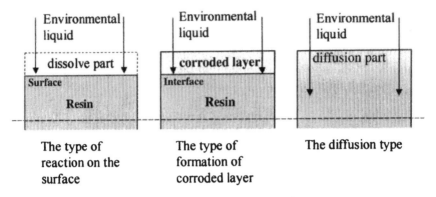

| The type of reaction on the surface | The type of formation of corroded layer | The diffusion type |

Figure I. Three types of degradation

(a) E glass fiber immersed in 5% acid at 50 °C for 1 day

(b) ECR glass fiber immersed in 5% acid at 50 °C for 30 days

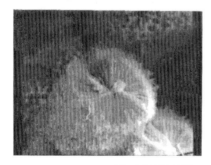

(c) E glass immersed in 16% NaOH at 50 °C for 15 days

(b) E glass fiber immersed in 16% NaOH at 50 °C for 15 days

Figure II. Stress corrosion of E and ECR glass fibers

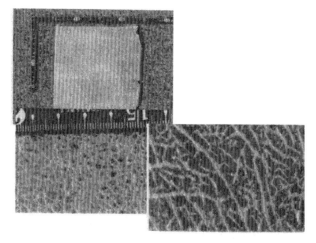

(a) GFRP tank
   leak out solution
   from the side wall

(b) The inner wall of tank

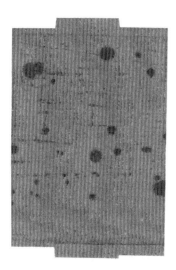

(c) The cross-section of tank wall

Figure III. Observation of tank wall

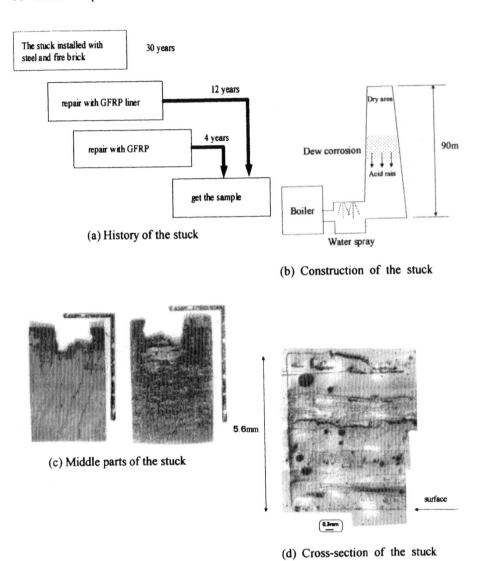

(a) History of the stuck

(b) Construction of the stuck

(c) Middle parts of the stuck

(d) Cross-section of the stuck

Figure IV. Degraded condition of the stuck

Applications of C/C composites to an engine for a future space plane

Hiroshi Hatta, Ken Goto
The Institute of Space and Astronautical Science (ISAS)
3-1-1 Yoshinodai, Sagamihara-shi,
Kanagawa-ken 229-8510, Japan
hatta@pub.isas.ac.jp

ABSTRACT

Feasibility studies were carried out regarding the application of carbon/carbon (C/C) composites to a turbine disk and heat exchangers for an engine used in a future space vehicle. In these applications, the maximum temperature is estimated to be about 1500°C. In order to overcome this high temperature, attempts were made to apply three-dimensionally reinforced C/C composites to both the structures. The most serious problem encountered in the application to the turbine disk was losing fragments of the material located near the outer periphery of the disk due to strong centrifugal force, which induced severe vibration due to rotational unbalance. The heat exchangers have complex shapes in order to realize a large heat exchanging area, so that joined structures were explored. In this application, our main effort has been focused on finding structures requiring minimum joining strength and materials with the lowest gas leakage.

INTRODUCTION

Carbon-carbon (C/C) composites are attractive materials because of their superior specific strength and specific elastic modulus at elevated temperatures over 2000°C [1,2]. In light of these advantages, attempts have been made to apply C/C composites to high temperature structures such as space vehicles, hypersonic airplanes, or fusion reactors etc [1-4]. However, C/C composites have seldom been applied to primary structures requiring high load bearing capability, mainly due to the composite's weakness against high temperature oxidation. Thus the improvement of oxidation resistance has been a primary research topic for C/C composites [1-3,5-7]. As a result, reliable short term oxidation protection for several hours, and up to about 1500°C is now guaranteed by the combination of a SiC coating and glass crack sealant, but long term and high temperature resistance is still under active study [3,5-7]. In addition to this serious shortcoming, few reliable mechanical design data are available regarding applications of C/C composites in load bearing structures [1-4]. In spite of the above difficulties, the authors have been making application studies of C/C composites to a turbine disk and heat exchangers which are to be used for a future spacecraft engine, the ATREX [8]. The ATREX engine, an air-turbo-ram-jet with an expander cycle, has been studied about 15 years at the Institute of Space and Astronautical Science (Japan). This engine is regarded as a suitable candidate for the 1st stage engine of a two-stage-to-orbit, TSTO, space plane. It is a combined cycle engine performing like a turbojet at subsonic to Mach 2 flight and a fan-boost ramjet at speeds of Mach 2 to 6. Figure 1 shows a schematic diagram of the ATREX engine under development, its main features including the utilization of hydrogen fuel, an expander cycle adopted to improve high speed capability, and the employment of C/C composites.

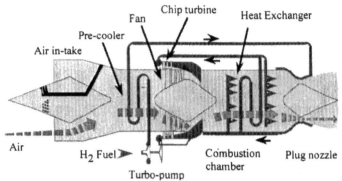

Fig.1 Schematic drawing illustrating the mechanism of an air-turbo-ram-jet engine for use in a future Japanese space vehicle.

CARBON-CARBON COMPOSITES

Mechanical properties

The tensile strength of C/C composites generally increase with increasing temperatures as shown in Fig.2. The similar tendency is obtained for other strength and toughness, so that we can conservatively design C/C composite structures based on room temperature properties.In the authors' experiment, 800 MPa has been obtained for room temperature strength of three-dimensionally reinforced C/C composites [9]. This tensile strength is sufficient for the ATREX applications. However, this improvement comes at the expense of interfacial strength between the fibers and matrix [10]. This interfacial degradation results in low compressive [11] and shear strength. In the applications in the turbine disk and heat exchangers, tri-axial stress distributions are anticipated. To avoid premature shear damage, three-dimensional, 3D, reinforcement is considered to be indispensable in the ATREX applications [8].

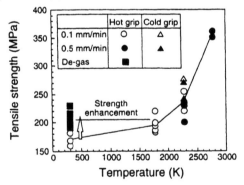

Fig.2 Tensile strength of a cross-ply-laminated C/C composites.

Thermal properties

The advantages of C/C composites also reside in their high thermal conductivity and low coefficients of thermal expansion, CTEs [12,13]. This low CTE of about 0-1 x 10^{-6} 1/K is maintained at temperatures over 2000°C However, owing to this extremely low CTEs of C/C composites, when the C/C composites are joined with other materials, such as metals, some measures must be taken to relief thermal mismatch strain. The thermal conductivity of C/C composites strongly depends on the extent of graphitization in the carbon fibers. The highest conductivity can be obtained using pitch based fibers heat-treated at high temperatures. Conventional C/Cs have intermediate conductivity. However, the highest conductivity of C/C is much higher than that of high conductivity metals [12]. These excellent properties make C/C composites highly resistant to thermal shock.

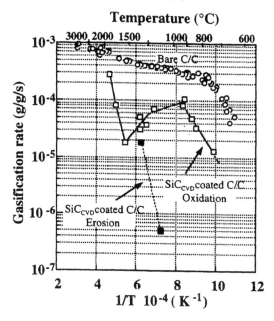

Fig.3 Mass loss rates due to oxidation for bare and SiC-coated C/C composites.

High temperature oxidation

To improve oxidation resistance, an oxidation protection coating is usually applied on the surfaces of C/C composites. For example SiC coating lowers the mass loss rate as indicated in Fig.3. However, there are still problems. As mentioned above, the CTEs of C/C composites are low and the coating is ordinarily applied at an elevated temperature, exceeding 1000°C. Thus, during cooling from coating temperature, many cracks appear in the coating due to thermal mismatch strains between a coating material and C/C composite. Thus, oxidation still proceeds from the coating cracks. Glass-based sealant is usually over-coated on a ceramic coating. A silicon carbide, SiC, coating over-coated with vitreous crack sealant is now believed to be most effective. However, it is useful only over short periods in a temperature range up to 1700°C [5,6] due to evaporation of the glass sealant. The glass crack sealant should be liquid in a temperature range for oxidation protection, so as to comply the variation of coating cracks width with temperature. Due to this requirement, the evaporation temperature of the glass sealant is inevitably low, e.g., B_2O_3 evaporates from 800°C. Thus complete suppression of oxygen by diffusion through coating cracks in a long term is difficult. In addition, over 1700°C, "active oxidation" occurs in the SiC coating, as shown in the left hand side in Fig.6, where the SiC coating continuously evaporates [14,15].

TURBINE DISK

The tip-turbine structure has a shape shown in Fig.4. Since the C/C composites are the only materials that have superior mechanical properties at elevated temperatures more than 1500°C [1-3], the utilization of the C/Cs is an indispensable factor to attain a flight speed of Mach 6. The turbine disk rotates up to 440m/s and the strength required of the C/Cs at this maximum speed was estimated to be 350 MPa. A tensile strength higher than 700MPa was shown to be attainable by 3D C/C. Thus using a r-θ-z reinforced 3D C/C, we have experimentally verified that we can clear this strength requirement.

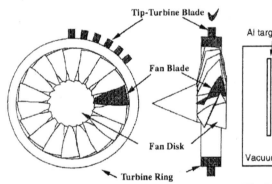
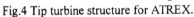
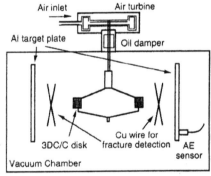

Fig.4 Tip turbine structure for ATREX.   Fig.5 Schematics of a spin burst tester.

High speed rotation tests

High speed rotation tests have been carried out. Figure 6 represents a schematic illustration of the utilized rotation burst tester driven by air turbines at a maximum rotation speed of 100,000 rpm. This tester equipped with telemeters and cameras to measure strain distributions and to take photographs of fracture initiation, respectively. The rotational tests were performed on various annular C/C disks with different reinforcing patterns, 2D or 3D reinforcement, varying ratio of inner and outer diameter, and changing notch length. These tests verified that fracture criteria established by static tests are applicable to the disk in rotational burst tests, and that the required rotational speed can be attained.

During the rotation burst tests of the 3D C/C disks, we found micro-fractures resulted in many fragment fly-outs, which occurred much earlier than did the total fracture. As shown in Fig.6, many local-fiber-bundle-fly-outs started at a peripheral rotation speed of 120 m/s, though the total burst occurred at 516 m/s. A large number of the fragment fly-outs induced vibration, which is fatally harmful to the ATREX engine, because the blades contact with the shroud. As illustrated in Fig.7, this phenomenon appeared at first in the form of debonding of outer-most fiber bundles due to stress concentrations at the ends of them by the hoop stress. The debonded bundles are subjected to bending force by centrifugal force. When the debonding length becomes sufficiently long, the fiber bundles fracture by bending and thy then fly-out. Sever micro-fractures occurred especially for the 3D C/Cs. It is due to that 3D C/Cs are reinforced by thicker fiber bundles than are 2D C/Cs and tend to possess weaker interface bonding. The micro-fly-out phenomenon can be predicted by the crack extension model shown in Fig.7. To prevent micro-fly-out, the interfacial strength was improved by the conversion of the interfacial carbon into SiC by Si impregnation only near the outer radius of a C/C disk. This treatment was confirmed to double the initial fly-out speed.

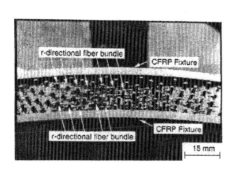

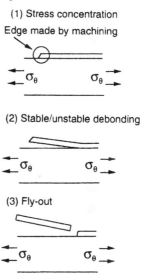

Fig.6 Outer periphery of a 3D C/C disk after high speed rotation test. After rotation of 200 m/s at the outer periphery, protruding needles are observed. These needles are the remainder of flown-out circumferential fiber bundles.

Fig.7 A model explaining fly-out of circumferential fiber bundles.

HEAT EXCHANGER

According to a preliminary calculation, a heat exchange area of 10 m² is required to effectively operate the expander cycle in a 40 mm diameter model. To realize this large heat exchange area, in addition to an conventional insert-type exchanger, the walls of the combustion chamber and plug nozzle should be utilized. Heat exchangers usually have complex shapes. The primary difficulties in applying of C/C composites to the heat exchangers are how to form them into such complex shapes. C/C composites can be bonded together by carbon or metal carbide. However, the bonding strength is extremely low as shown in Figs.8 (a) and (b). Hence we must consider structures requiring low bonding strength.

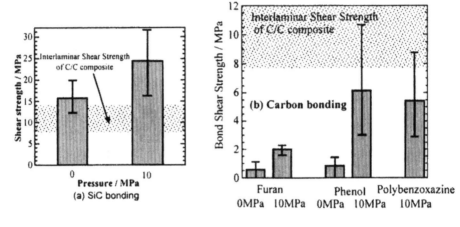

Fig.8 Fracture shear stresses of SiC- (a) and carbon- (b) bonded C/C composites.

For the ATREX application, two types of structures as shown in Figs.9 (a) and (b), the walls of combustion chamber and plug nozzle are to be used for heat exchangers. In the ducts of the heat exchangers, hydrogen flows at a maximum pressure of 40 atm. On the other hand, outside of the heat exchangers, combustion gas runs at a maximum of 10 atm. Hence, the duct wall sustains a maximum pressure difference of 30 atm at a temperature of 1500°C. The wall type heat exchanger shown in Fig.9 (a) consists of a r-θ-z 3D reinforced C/C composite. In the preform of this main tube, copper tubes are inserted to form ducts for the hydrogen flow. Copper is to be used because it is chemically stable with carbon even at elevated temperature. This preform is infiltrated with phenolic resin followed by carbonization of the resin. This cycle is repeated until a sufficient density is attained, about 1.7 g/cm³. Then, the copper tubes are chemically dissolved with nitric acid and resulting C/C is to be bonded to the manifold.

The insert type heat exchanger has the similar shape to the metal one, composed of thin curved tubes. The primary problem regarding this type lies in the formation of thin-walled and small diameter tubes. This type of tube is to be formed from a CFRP tube formed on a thin copper tube. Precursor resin infiltration and carbonization process are then repeated at around 600 to 800 °C to form a C/C on the surface of the metal tube. After that copper tube is dissolved with nitric acid. Finally the graphitization process is performed at over 2000 °C.

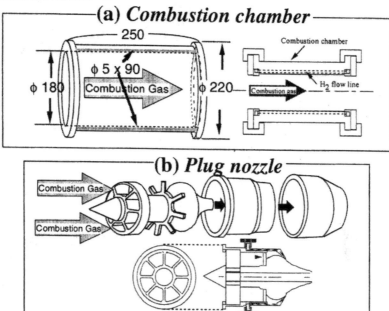

Fig.9 Heat exchangers made of C/C composites under development. Cylindrical heat exchangers utilizing the wall of a combustion chamber (a) and plug npzzle.

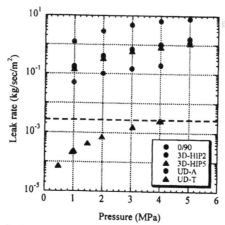

Fig.10 Nitrogen gas leak rates of various C/Cs as a function of gas pressure. 0/90; bare

cross ply laminate, 3D-HIP2; 3D C/C with 2cycles of HIP-treatment, 3D-HIP5; 3D C/C with 5 cycles of HIP-treatment.

The final problem is leakage of hydrogen gas from the C/C wall and bonded regions. The C/C composites usually include plenty of pores and cracks. Thus, gas passes easily through C/Cs if no measures are adopted. The engine system requires a hydrogen gas leakage under 3 %. However, non-treated C/Cs leaks gasses more than that as shown in Fig.10 in which the 3% gas leakage level is indicated by the dashed line. To prevent the leakage, Si was infiltrated into pores and cracks in the C/C composites. By this treatment performed at 1500°C, most of the impregnated Si was converted to SiC with volumetric expansion [16, 17]. The pores and cracks are then almost sealed. Thus, the leakage was confirmed to lowered 3 orders of magnitude of 10^{-3}, which is sufficient for the ATREX application. Similar leakage problems have been recently discussed by several authors but in different situations [18, 19]. One report dealt with the development of a hydrogen rocket tank for use at cryogenic temperatures.

REFERENCES
[1] G. Savage, "Carbon-Carbon Composites", Chapmanhall, 1993.
[2] E. Fitzer, L.M. Manocha, "Carbon Reinforcements and Carbon/Carbon Composites", Springer, 1998.
[3] J. D. Buckley, D. D. Edie eds., "Carbon-Carbon Materials and Composites", NASA Reference Pub. 1254, 1992-2.
[4] D.L. Schmidt, K.E. Davidson, L.S. Theibert, SAMPE J., 35 (1999) 27-39, 51-63, 47-55.
[5] J. R. Strife, J. E. Sheehan, Am. Ceram. Soc. Bul., 67(1988) 369-374.
[6] T. Aoki, H. Hatta, Y. Kogo, H. Fukuda, Y. Goto, T. Yarii , Trans. Japan Soc. Metal (In Japanese) 62(1998) 404-412.
[7] H. Hatta, T. Aoki, Y. Kogo, T. Yarii, Composites, Part A, 30(1999) 515-520.
[8] N. Tanatsugu et.al., 19th ISTS, 1994-5 Yokohama.
[9] T. Ikegaki, Master's thesis, Kyoto Inst. Tech. 1999.
[10] H. Hatta, K. Suzuki, T. Shigei, S. Somiya, Y. Sawada, Carbon, 39(2001) 83-90.
[11] P.S. Steif, J. Compos. Mater., 22-September (1988) 818-828.
[12] T. Arai, K. Kodama, H. Horiike, Tanso 1992(1992) 120-127.
[13] H. Hatta, Y. Kogo, et.al., Materials System, 14(1995) 15-24.
[14] W.L. Vaughn, H.G. Maahs, J. Am. Ceram. Soc., 73 (1990) 1540.
[15] M. Balat, G. Flamant, G. Male, G. Pichelin, J. Mater. Sci., 26(1991) 1093.
[16] H. Hatta, E. Sudo, Y. Kogo, I. Shiota, J. Jpn. Inst. Metals, 62(1998) 861-867.
[17] H. Mucha, A. Kamiya, W. Wielage, Proc. ICCM 11, Gold Coast, Australia, 717 (1997).
[18] H. Kumazawa, Y. Hayashi, T. Ishikawa, T. Aoki, Y. Morino, 2000-7 Mishima, pp.705-712.
[19] M.P. Bacos, J.M. Dorvaux, O. Lavingne, Y. Renollet,, Carbon, 38(2000) 77-92.

Overview of Polymer Composite Research Activities with Textile Reinforcements (Stitching and 3-D Fabric) Conducted in National Aerospace Laboratory of Japan

Yutaka Iwahori and Takashi Ishikawa
National Aerospace Laboratory of Japan
6-13-1, Ohosawa, Mitaka, Tokyo, 181-0015, Japan
iwahori@nal.go.jp and isikawa@nal.go.jp

ABSTRACT

National Aerospace Laboratory of Japan (NAL) has been conducting research of stitching and 3-D textile composite materials for aerospace applications, targeting more light weight and low cost structures with RTM or RFI resin infusion techniques. NAL clarified that compression after impact (CAI) strength of stitched laminate by RTM is 20% higher than non-stitched laminate. They also found that relationships between stitch density and mode I interlaminar fracture toughness (G_{Ic}) is linearly increasing by conducting double cantilever beam (DCB) tests and FEM analysis. Interlaminar fracture toughness of 3-D fabric in various Z-Fibers (Carbon or Kevlar) were also evaluated by DCB tests and analysis that include through-the-thickness of fiber slack and friction of broken fiber effects. These researches were co-operated with NIPPI for stitching and TMIT (fabricated by MHI) for 3-D textile composite.

25

INTRODUCTION

Composite materials, especially CF/epoxy 2-D laminates, have capability of light weight design for in-plane high loading structures. On the other hand, interlaminar strength is not strong against through-the-thickness direction load (laminate peeling to out of plane). In some aircraft structural design, weakness of the CFRP properties is critical and CAI (compression after impact), OHC(open hole compression) or resistance of interlaminar peel load such as interface of skin-stringer structure caused by buckled skin panels are typical examples. For these reasons, improvement of the CFRP's interlaminar strength gives a possibility to more light weight structures by damage tolerance or post buckling design. One of the CFRP research activities which through-the-thickness characteristics improvement for aircraft structures, stitching and 3-D fabric with low cost consolidation technique such as RTM (resin transfer molding) and RFI (resin film infusion) are evaluated in US and other countries. In Japan in 1980's, NAL reported that stitched CFRP laminates manufactured by prepreg stitch/autoclave consolidation was evaluated(1). The results were confirmed in ability of crack arresting and to postpone final fracture in static strength. On the other hand, disadvantage such as decline of in-plane and bending strength were admitted. The reason for lower strength was that stitch needle gave a damage to CF bundles in prepreg or voids were formed at through-the-thickness thread holes. Recently, however, RTM or RFI technique such as dry-preform consolidation system has been matured and the preform damage by stitching could be reduced by using these methods. The authors' group carried out re-evaluation of stitched CFRP specimens by RTM processes. 3-D fabric composite interlaminar strength is also evaluated.

OVERVEW OF STITCHING AND 3D FABRIC

Effect of CAI (compression after impact) strength

NAL conducted CAI test about stitched CFRP laminates that various impact damage levels(2). The impacts was given by using an impact machine (Dynatup:GRC 8250) and damaged specimen was tested INSTRON 1128 screw driven testing machine. These results were shown Figure 1. As the results, stitched CFRP laminates have higher compressive strength than unstitched laminates. The CAI strength improvement was about 20% at the tested impact levels.

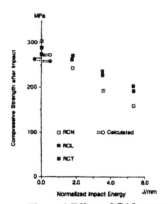

Figure 1 Effect of CAI strength

Stitch effect for open hole fatigue

Stitch effect for open hole tension and compression fatigue strength tests were conducted(3). Figure 2 shows tension fatigue test specimen with open hole. The specimens were made from T300 plane woven fabric stitched by Kevlar® 29. The preform was impregnated by resin transfer molding after stitch process.

The stitch position effects were evaluated for open hole specimen, fatigue test carried out under tension-tension (T-T, R=0.1) cyclic load condition. In summary, stitching position

Figure 2 T-T fatigue test

affects fatigue life. If stitching is too close to the open hole, static strength and fatigue strength were reduced by stitching. It was suggested that the stitching must be avoided in the vicinity to the open hole where stress concentration occurs for static and fatigue loads. Figure 3 shows ultrasonic inspection results after 2000000 cycles for parallel stitch and unstitch specimen. No low cycle failure happened in this stitch pattern. In fatigue tests, the damage area propagated from open hole to the edge similarly to unstitch cases. In parallel stitch, after damage area reached to the stitch lines, it was arrested by the stitch lines.

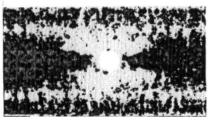   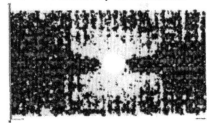

(a) Stitch after 2000000 cycles   (b) Unstitch after 2000000 cycles
291.9MPa ⇔ 29.2MPa   286MPa ⇔ 28.6MPa
Figure 3 Ultrasonic inspection results(T-T)

Compression-compression (C-C) open hole fatigue test and ultrasonic inspection were conducted to determine the fatigue life of CFRP laminates and the effect of stitching on the delamination damage propagation. During C-C fatigue test, stress ratio (R=min./max.) was 10. Figure 5 depicts S-N plot of C-C fatigue test. It is concluded that stitched specimens have equivalent or slightly better fatigue resistance than unstitched specimens and damage propagation arrest is observed on some specimens by ultrasonic inspection.

Figure 4 Picture of
fatigue test

Figure 6 shows the data plots of delamination area propagation in the specimens measured from the ultrasonic inspection results. There is a trend that increments in damage area for stitched specimen was lager than those of unstitched specimen. This phenomenon is roughly explained as that the damage grows vertically after arrest by stitch lines. It should be noted that simple damage area can not be correlated with fatigue life.

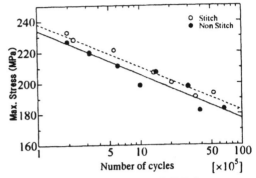

Figure 5 S-N Plotted of C-C Fatigue test

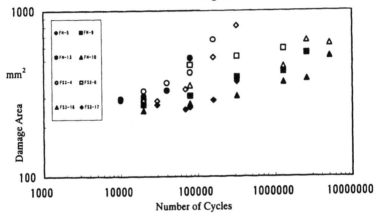

Figure 6 Damage Area Propagation for C-C fatigue

Interlaminar fracture toughness evaluation for textile composites

NAL evaluated interlaminar fracture toughness of stitched laminates and 3-D fabric composites(4,5). These CFRP's have through-the-thickness fiber such Kevlar or carbon fibers. Measurements of interlaminar fracture toughness value (G_{IC}: Mode I crack energy release rate) in through-the-thickness materials require high peel loading on DCB (double cantilever beam) test specimen. Stitching specimens were manufactured by NIPPI (Japan Aircraft Mfg. Co. Ltd), 3-D fabric research co-operated with TMIT (Tokyo Metropolitan Institute of Technology) and 3-D fabric were manufactured by MHI (Mitsubishi Heavy Industry Co. Ltd).

NAL developed insert tongue type loading fixture applied to stitching laminates or 3-D fabric that is shown Figure 7. As another aspect, DCB specimen requires bending strength caused by high peel loading as crack can be propagated. Therefore, unidirectional CFRP tabs bonded to top and bottom of DCB specimen were employed. These tabs can relief bending stress and avoid specimen arms to fail before crack propagation. A picture of DCB tests is shown in Figure 8. UD tabbed stitch specimen was installed in loading fixture. Typical load-COD (crack opening displacement) curves are shown in Figure 9 (a) for stitching and (b) for 3-D fabric.

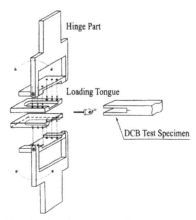

Figure 7 Insert tongue loading fixture

Figure 10 indicates relationships of stitch density and G_{IC}. From DCB test results and FEM analysis using NASTRAN, linear relationships in improvement in G_{IC} about stitch density are obtained. Figure 11 shows R-curve in 3-D fabric of different content of Z-fiber: C0x has no Z fiber, C1x is 0.27%, and C3x has 0.77%. NAL also evaluated interlaminar fracture toughness of 3-D fabric composites with different Z-fiber slack. Figure 12 shows schematic of low vertical tension (full slack) and normal vertical tension load (slight slack) in Z fiber. These DCB test data are shown in Figure 13. It was observed that strain energy release rate gradually increase for full slack case (C'1x) when crack length is propagated.

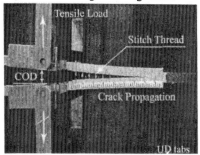

Figure 8 Stitched specimen under DCB test

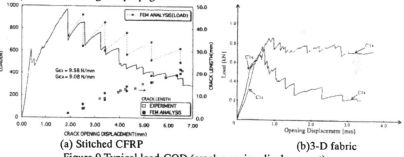

(a) Stitched CFRP   (b)3-D fabric

Figure 9 Typical load-COD (crack opening displacement) curve

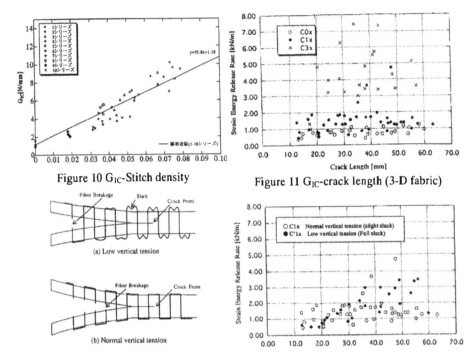

Figure 10 G$_{IC}$-Stitch density

Figure 11 G$_{IC}$-crack length (3-D fabric)

Figure 12 Schematic of vertical tension
load in Z fiber

Figure 13 Effect o f vertical tension in Z
fiber(3-D fabric)

FUTURE SCOPE

Double open hole fatigue test for stitched CFRP

NAL is ready to evaluate for double open
hole fatigue test for stitched CFRP. Static tension
strength test specimen installed in a testing
machine is shown in Figure 14. Tension-tension
fatigue test and non-destructive inspection by UT
are planned. The point of interest of these tests
will be observed for propagation of damage area
and interference in damage propagating from each
open hole.

Figure 14 Double open holes specimen

Stiffened panel test

For evaluation of interface strength between skin and stiffener under cyclic load, stiffened shear panel tests are planned. Figure 15 shows a stiffened panel test specimen. Interface of stiffener and skin panel is stitched and co-cured by RTM. Unstitched stiffened panel is also prepared. There specimens will be installed into picture frame type of shear loading fixture and tested under cyclic load reaching a buckled level of skin-panels.

Figure 15 Stiffened panel specimen

REFERENCE

(1) Y.Tada, T.Ishikawa, 1989. "Experimental Evaluation of the Effects of Stitching on CFRP Laminate Specimens with Various Shapes and Loading" Key Engineering Materials, vol.37, pp.305-316.

(2) Y.Tada,M.Matusima et al, 1994. "Stitching Effect upon Strength of Composite Laminates." Proc. 19th Symposium on Composite Materials, pp.57-60.

(3) Y.Iwahori, T.Ishikawa et al, 2000. "Open Hole Compression Fatigue Test of Stitched CFRP laminates" Proc. 9th US-Japan Conference on composite Materials, pp.761-768.

(4) Y.Tanzawa, N.Watanabe & T.Ishikawa, 1999. "Interlaminar Fracture Toughness of 3-D Orthogonal Interlocked Fabric Composite" Composite Science Technology, 59, 8, pp.1261-1270.

(5) Y.iwahori, T.Ishikawa, 1999. "Stitch Parameter Effects on Interlaminar Fracture Toughness of CFRP Laminates" Proc. 44th SAMPE Symposium and Exhibition, pp.2289-2300.

X-ray Diffraction Studies on Stress Transfer of Fiber Reinforced Polymer Composites

Takashi Nishino

Department of Chemical Science and Engineering, Faculty of Engineering,
Kobe University, Rokko, Nada, Kobe 657-8501, JAPAN
TEL: +81-78-803-6164, FAX +81-78-803-6205, E-Mail: tnishino@kobe-u.ac.jp

ABSTRACT

For fiber reinforced composites, it is important to clarify the reinforcement effect of the incorporated fiber. Here, I would like to propose a new technique "X-ray diffraction Method" to detect the stress on the incorporated fibers in the composites under load, and discussed the stress transfer and the reinforcement behavior in terms of microscopic point of view.

Stress σ_o was applied in the direction perpendicular to the fiber direction for the rectangular sample of one dimensional reinforced composite on the X-ray goniometer. The relationship between σ_o and σ_c, the stress σ_c on the incorporated fiber, was investigated for the composites with different combinations under various environments. We have applied this technique to various fiber reinforced (gel-spun polyethylene fiber (HP-PE) [1,2], carbon fiber [3,4], poly(p-phenylene benzobisoxazole) (PBO) fiber [5]), and particulated (Aluminum [6-8], Silica [9]) composites.

In this study, our results on the carbon fiber composite will be focused. The effects of the fiber surface treatment, the mechanical, thermal fatigues on the stress transfer was also investigated by X-ray diffraction, and a novel method for the repairing of the deteriorated composite will be proposed.

INTRODUCTION

Recently high performance materials are required in a lot of important fields, while various demands in each field could not be always fulfilled with materials consisting of only one kind of substance. Thus, materials with two or more kinds of substances, *that is*, composite materials have been designed for complying with each demand. Nowadays high performance composite materials are available with a whole range of properties, and they are used in applications varying from aerospace to electronics and sports equipments [10]. Continuous carbon fibers are often introduced unidirectionally in the composite for reinforcement, which possesses high elastic modulus and strength longitudinally. However, pronounced anisotropy results in inferior inter fibrillar strength, and transverse cracking of the composite. Thus for the unidirectional fiber reinforced composites, it is important to clarify the transverse reinforcement effect and stress transfer. It has received detailed analysis, because of its significance for the prediction of mechanical performance of composite, by using photoelasticity [11], Raman spectroscopy [12], mechanical modeling [13], infinite elemental analysis [14] and so on.

In this study, unidirectional continuous carbon fiber reinforced epoxy resin composite (CFRP) was subjected to the stress in the direction perpendicular to the fiber direction. This stress was transferred to the incorporated fibers through the matrix, which was appeared as the shift of the equatorial X-ray diffraction peak of carbon fibers. By detecting the peak shift quantitatively, we could get information how much stress the incorporated fiber was subjected to. The relation between X-ray beam and composite has been limited to the X-ray radiography, where defects, fiber alignment could be examined [15]. On the other hand, we here proposed a novel technique "X-ray Diffraction Method" to detect the stress on the incorporated fiber in the composite under transverse load *in situ* and non-destructively, and discussed the stress transfer in terms of microscopic point of view. The effects of the fiber surface treatment, the mechanical, thermal fatigues on the stress transfer was also investigated by X-ray diffraction

EXPERIMENTAL

Materials

A liquid diglycidyl ether of bisphenol A type epoxy resin (Epikote 828; Shell Chemical Co.), methyl tetrahydrophthalic anhydride (Ciba Geigy) as a curing agent, and 1-methyl imidazole (Aldrich) as an accelerator were chosen for the resin system because of low viscosity before curing and high mechanical performance after curing. Their chemical formulas are shown in Figure 1. Resin composition with epoxy resin : curing agent : accelerator = 100 : 87.5 : 1.5 phr was determined from the preliminary experiments.   Glass transition of the cured resin was 140℃ measured by a differential scanning calorimeter (Seiko Instruments, DSC-220U) at the heating rate of 10℃/min and the sample weight of 20 mg.

Table 1 Chemical structure of epoxy resin, curing agent and accelerator.

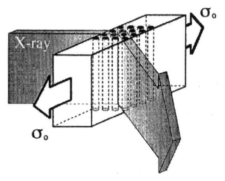

High   modulus   carbon   fibers (TORAYCA M40B, Toray industries, Inc.) were used as the reinforced fibers.   Carbon fibers were impregnated into degased epoxy resin, followed by pultrusion, curing at 80˚C, 6h, and 180˚C, 12h., and finally unidirectional CFRP was obtained.   The fiber weight fraction was 40% determined from thermo-gravimetry trace.

X-ray Diffraction

Rectangular   strip of unidirectional CFRP was clamped and set on the X-ray goniometer.   Stress σ_o was applied to the whole composite in the direction perpendicular to the fiber axis, where equatorial reflection of carbon fiber can be detected as shown in Figure 2.

Several sharp reflections from the carbon fibers could be observed in the equatorial X-ray diffraction profile of CFRP with overlapping with the amorphous halo (2θ =ca. 18 °) of the epoxy resin.   The 004 reflection (2θ =53.5° for CuK α radiation) was employed for the measurement of the lattice spacing,   because   it   located   at   higher diffraction angle and did not disturbed by the presence of the amorphous halo.

The crystal strain ε was measured by detecting the change in the peak position.

$$\varepsilon = \Delta d/d_0 \qquad (1)$$

where, d_0 shows the initial lattice spacing for the 004 reflection determined with the Bragg's equation, and Δd is the difference in the lattice spacing induced by a constant stress.   The experimental error in measuring the peak shift due to the lattice extension was evaluated to be generally less than ±

Fig. 2 Schematic model for the measurement of the strain on carbon fibre incorporated into the epoxy resin by X-ray diffraction.

$1/100°$ in 2θ, which corresponds to the strain of 0.013%.

The stress σ_c on the incorporated carbon fiber was calculated as follows:

$$\sigma_c = \varepsilon \cdot E_t \tag{2}$$

where, E_t is the elastic modulus (36.5GPa) [16] of the inter-graphite networks, which is perpendicular to the direction of covalent bonds.

RESULTS AND DISCUSSION

Stress Transfer

Figure 3 shows the relationship between the stress σ_o applied to the whole CFRP and the stress σ_c on the embedded carbon fiber. As-supplied carbon fibers were incorporated in this case. Several symbols are the results from different specimens, and they coincided one another within experimental error. The crystal strain ε was converted into the stress σ_c on the incorporated carbon fiber by using equation (2). Here, the homogeneous stress distribution within each carbon fiber was assumed. In the initial stage, σ_c increased

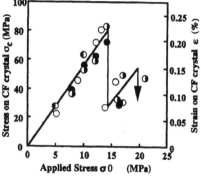

Fig.3 Relationship between the stress σ_o applied to the carbon fiber/epoxy resin composite and the stress σ_c on the crystalline regions of embedded carbon fiber.

linearly with increasing σ_o through the origin. This indicates that the applied stress was transferred from the matrix to the fibers certainly. So the "X-Ray Diffraction Method" is a powerful tool for detecting the stress on the incorporated fiber in the composite under load *in situ* and non-destructively. The initial slope of the curve can be correlated to the efficiency of the stress transfer to the incorporated fibers. When 10 MPa was subjected to the whole composite, .ca.69 MPa was concentrated on the incorporated fibers. This implies that the carbon fibers show reinforcement even in the perpendicular direction of the fiber alignment of CFRP. Sudden breakage occurred around 14 MPa, which will be due to the interfacial failure between the matrix and the incorporated fibers. However, the remained un-debonded fibers seemed to be still stressed even after the macroscopic failure. This is quite important phenomenon in the view point of damage torelance for CFRP after undergoing the stress more than the tolerance limits.

Fatigue

Next, in order to investigate the environmental deterioration of the composite, the composite was immersed into the boiling water for 12 hr, then dried at 110℃ for 12 hr. The stress transfer was examined by X-ray diffraction at each stage.

Figure 4 shows the relationship between the stress σ_o applied to the CFRP and the stress σ_c on the embedded carbon fiber (○) after immersing in boiling water, (●) followed by drying at 110℃. The σ_o - σ_c curve for the control CFRP (The results of Fig.3) was also superimposed on the figure with a half-tone line.   Comparing with the results for the control, the maximum attainable stress decreased, and the initial slope also decreased after immersing into boiling water.   These indicate that the stress transfer from the matrix to the fiber went worse by the water absorption.   Further , after drying at 110℃, the maximum attainable stress rather recovered, however, the initial slope remained low.   The temperature of 110℃ is below the glass transition temperature (140℃) of the matrix resin. Even after drying at 150℃ for 12 hr, the initial slope remained low .

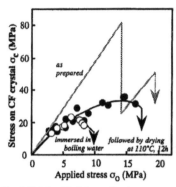

Fig. 4 Relationship between the stress σ_o applied to the carbon fiber/epoxy resin composite (○) immersed in boiling water (●) followed by drying and the stress σ_c on the crystalline regions of embedded carbon fiber.

Figure 5 shows the relationships between the immersing time into the boiling water and the tensile strength, the amount of the absorbed water of CFRP.   The composite absorbed water much more than the epoxy resin itself.   Carbon fiber does not adsorb water, so the excess absorption for the composite compared with that of the matrix is considered to be attribute to that at the carbon fiber / matrix interface.   The averaged thickness of the

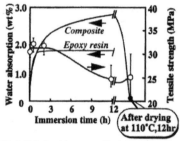

Fig.5, Relationships among the immersion time in boiling water and the amount of the water absorption, and the tensile strength of the carbon fiber / epoxy resin composite.

absorbed water layer at the interface could be evaluated as 18.5 nm around each fiber when the excess amount of the absorbed water after 12 hr equally locates at the interface. The initial increase of the absorbed water obeyed the Fick's low, and the apparent diffusion coefficient of water into the CFRP was 0.53×10^{-11} m^2/s. This value is one third of the matrix resin.   The tensile strength of the composite decreased by the water absorption.   After drying at 110℃ for 12 hr, the amount of the absorbed water did not become zero, and the tensile strength of the composite did not recover and remained low, which are in contrast with the cases of the matrix resin, where the mechanical properties recovered to their initial values completely.   These suggest that the chemical and / or physical bonds were destructed at the composite interface.   This is reflected for the irreversible decrease of the stress transfer by X-ray diffraction as shown in Fig.4.

Next, a cyclic tensile stress of 10 MPa, corresponding to 80% of the macroscopic failure stress, was applied to the composite in the direction perpendicular to the fiber direction as schematically shown in Figure 6, and the stress transfer was examined by X-ray diffraction.

Figure 7 shows the relationship between the stress σ_c on the embedded carbon fiber and the stress σ_o applied to the whole CFRP after applying the cyclic stress.   The $\sigma_c - \sigma_o$ curve for the control CFRP was again superimposed on the figure with a half-tone line.   With increasing the cyclic number, the initial slope decreased.   This indicated that the stress transfer coefficient from the matrix to the incorporated fiber went worse by the mechanical fatigue.   The interfacial debonding progressed during the cyclic stress, even below the failure stress.   Subsequently, the crack propagation at the interface is considered to bring about the decrease in the stress transfer at the interface of the composite.   This reveals that the fatigue behavior can be also clearly monitored by the "X-ray diffraction method".

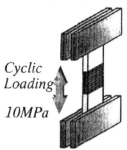

CF/Epoxy Composite

Cyclic Loading

10MPa

Fig. 6 Schematic representation for applying cyclic tensile stress to the carbon fiber / epoxy resin composite.

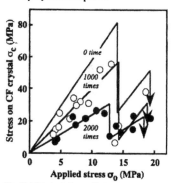

Fig. 7 Effect of the cyclic loading upon the stress transfer system of carbon fiber/ epoxy resin composite.

Repair

The mechanically deteriorated CFRP was tried to be repaired by immersing the specimen into monomer, then polymerized.   By this method, the monomer will be penetrated into the crack portion, and polymerization / solidification will be expected to heal the crack.

Figure 8 shows the relationship between the immersing time in four kinds of monomers (AN; acrylonitrile, AAm; acrylamide, MMA; methyl methacrylate, St; styrene, and Bisphenol A DGE shown in Fig.1) at 30°C and the degree of swelling of the cured epoxy resin.   Here, AAm and bisphenol A DGE were dissolved in acetone.   The resin was swollen by adsorbing each monomer, especially the degree of swelling was very high for AN.   This suggests high affinity of AN to the cured epoxy resin.   In this study,

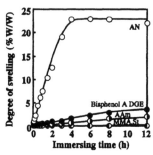

Fig. 8 Relationship between the immersing time in four kinds of monomer (AN , AA , MMA and bisphenol A DGE) at 30°C and the degree of swelling of cured epoxy resin.

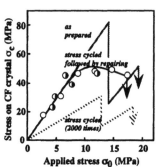

Fig. 9 Relationship between the stress σc on the crystalline regions of embedded carbon fiber and the stress σ0 applied to the stress cycled (10MPa, 2000 times) carbon fiber/epoxy resin composite immersed in AN followed by annealing at 150°C.

the mechanically deteriorated CFRP was immersed into AN monomer for 6 h at 30 °C, then polymerized with t-butyl hydroperoxide for 12 h at 150°C.

Figure 9 shows the the relationship between the stress σ_c on the embedded carbon fiber and the stress σ_0 applied to the CFRP deteriorated (10 MPa, 2000 times), then repaired with the method shown above.   The σ_c - σ_0 curve for the control and the deteriorated CFRP were superimposed on the figure with a half-tone and the broken line, respectively.   The initial inclination of the curve recovered to its initial value after repair, which could not be attained only by heating at 150°C without monomer incorporation. The polymerized AN, $i.e.$, polyacrylonitrile (PAN) is a polar polymer.   Thus, PAN adhered to the resin and healed the crack.   It has been revealed that this method is effective for th repair of the deteriorated composite.

REFERENCES

1. T.Nishino, Y.Tanaka, K.Nakamae, X-ray Studies on the Stress Transfer in High Performance Polyethylene Reinforced Composite -In the Direction Perpendicular to the Fiber Axis-, Jpn. J.Materials Sci.,vol.47, 1998, 293-298.

2. T.Nishino, Y.Tanaka, K.Nakamae, X-ray Studies on the Stress Transfer in High Performance Polyethylene Reinforced Composite -In the Direction Parallel to the Fiber Axis-, Jpn.J.Materials Sci.,vol.47, 1998, 1083-1087.

3. T. Nishino, H. Naito, K. Nakamura, and K. Nakamae, X-ray Diffraction Studies on the Stress Transfer of Transversely Loaded Carbon Fibre Reinforced Composite, Composites, Part A, vol.31, 2000, 1225-1260.

4. T. Nishino, D.Hirokane, K. Nakamae, X-ray Diffraction Studies on the Environmental Deterioration of Transversely Loaded Carbon Fibre Reinforced Composite, Composite Science and Technology, vol.61, 2001, 2455-2459.
5. T. Nishino, K. Nakamura, and K. Nakamae, Fatigue Behaviour of Poly(p-phenylene benzobisoxazole) Fibre Reinforced Composite by X-ray Diffraction, Proceedings of the 15 the Annual Meeting of the Society of Polymer Processing, Published in CD-ROM (8 pags), 1999, Den Bosch, The Netherlands
6. K.Nakamae, T.Nishino, Xu Airu, T.Matsumoto, and T.Miyamoto Studies on Mechanical Properties of Polymer Composites by X-ray Diffraction I.Residual Stress in Epoxy Resin by X-ray Diffraction., *J.Appl.Polym.Sci.*, vol.40, 1990, 2231-2238 .
7. T.Nishino, Xu Airu, T.Matsumoto, K.Matsumoto, and K.Nakamae, Residual Stress in Particulate Epoxy Resin by X-ray Diffraction, *J.Appl.Polym.Sci.*, vol.45, 1992, 1239-1244 .
8. K.Nakamae, T.Nishino, and Xu Airu, Studies on Mechanical Properties of Polymer Composites by X-ray Diffraction:3.Mechanism of Stress Transmission in Particulate Epoxy Composite by X-ray Diffraction., *Polymer*, vol.33, 1992, 2720-2724 .
9. Xu Airu, T.Nishino, and K.Nakamae, Stress Transmission in Silica Particulate Epoxy Composite by X-ray Diffraction. *Polymer*, vol.33, 1992, 5167-5172.
10. R.Talreja, J.-A. E.Manson, Polymer Matrix Composites, Elsevier, Oxford, UK, 2001
11. W.R.Tyson, , G.J.Davies, A photoelastic study of the shear stresses associated with the transfer of stress during fiber reinforcement. *Brit.J.Appl.Phys.*, 16 ,1965, 199-205.
12. R.J.Young, Characterization of interfaces in polymers and composites using Raman spectroscopy, In *Polymer Surfaces and Interfaces II*, eds. W.J.Feast, H.SMunro, R.W. Richards, Wiley, Chichester,1993, 131-160.
13. E.S.Folias, On the prediction of failure at a fiber/matrix interface in a composite subjected to a transverse tensile load, *J.Comp.Mater.*, 25 , 1991, 869-886.
14. D.F.Adams, Inelastic analysis of a unidirectional composite subjected to transverse normal loading, *J.Comp.Mater.*, 4 , 1970, 310-328.
15. L.A.Pilato, M.J.Michno, , In *Advanced Composite Materials*, Springer-Verlag, Berlin, 1994.97-107.

MANUFACTURING METHODS

Investigations into the mechanism of electron-beam curing of an epoxy resin

Kenneth C. Cole and Minh-Tan Ton-That
National Research Council Canada, Industrial Materials Institute
75 De Mortagne Blvd., Boucherville, Quebec, Canada J4B 6Y4
kenneth.cole@nrc.ca; minh-tan.ton-that@nrc.ca

Andrew Johnston, Mehdi Hojjati, and Kim Valcourt
National Research Council Canada, Institute for Aerospace Research
Montreal Road, Ottawa, Ontario, Canada K1A 0R6
andrew.johnston@nrc.ca; mehdi.hojjati@nrc.ca

Vince J. Lopata
Acsion Industries Inc., Pinawa, Manitoba, Canada R0E 1L0
lopata@acsion.com

ABSTRACT

Electron-beam (E-beam) curing of composite materials has a number of advantages over conventional thermal curing, one of which is the fact that it is not necessary to heat the mould in order to achieve the cure. However, since the crosslinking process involves exothermic reactions governed by chemical kinetics, it is clear that temperature does play an important role. The object of this study was to develop a better understanding of the mechanism of the E-beam curing process and the effect of temperature. Samples of a typical E-beam epoxy resin system (Tactix 123 + 3 phr CD1012) were partially or fully cured under varying conditions and characterized by different techniques, including the monitoring of temperature during the irradiation, differential scanning calorimetry, infrared spectroscopy, and dynamic mechanical thermal analysis. The results indicate that the main E-beam reactions proceed reasonably fast at moderate temperatures (50–150°C), but that the process parameters like dose rate and total dose should be chosen in such a way as to ensure that the temperature developed internally is high enough to achieve sufficient curing in the time allowed. The results also suggest that E-beam irradiated samples contain "trapped" active sites that remain stable for a long time at room temperature. In the case of partially cured samples, these sites promote further crosslinking when the sample is subsequently heated above room temperature. With the proper choice of conditions, E-beam curing of the system investigated (Tactix 123 + CD1012) can give a product with a glass transition temperature around 190°C.

43

INTRODUCTION

Over the last several years electron-beam (E-beam, EB) curing of composite materials has received increasing attention as an interesting new process technology. Its main advantage over conventional thermal curing lies in the fact that crosslinking of the polymer matrix can be achieved in a much shorter time without the introduction of any external heat. This means that the potential for development of residual stresses upon cooling, as well as the cost of tooling and consumables, is considerably reduced. If desired, the initial temperature can be varied and the cure rate can be controlled by varying the dose rate. Furthermore, EB curing can be used to selectively cure specific sections of a part. As in all manufacturing processes, however, a good understanding of the chemical and physical principles involved is required in order to assure proper control and the production of consistently high-quality parts. In the case of epoxy resins, this understanding is not yet as highly developed for EB curing as it is for thermal curing. The object of the present study is therefore to advance our state of knowledge of EB curing, in particular the mechanism of the curing reaction. Such information is crucial to our work on the development of a model to treat the kinetics of the curing process (1).

Thermal cure of epoxy resin matrices is usually achieved by incorporating a "hardener" containing reactive groups like amine or anhydride that form crosslinks through chemical reaction with the epoxide rings. In the case of EB curing, the epoxy resins are in general the same ones used for thermal cure but the hardener is replaced by an initiator that is activated by the electron beam and initiates cationic homo-polymerization of the epoxy resin (2). Cationic curing of epoxies can also be achieved thermally (3-5). Although the mechanism is complex, the most important reactions are shown below. A typical initiator is a salt of a diaryliodonium cation Ar_2I^+ and a hexa-fluoroantimonate anion SbF_6^-. An electron beam or ultraviolet light excites it and causes it to react with neighbouring molecules to produce a superacid (e.g. $HSbF_6$):

$$In^+ \ X^- \xrightarrow{\ h\nu \ or \ e^-\ } H^+ \ X^- \ + \ \text{other products} \tag{1}$$

Initiator   *Superacid*

The superacid then protonates a neutral epoxide ring to produce activated monomer:

$$ \tag{2}$$

Epoxy Monomer   *Activated Monomer*

which in turn reacts with another epoxide ring to give an activated chain end:

$$ \tag{3}$$

The reaction then propagates through further addition of epoxide rings in this "activated chain end" mechanism. Other reactions can also occur, the most important of which involve reaction between an activated monomer molecule or an activated chain end and a hydroxyl group. It is obvious that the kinetics of these reactions and the structure of the molecular network formed can be quite different from those associated with thermal cure. Furthermore, since the epoxide ring opening is exothermic for EB cure as it is for thermal cure, the heat generated will produce a rise in temperature that will affect the propagation reaction rate. A reliable process model must take into account these effects. Since the information available is still rather limited, in order to obtain a better insight into the nature of the cure process we have undertaken a study of a number of samples of an E-beam epoxy resin system partially or fully cured under different conditions. The techniques used include monitoring of temperature during E-beam cure, differential scanning calorimetry, infrared spectroscopy, and dynamic mechanical thermal analysis. This paper describes the results and their interpretation.

EXPERIMENTAL

The resin system studied was based on Dow Chemical's epoxy resin Tactix 123, which is a diglycidyl ether of bisphenol A. To this was added 3 phr of Sartomer's SarCat CD1012 (a diaryliodonium hexafluoroantimonate) as initiator. For irradiation (done at Acsion Industries Inc.), the resin mixture was introduced into 1 mL polypropylene syringes (4.7 mm ID, 58 mm long). EB irradiation was done on Acsion's IMPELA 10 MeV linear accelerator. The main experiments involved a sort of "on-line calorimetry". A thermocouple was inserted into the resin in each syringe and the syringes were embedded in holes in a foam block designed to act as a thermal insulator so as to minimize heat losses and inasmuch as possible attain adiabatic conditions. The block and samples were irradiated in a single pass in which the dose rate and total dose were varied by changing the velocity of the conveyor belt used to pass the samples under the beam. In order to obtain as much information as possible, these experiments were done with deliberately low doses (nominally 4, 5, 6, and 8 kGy) so as to achieve only partial cure. A typical experiment involved eight syringes, four containing uncured resin and four in which the resin had been fully cured by E-beam previously. These precured specimens served as reference samples to monitor temperature changes arising solely from absorption of the high-energy electrons. In addition to these "calorimetry" samples, which are designated "Set 1" and included a nonirradiated specimen, some highly cured specimens ("Set 2") were prepared with (i) 6 passes of 25 kGy E-beam, (ii) 100 kGy of X-ray, and (iii) 150 kGy of γ-ray irradiation.

The samples were analyzed by Fourier transform infrared (FT-IR) spectroscopy, differential scanning calorimetry (DSC), and dynamic mechanical thermal analysis (DMTA). The FT-IR spectra were acquired on a Nicolet Magna 860 instrument with a resolution of 4 cm^{-1} by means of the attenuated total reflection (ATR) technique. The DSC curves were measured on a Perkin-Elmer DSC7 instrument with the following temperature program: heat from 30°C to 280°C at 20°C·min^{-1} ("first scan"), hold at

280°C for 5 min, cool to 50°C at 20°C·min⁻¹, reheat to 280°C at 20°C·min⁻¹ ("second scan"). DMTA data were obtained on a Polymer Laboratories instrument in single cantilever bending mode at a frequency of 1 Hz, with specimens about 25 mm x 4 mm x 1 mm. The temperature program used was: heat from 30°C to 280°C at 2°C·min⁻¹ ("first scan"), cool to 30°, reheat same specimen to 280°C at 2°C·min⁻¹ ("second scan").

RESULTS AND DISCUSSION

The designation of the different samples is given in Table 1 along with some other information to be discussed later.

Table 1 – Sample Description and Relevant Data

Sample	Description	Conveyor Velocity (cm·s⁻¹)	Measured Dose* (kGy)	Temp. Rise** (°C)
EB-0	Nonirradiated	—	—	—
EB-4	E-beam ~ 4 kGy	0.48	4.25	3.4
EB-5	E-beam ~ 5 kGy	0.38	4.68	4.0
EB-6	E-beam ~ 6 kGy	0.30	5.72	4.6
EB-8	E-beam ~ 8 kGy	0.22	7.72	6.4
EB-150	E-beam 6 x 25 kGy	—	—	—
XR-100	X-ray 100 kGy	—	—	—
GR-150	γ-ray 150 kGy	—	—	—

*Dosimeters placed on foam block and later read with FWT-100 Radiochromic Reader

**Temperature rise observed for precured reference specimens, corresponding to thermal electron absorption only.

Figure 1a shows the temperature evolution during the E-beam cure of the uncured "calorimetry" samples. The time scale has been adjusted so that $t = 5$ min corresponds to the point at which the samples are directly under the electron beam. The temperature evolution appears to involve three stages. First there is an initial rise of about 10-20°C while the sample is being irradiated (between $t = 4$ and $t = 6$ min). This is followed by a second rise whose rapidity and magnitude are very much dependent on the dose rate. (At 8 kGy the second stage cannot be distinguished from the first.) Finally there is the third stage, which corresponds to cooling as the resin loses heat to its surroundings more quickly than heat is evolved. Figure 1b shows an expanded view of the first stage and also includes data for the precured reference specimens. The latter show a rise in temperature as they pass under the beam but little change afterwards. The temperature rise can be fitted quite well by an error function, which is the integral of a Gaussian curve. This indicates that, as expected, the E-beam energy is spatially distributed about its centre with a Gaussian distribution. Despite noise in the data, by means of the curve fits it is possible to determine reasonably accurate values of the temperature rise ΔT for the different precured reference specimens. These are included in Table 1. The ΔT values correlate very well with the total dose as measured by dosimetry, and the proportionality constant corresponds to a specific heat capacity value of $C_p = 1.22$ $J \cdot g^{-1} \cdot K^{-1}$. This value is in excellent agreement with the value determined by DSC for the cured resin around 25°C. Thus the rise in temperature for the precured specimens can be explained entirely in terms of the energy contributed by the absorbed electrons, as expected. However the rise observed for the uncured specimens (in the first stage) is significantly higher than for the precured ones. The difference cannot be explained in terms of heat capacity, because uncured resin would be expected to have a higher C_p, which would result in a lower ΔT. Thus it appears that the first stage in Figure 1a involves some exothermic reaction, but that at lower dose rates the reaction slows somewhat after the sample leaves the beam only to resume speed as the degree of cure advances. One possible explanation is that the first stage involves mainly the production of superacid and the protonation of epoxide rings (Eqs. 1 and 2), but the chain extension reaction (Eq. 3) is kinetically slower and occurs mainly in the second stage.

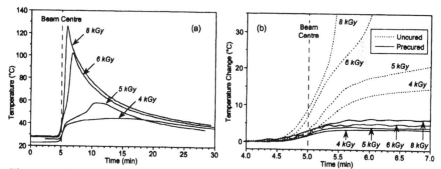

Figure 1 – Temperature evolution during E-beam cure at different doses: (a) uncured samples; (b) uncured and precured samples, during irradiation period only.

Figure 2 shows the infrared spectra of the different samples, including the nonirradiated resin mixture EB-0, after normalization with respect to the intensity of the aromatic ring band at 1509 cm^{-1}. The most significant change is the decrease of the epoxide ring band at 915 cm^{-1} upon curing. This decrease is accompanied by the growth of a broad band around 1170–960 cm^{-1} that can be assigned to stretching of the C–O–C ether linkages formed by reaction of the epoxide rings. By comparing the area of the epoxide peak to its area in the uncured resin spectrum it is possible to calculate the degree of cure. The spectra show that the "calorimetry" samples are significantly undercured, as anticipated because of the low dose. However the samples that received a higher dose show only a very weak epoxide peak, indicating a high degree of cure.

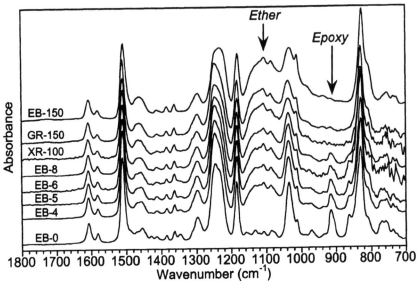

Figure 2 – Infrared spectra of the various samples.

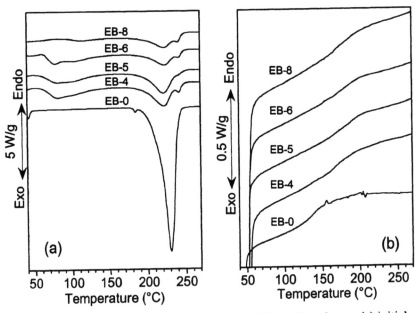

Figure 3 – DSC results for specimens subjected to different low doses: (a) initial scan; (b) repeat scan on same specimen after cooling.

The DSC results for Sample Set 1, given in Figure 3, show a number of interesting features. In the case of nonirradiated resin (EB-0), the initial scan shows a strong exothermic curing peak at 231°C with enthalpy $\Delta H = 574 \, J \cdot g^{-1}$. This is preceded by a small exothermic peak at 184°C. It is important to note that a DSC scan of neat Tactix 123 (i.e. without initiator) from 30 to 290°C showed no significant exotherm. Thus the most likely explanation of the nonirradiated resin result is that the small peak at 184°C corresponds to thermal decomposition of the initiator (Eq. 1), which then leads to epoxy polymerization (Eqs. 2 and 3). The irradiated (partially cured) samples show much more complex DSC curves, with four apparent exothermic peaks: two at rather low temperature (maxima around 81°C and 102°C) and two at high temperature (maxima at 222°C and 242°C). Infrared spectra on specimens removed from the DSC at the 160°C point and at the 280°C point confirmed that both the low-temperature and high-temperature exotherms correspond to loss of epoxide rings. It appears that in these low-dose experiments, the temperature-time profile during the EB cure process results in only partial cure. However, active cationic species remain trapped in the sample and on further heating to only 50–120°C in the DSC the reaction continues. The two peaks in this range could correspond to reaction of activated chain ends with epoxy and hydroxyl groups respectively. The activated species appear to be quite stable, given that the samples passed several weeks at room temperature between the irradiation and the DSC experiments. It also appears, however, that these species are not sufficient to produce complete cure at low temperature, because all the samples show a significant high-temperature exotherm as well. This could be due to further reaction induced by thermal decomposition of residual initiator, and again the two peaks observed could be due to reactions with epoxy and hydroxyl groups respectively. It is interesting to note that for the 8 kGy sample, where the temperature briefly reached almost 130°C during the irradiation experiment, the low-temperature exotherm is quite weak in the DSC curve. The degree of cure was calculated for the different samples by comparing the total area under the exothermic peaks to that for the nonirradiated sample. The results are given in Table 2 along with values calculated from the IR spectra, and the two sets are seen to be in reasonably good agreement. The degree of cure ranges from about 0.5 at a dose of 4 kGy to about 0.7 at 8 kGy. It is interesting to note that for all the irradiated samples the high-temperature exotherm in the DSC accounts for 20–30% of the overall cure. After the samples are fully cured in the first DSC scan, the second scan (Figure 3b) shows only the glass transition. Here again an interesting effect is observed. For the irradiated samples the ultimate glass transition temperature T_g is always near 185°C, but for the nonirradiated sample it is only 138°C. It may be that when the cure is rapidly achieved entirely thermally above 200°C the network structure is different.

Figure 4 shows the DMTA tan δ results for Sample Set 1. In general they corroborate the DSC results. The first scan (Figure 4a) shows two distinct loss tangent peaks that correlate with the low-temperature and high-temperature curing exotherms in the DSC curves. The correlation is not exact because of the quite different scan rates ($2°C \cdot min^{-1}$ for DMTA, $20°C \cdot min^{-1}$ for DSC) and the fact that DSC measures thermal response while DMTA measures mechanical response. The first loss peak moves to higher temperature and decreases in magnitude as the dose is increased. The second is fairly constant at about 200°C. After complete cure during the first DMTA scan, the second scan gives a single tan δ peak corresponding to the glass transition, which occurs at about 190°C in all cases (Figure 4b).

Table 2 – Degree of Cure for the Various Samples

Sample	Description	Degree of Cure (FT-IR)	Degree of Cure (DSC)
EB-0	Nonirradiated	0	0
EB-4	E-beam ~ 4 kGy	0.49	0.48, 0.55*
EB-5	E-beam ~ 5 kGy	0.49	0.55, 0.55*
EB-6	E-beam ~ 6 kGy	0.58	0.57, 0.60*
EB-8	E-beam ~ 8 kGy	0.66	0.73, 0.68*
EB-150	E-beam 6 x 25 kGy	~ 1	0.94
XR-100	X-ray 100 kGy	~ 1	0.96
GR-150	γ-ray 150 kGy	0.94	0.89

*Results for two different specimens.

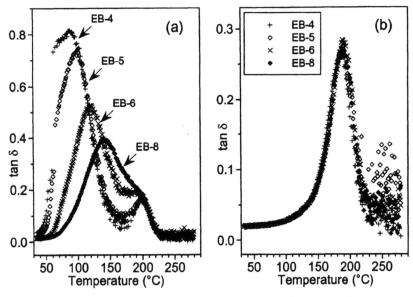

Figure 4 – DMTA tan δ results for specimens subjected to different low doses: (a) initial scan; (b) repeat scan on same specimen after cooling.

Figure 5 gives the DSC and DMTA results for the high-dose Sample Set 2. These are all highly but not completely cured. The initial DSC scans show weak low-temperature exotherms, matched by weak loss tangent peaks (indicated by an asterisk) in the DMTA scans, that are absent in the second scan. A very small amount of cure may also be seen near 200°C in the DSC scan. The initial degree of cure (Table 2) is greater than 90% in all cases. Again, the ultimate T_g appears to be somewhat sensitive to the nature of the curing, since it ranges from 177°C for the γ-ray sample to 183°C for the X-ray sample to 195°C for the E-beam sample. (These values correspond to the tan δ maximum in the second DMTA scans.) Another interesting point is that low-temperature curing is observed in spite of the fact that these samples are already highly cured and would thus be expected to have a high T_g. This suggests that the trapped active species are close enough to epoxy rings so that diffusion control does not prevent them from reacting once the temperature is kinetically favourable. There are thus several questions raised by these results that will require further clarification.

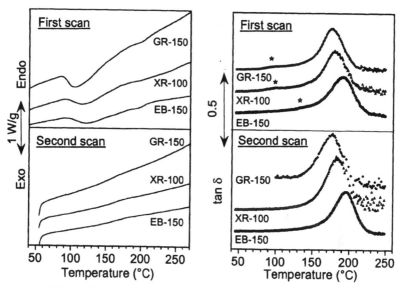

Figure 5 – DSC results (left) and DMTA results (right) for more highly cured samples.

CONCLUSIONS

The results obtained provide some new insight into the complex nature of the E-beam curing of Tactix 123 containing 3 phr CD1012 initiator. In particular, several specific points can be made. Firstly, although the electrons are responsible for exciting the initiator, the subsequent cationic epoxy reactions nevertheless appear to be governed by normal chemical kinetics. Hence temperature plays an important role and the rate of reaction advancement or degree of cure will depend on the particular combination of dose rate, temperature, and time. Secondly, when samples only partially cured by EB radiation are subsequently reheated, as in a DSC experiment, completion of the curing is rather complex, apparently involving at least four reactions. Thirdly, two of these reactions occur at rather low temperature (in the range 50 to 150°C) and are believed to involve cationic species "trapped" in the sample. These species appear to be stable for long periods of time at room temperature. Since these reactions occur even in highly cured samples, they do not appear to be strongly hindered by diffusion control limitations. This may be because at least some of the trapped species are relatively close to unreacted epoxy groups. Fourthly, these "low-temperature" reactions are not always sufficient to produce complete cure. In such cases this can be achieved by going to higher temperatures (above 200°C) where the remaining two reactions become important. The most likely explanation for this is that the higher temperature thermally activates residual initiator that has not already been activated by electrons. Another possibility is that it is necessary to exceed the ultimate glass transition temperature of about 190°C in order to overcome diffusion control limitations for trapped species that are not close to epoxy groups. Fifthly and finally, the glass transition temperature of the fully cured resin depends on the means used to achieve that cure. This is believed to be related to the different reactions involved in building up the molecular network. Samples that have been at least partly cured by E-beam tend to give the highest T_g values, around 190°C.

Further studies are required to obtain a clearer understanding of these phenomena.

REFERENCES

1. A. Johnston, K. C. Cole, M. Hojjati, M. Mascioni, G. R. Palmese, and V. J. Lopata, "A Basic Process Model for EB Curing of Composite Materials", Int. SAMPE Symp./Exhib., **47**, 874-888 (2002).

2. J. V. Crivello, M.-X. Fan, and D.-S. Bi, "The Electron-Beam-Induced Cationic Polymerization of Epoxy Resins", J. Appl. Polym. Sci., 1992, **44**, 9-16.

3. L. Matejka, P. Chabanne, L. Tighzert, and J.-P. Pascault, "Cationic Polymerization of Diglycidyl Ether of Bisphenol A", J. Polym. Sci. A: Polym. Chem., 1994, **32**, 1447-1458.

4. L. Matejka, K. Dušek, P. Chabanne, and J.-P. Pascault, "Cationic Polymerization of Diglycidyl Ether of Bisphenol A. II. Theory", J. Polym. Sci. A: Polym. Chem., 1997, **35**, 651-663.

5. L. Matejka, K. Dušek, P. Chabanne, and J.-P. Pascault, "Cationic Polymerization of Diglycidyl Ether of Bisphenol A. III. Comparison of the Theory with Experiment", J. Polym. Sci. A: Polym. Chem., 1997, **35**, 665-672.

Numerical simulation of resin flow under flexible cover in vacuum assisted resin infusion

Laurent Joubaud and François Trochu
Department of Mechanical Engineering, Applied Research Centre on Polymers (CRASP), Ecole Polytechnique, C.P. 6079, Station « Centre-Ville », Montreal, H3C 3A7, Quebec.

Jérôme Le Corvec
Kaizen Technologies, 75 de Mortagne blvd., Boucherville, J4B 6Y4, Quebec.

ABSTRACT

In this paper, techniques to simulate resin infusion using classical RTM simulation software have been investigated. The difference in the filling behavior between "rigid" and "flexible" moulds is evaluated and explained. A model describing the evolution of permeability with pressure is developed for flexible moulds. This model takes into account the changes in thickness of the cavity following deformations of the mould cover, as well as the compressibility of the reinforcement. The model will first be validated by simulations for a complex industrial part and numerical results will be discussed.

INTRODUCTION

Resin Transfer Moulding (RTM) has become a widely used process to manufacture glass-reinforced composites. A good technical description of the basic issues related to RTM manufacturing can be found in the reference book of Cauchois [1]. In this process, a stack of dry fibrous reinforcements is placed in the cavity of a rigid mould and the resin is injected at low pressure. The stiffness of the mould is often a concern when manufacturing large parts with a high fiber volume content. The filling time is increased significantly and sometimes it is not possible to inject the part.

The liquid composite moulding process known as *Vacuum Assisted Resin Infusion* (VARI), first introduced by Marco [2], enables to manufacture with success large parts at a relatively low cost. In this process, the stack of dry fibrous reinforcement is placed between a stiff mould half and a plastic bag. The resin is injected by gravity after partial or total vacuum has been achieved in the cavity containing the reinforcement. Williams et al. [3] gives an interesting review on the main developments in this field. Many variants of this method have been developed, the *Seeman's Composite Resin infusion Moulding Process* (SCRIMP) [4] being the most famous. The special optimized approach that will be studied here was developed by Kaizen Technologies under the name *Kaizen Infusion system* (K.I.S.). These techniques offer the advantages of low cost tooling and allow to reduce significantly the filling time compared to classical RTM.

However, the behavior of the resin when it impregnates the fabric is not fully understood yet and the injection strategy used in many applications is not always optimal. Problems of micro-porosity, irregular thickness and defects have been observed locally in parts manufactured by resin infusion. Although this is not a concern in the marine industry, the main field today of application of resin infusion, it is an important issue in aerospace applications for certification. Numerical simulation of resin injection can assist in positioning the inlet ports and vacuum intakes, especially for large and complex parts. Optimal injection strategies can be studied on the virtual model prior to prototype testing, hence helping to reduce process set-up costs.

The goal of this investigation is to verify if sufficiently accurate simulations of resin infusion can be performed using numerical tools originally developed for classical RTM, i.e., to simulate the injection of resin in rigid moulds. Corden and Rudd [9] have also simulated resin infusion by using, in order to predict the resin flow, a permeability measured in a flexible mould that reproduces the actual processing conditions. This paper follows a different approach: a model based on the compressibility of the fibrous reinforcement predicts the permeability that will be used in the numerical simulation. We begin by recalling Darcy's equation which governs the resin flow. Then we present the experimental results on permeability and compressibility, that are used in the prediction of permeability. Finally, numerical results for an ambulance roof are compared with the industrial process.

Analysis of resin infusion using RTM flow simulation

Recall on RTM simulation

The impregnation of a fibrous preform is usually modeled as a flow through porous media. The low resin velocity allows to use Darcy's law, which gives the flow rate per unit area -often called Darcy's velocity - in function of the pressure gradient:

$$v = -\frac{K}{\mu} \nabla P \qquad (1)$$

where K is the permeability tensor, μ the fluid viscosity, and P the pore pressure. Combining Eq. 1 with the continuity equation gives the equation that governs the porous flow:

$$\nabla \cdot \left(\frac{K}{\mu} \nabla P \right) = 0 \qquad (2)$$

The numerical software LCMFlot will be used here to solve the above equation at each time step by the finite element method [5]. Then a filling algorithm is used to displace the flow front to its new position at the next time step.

The permeability of a preform depends of several factors, the main one being its porosity ω. Usually, experimental measurements are required to obtain the values of this key parameter [6]. The most commonly used empirical model to describe permeability in function of fiber volume fraction is exponential:

$$K = A_1 V_f^{b_1} \qquad (3)$$

where $V_f = 1 - \omega$ is the fiber volume content. The parameters A_1 and b_1 are determined experimentally for each reinforcement.

Main differences between resin infusion and RTM

Because of the flexibility of the plastic bag used in resin infusion, the porosity of the reinforcement and hence permeability, depends on the level of vacuum achieved in the cavity of the mould. For this reason, the permeability measured in a stiff mould can no longer be used to simulate this process. A way to avoid this problem consists in using a modified value of permeability in order to reflect the flexibility of the mould. The main idea of this investigation is to model a "*flexible mould*" permeability.

EXPERIMENTAL

In order to modify the permeability according to the flexibility of the plastic bag and the vacuum level achieved in the cavity, a simple approach is to derive the "flexible mould" permeability of the fabric from "rigid mould" measurements. In this work, this method has been used to simulate the infusion of a part of an ambulance roof. The fabric used was the "Multimat" from Syncoglass, Belgium, which is a knitted fabric supposed to have a high permeability together with a good drapability. Luo [7] studied the permeability characteristics of the Multimat. The results of Table 1 show that the fabric is almost isotropic.   Therefore, additional permeability measurements were only performed in the stitch direction. Firstly, the "rigid mould" permeability was measured using a rigid rectangular mould and following the unidirectional method proposed by Ferland et al. [6].

Luo [7] employed a radial flow test method to measure the permeability of the Multimat. The permeability K_1 and K_2 in the principal material directions (stitch and cross direction) were measured for various fibre volume contents. The angle θ between the principal direction and the laboratory axis was also measured. The results can be found in Tables 1 and 2:

V_f (%)	K_1 (10^{-10} m^2)	K_2 (10^{-10} m^2)	θ (°)	K (10^{-10} m^2)
17.1	18.70	17.30	-12	18.00
26.0	4.33	3.99	-5	4.16
34.2	1.66	1.39	10	1.52
38.3	1.11	1.00	-15	1.05

Table 1: "Rigid-mould" permeability of the Multimat measured by Luo [7].

where K is an average of K_1 and K_2 and K_{int} is an interpolated value obtained from Luo's measurements.

V_f (%)	K (10^{-10} m^2)	K_{int} (10^{-10} m^2)
20.6	10.6	9.33
25.2	4.7	4.57
31.3	1.59	2.12

Table 2: Comparison between unidirectional measurements
and interpolated values from Luo's results.

The interpolated results (Fig. 1) show a good agreement between the rigid-mould permeability measured by Luo using radial injections method and the results obtained by the unidirectional method.

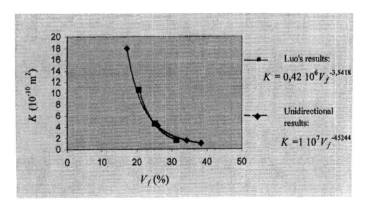

Figure 1: Experimental curves of permeability in function of fibre volume fraction.

A simple model for a "flexible mould" permeability

We assume that resin injection through fibrous reinforcement is a slow process that can be modeled by a quasi static approximation, i.e., at each time step the plastic bag is in equilibrium between the atmospheric pressure P_0 and the effective stress of the compressed fabric and the fluid in the cavity (Fig. 2).

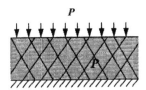

Figure 2: The equilibrium state of the plastic bag.

A porous medium follows Terzaghi's law, which states that the total stress σ is decomposed into the effective stress σ' that acts on the fabric and the pore pressure of the fluid:

$$\sigma_{ij} = \sigma_{ij}' - P \delta_{ij} \tag{4}$$

The compressibility of fabrics was studied by many authors and by Robitaille and Gauvin [8], who proposed an empirical model for the compaction pressure in function of the fibre volume content:

$$V_f = A_2 P^{b_2} \tag{5}$$

where P is the pressure applied on a fabric sample, A_2 and b_2 are experimental parameters depending on the particular fabric used. Combining Eq. 4 and Eq. 5 leads to:

$$V_f = A_2 \left(P_0 - P_{vac} \right)^{b_2} \qquad (6)$$

where P_{vac} is the vacuum level achieved in the cavity and P_0 is the atmospheric pressure. Introducing Eq. 6 in Eq. 3, we can express the flexible-mould permeability in function of the vacuum level achieved in the mould:

$$K_{eff} = A_1 A_2^{b_1} \left(P_0 - P_{vac} \right)^{b_2 b_1} \qquad (7)$$

Luo [7] measured the compressibility of the Multimat and found $A_2 = 0.157$, $b_2 = 0.162$ when the compaction pressure P is expressed in kPa.

NUMERICAL SIMULATIONS

Numerical simulations using the flexible-mould permeability defined previously will now be carried out for an ambulance roof.   The ambulance roof of figure 3 is manufactured by the company Fibres Design, Chambly, Quebec, using a vacuum assisted resin infusion technique.

Description of the process

One layer of a "Multimat" reinforcement, is placed between a stiff mould half and a plastic bag. The resin is injected through an injection line by gravity after a vacuum level of around 50 kPa has been achieved in the cavity containing the reinforcement. Note that this process, known as *Kaizen Infusion System* (K.I.S.), does not use any flow-enhancing layer, which is a main difference compared to classical S.C.R.I.M.P [4]. The resin used is a low viscosity polyester.

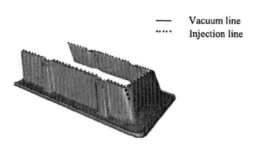

—   Vacuum line
····   Injection line

Figure 3: A part of an ambulance roof.

Numerical simulation of the process

A numerical simulation of the injection process using has been performed with LCMFlot. We used the "flexible-mould" permeability defined in the previous section. The numerical values of the parameters used for the simulations are available in Table 3:

K (m^2)	μ (cp)	V_f (%)	P_0 (kPa)	P_{vac} (kPa)	h (mm)
$2.17.10^{-10}$	250	29.7	101	50	2.5

Table 3: Numerical values of the parameters used in the simulations.

where h is the thickness of the part, T the temperature and the other parameters have been defined previously. The advancement of the flow front in time is displayed in Figure 4.

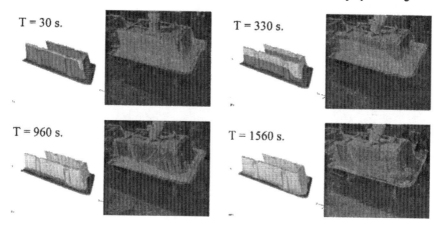

T = 30 s.   T = 330 s.

T = 960 s.   T = 1560 s.

Figure 4: Comparison between the simulation and the infusion process for the ambulance roof.

Figure 4 shows a good agreement between the simulation and the infusion process for the ambulance roof. Until around 1000 s, the flow fronts in the simulation and in the infusion are very close. After that time, the bottom right corner takes a long time to be filled because of the low pressure gradient driving the flow. As a matter of fact, the resin flows through the vent line located along the bottom edge of the part, instead of filling the reinforcement. Note the difficulties observed in Figure 4 to impregnate the corners of the part both in the simulation and in the industrial process.

CONCLUSION

A simple approach to simulate resin infusion was presented, based on the use of an equivalent "flexible mould" permeability to account for the deformation of the mould cover. This permeability can be derived from the real permeability and compressibility of the fabric. Numerical simulations for a complex industrial part were performed. The results showed reasonable agreement between the simulation and the real injection. Therefore, it is possible to simulate with a reasonable accuracy vacuum assisted resin infusion with a RTM software solving Darcy's equation at each calculation step. It is now necessary to perform flexible mould experiments for various vacuum levels and different fabrics, in order to establish the validity and limitations of this approach. For this purpose, an experimental procedure will be set up to measure the "flexible mould" permeability. On the other hand, during the vacuum infusion process, the thickness of the cavity varies, so permeability changes during the infusion process. This variation of permeability needs to be taken into account in order to further refine the model.

REFERENCES

1-   J.-P. Cauchois, R.T.M. Process, Éditions Syntech, 228 (1997).

2-   Marco Method, US Patent No 2495640 (1950).

3-   C. Williams, J. Summerscales and S. Grove, "Resin Infusion under Flexible Tooling: a Review.", Composites, **27A**, pp 517-524 (1996).

4-   W. H. Seaman, "Plastic Transfer Molding techniques for the production of fibre Reinforced plastic structures", US Patent No. 4902215, filed 30th March 1989.

5-   F. Trochu, R. Gauvin and D.-M. Gao, "Numerical Analysis of the Resin Transfer Molding Process by the Finite Element Method", Advances in Polymer Technology, **12**, pp 329-342 (1993).

6-   P. Ferland, D. Guittard and F. Trochu, "Concurrent Methods for Permeability Measurements in Resin Transfer Molding", Polymer Composites, **17**, pp 149-158 (1996).

7-   Y. Luo, "Resin Transfer Moulding of Knitted Fabric Reinforced Composites", PhD Thesis, Katholieke Univeristeit Leuven Belgium (2001).

8-   F. Robitaille and R. Gauvin, "Compaction of Textile Reinforcements for Composites Manufacturing. I: Review of Experimental Results", Polymer Composites, **19**, pp 198-216 (1998).

9-   T. J. Corden and C. D. Rudd, "Permeability Measurements And Modelling Techniques for Vacuum Infusion", Fifth International Conference on Automated Composites, ICAC 97, pp 231-242 (1997).

Die Design Procedure and its Application to SMC Moulding in Consideration of Rib Parts for Ground or Aquatic Vehicular Component

Prof. em. of Doshisya University   T. Hirai
3-9 Nanasegawa Miyamagi Kyo-tanabe shi   610-0313 Japan

ABSTRACT

When prepreg composite materials are used in compression moulding process, problems might arise due to heterogeneity of materials, caused by fibre orientation during flow. To overcome the problems occurring during the process, by improvement in moulding design and forming conditions, it should be necessary to investigate the flow characteristics of composite materials during unsteady forming process. Forming procedure in consideration on charging ratio might be selected by quantitative analyses de dependent on applied condition of components. For the flow of SMC Class-A, homogeneous orthogonal pseudo-plastic equations have been derived and the application on the surface and rigidity is developed by CAD system.

Keywords: SMC compression moulding, Low charge ratio, Sink mark, Snaking behaviour in rib parts , Symmetric or asymmetric flow pattern.

INTRODUCTION

Composite prepreg materials of SMC Class A used extensively today in manufacturing the outer panels and structural parts of transportation vehicles because of their lightweight and high degree of moulding freedom. Products moulded from fibre reinforced material are subject to several problems during the manufacturing process such as sink marks, warp, short shot shots irregularity of volume fraction caused by asymmetrical flow pattern and snaking orientation reinforcement at flange panel.

FORMULATION AND SOLUTION PROCEDURE

A typical Load-compression curve for SMC compression moulding is shown in Fig.1. To simplify the analysis, the filling process is considered   in three stages, that is, initial deformation, flow behaviour and final flow process as shown in Fig,2.

Composites prepreg materials are essentially heterogeneous structures and moulding them results in unstable deformation. The flow state during a press forming process can be analysed by prepreg materials are essentially heterogeneous

61

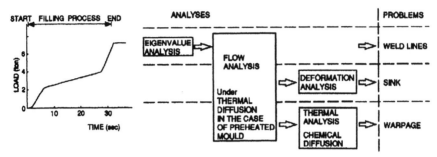

Fig.1 Typical al compressive

-load curve.   Fig.2. Analytical frame work. Composites
prepreg materials are essentially heterogeneous structures and moulding them results in
unstable deformation. The flow state during a press forming process can be analysed by
representing the material flow as being unsteady. For an unstable condition, a
Lagrangean description should be applied, but to simplify the analysis a progressive step-
by-step formulation method(1) is developed using an Eulerian

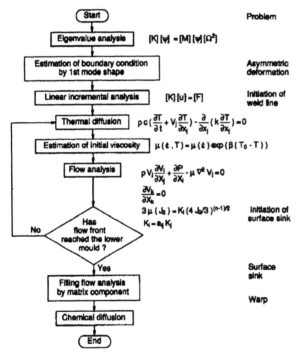

description. By minimising
energy, the eigenvalue and
linear incremental analysis
can be applied to predict
the progress of flow for
initial conditions.   The
numerical results reveal the
initiation of the weld line
which can lead to fracture
of composite structure. An
additional outcome of the
work is the development of
a hybrid moulding method,
which avoids this trouble
by employing laminates
composed of materials of
different stiffness.   The
progressive flow of a
composite during moulding
has been analyzed and

Fig. 3 Flow chart for compression moulding.

numerical results obtained here by the progressive step-by-step method are shown to be good agreement with experiment. The numerical procedure thus developed is shown in Fig.3.

A full-charge mode has been assumed (i.e., a 100% charge ratio) to simplify the analysis, and the dimensional configuration of rib is larger than in normal practice, due to the idealized homogenous characteristics of the material. In order to determine the flow characteristics of the composite, the ratio of the pseudo-plastic viscosities K_T / K_L is given a high value, to represent a highly anisotropy flow resistance. The practical case of forming a T-shaped component from blank is a typical practical case and a result is shown in Fig.4. The dots shown in the figure indicate where the reaction forces at the nodal points show stress-free conditions. The phenomenon indicates the possibility of micro-buckling in laminae at the surface.The asymmetry of dotted nodal points suggest a tendency for any subsequently formed weld line in the rib section to undulate or ' snake'. The method of weight residuals is used to solve the equations. Numerical results for flow states during at constant room temperature are shown in Fig. 4, as an example of unsteady flow. In the figures it is evident that the flow swing around the rib

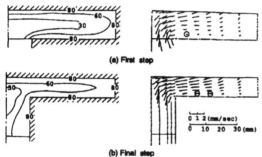

(a) First step

(b) Final step

Fig. 4 Temperature distribution and flow states
By the progressive step by step method.

Corner, as shown by ringed vectors. The flow state suggests that there is a resin rich region in the product. To prevent or minimise faults in the product, the thermal diffusion condition achieved by using preheated die reduced the tendency. (2)

A surface sink mark might be produced during the final stage of moulding as resin flows down into the lower corers of the rib through the oriented reinforcement composed by flow state, after the flow front reaches at the lower mould.

EFEFECT OF RIB POSITION AND CHARGE RATIO
ON THE FLOW   SYMMETRY INTO A RIB

SMC is used in the manufacture of automobile parts because of the advantage of light weight, corrosion resistance and design freedom. However, the problem of sink marks on a panel face on opposite side to the stiffening ribs must be overcome.

It might be possible to improve a sink mark fault by altering the charging mode and the position of SMC in the mould. In SMC moulding, the charge ratio is normally kept considerably lower than 100%, but this can lead to an asymmetric orientation of the reinforcement. Experimental results for an example of asymmetric filling with a 50% charge ratio are shown in Fig. 5. The volume fraction distribution of the reinforcement for symmetric and asymmetric flow into the rib channel are shown in Fig. 6. Since important requirement of a vehicle component are rigidity and strength, it might be more desirable to mould symmetrically, as indicated in Fig. 7 (3). The rigidity and strength of an asymmetrically moulded rib are very low, especially on the resin-rich side due to the stress concentration caused by asymmetric reinforcement.

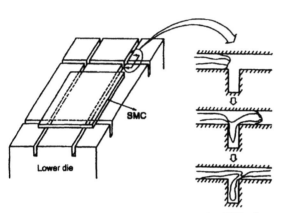

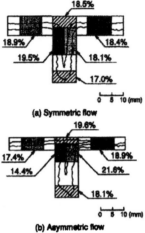

Fig.5. A symmetric flow into the rib part by 50% charge ratio

Fig.6. Volume fraction distribution

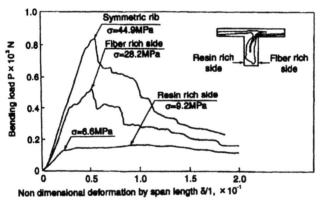

Fig.7   Relation between bending load   and deflection

UPPER BOUND APPROACH FOR THE BRANCHING FLOW AT RIBCHANNEL

The theoretical method should also be effective for dealing with composite prepreg   materials. It might have a tendency to apply a SMC compression moulding selecting smaller charge ratio of prepreg material in order to improve a surface quality overcoming   the faults in practice such as sink marks. The purpose of this investigation is to develop a design procedure obtaining high modulus without surface sink using CAD system to control charge patterns and design configurations of mould with ribs. When small charge ratio is applied in the process, the branching flow ought to rise up into sequent plain mould and perpendicular rib canals. It should be obliged to consider the kinematically admissible velocity by in-mould flow from the plane channel and distortion into rib parts. When the flow front states near the rib part, the solution for stays could be obtained by upper bound approach. Material flow behaves to resistance, considering the behaviour dependent on the incremental deformation caused by progressive punch strokes in the domain as shown in Fig.8. Considering admissible velocity field along the boundary,   the equilibrium equation   is supposed to hold the equilibrium   between the

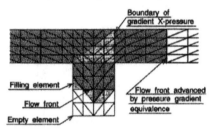

Fig.8 Analytical model at branching flow domain

surface traction of inlet from flow front and resistance into in-moulding plane channel and distributed to rib part in indicating the viscosity in both direction of the plane and orthogonal rib channels and the mechanical behaviour during the process in mouling as shown in Fig.9. The numerical results obtained the mode shaped by eigen value analysis for various mode shape of tangential flow

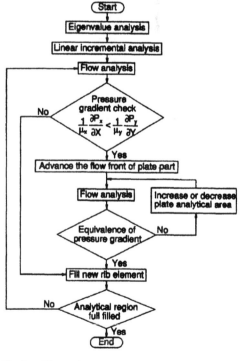

Fig.9.   Flow chart

behaviour into the rib channel suggest as similar experimental results on each adapted conditions. SMC Compression Moulding is used extensively today in manufacturing the panels and structural parts of transportation vehicles.

DESIGN PROCEDURE FOR THE RIB CONFIGURATION

The dependence of surface quality and rigidity on the flow pattern are shown in Fig.10. A poor surface quality is evident with symmetric flow, but the defects almost disappear with asymmetric flow. However, rigidity and strength are adversely affected by asymmetric flow, and it is necessary to devise a method which avoids this problem. Although surface quality can be improved by using a lower charging ratio, the rigidity and strength become very low due to the asymmetrical flow into the rib. Then die design procedure to SMC moulding using a full charge ratio might be took into service for an aquatic vehicular component like as water jet.

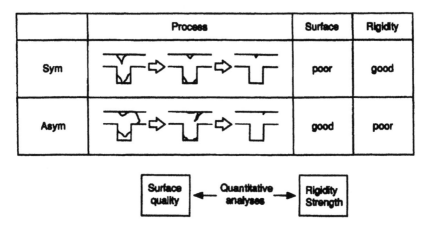

Fig. 10   Difference of quality by symmetric or asymmetric flow pattern

A theoretical optimum design to prevent sink marks with symmetric flow into the rib channel is achieved by installing a counter punch. This opposes the motion of flow front as shown in Fig. 11, and the technique is effective for preventing weld lines in the initial stage and sink marks in the final stage.

We developed a programmed control system for the back pressure as a function of the punch stroke. The restriction of a flow front by a counter punch depends on the filling state in the domain, and can be expressed by the reciprocal of the index of flow resistance, i.e., K_L/K_T.

The theoretical results obtained with flow-front moulding are shown in Fig .12.

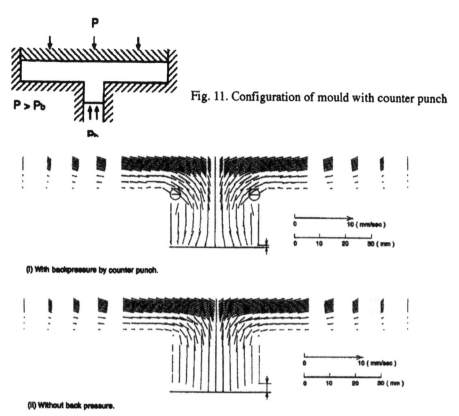

Fig. 11. Configuration of mould with counter punch

(I) With backpressure by counter punch.

(II) Without back pressure.

Fig. 12   Configuration of flow states with and without back pressure by a counter punch.

The results show the effect of optimising the control with a counter punch. The ringed horizontal velocity vectors in Fig. 12 at the entrance to the rib suggest that the flow swings around the rib corner.   The incidence of a sink mark and the effect   of swing around the rib corner can be expressed by the deviation factor of the flow front and the horizontal velocity vectors at the entrance to the rib, which is related to the potential value of the flow front. Typical results from an analysis using controlled back pressure are shown in Fig. 13. The factors are indicated by black and white bars in the figure as the deviation factor and the effect of swing, respectively, in relation to the relative values divided by the velocity of the punch stroke. The most satisfactory results were obtained for condition B2, which uses a stepped back pressure pattern, the surface roughness measurements across the face opposite the rib being shown in Fig. 14 (a) for this case. In

Fig.14 (b), which was the case of non-addition of back pressure, an apparent depth of sink mark can be measured. The results suggest that the application of back pressure is effective in improving surface quality.

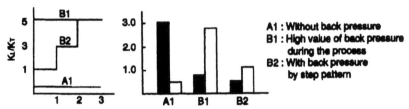

Fig. 13. Theoretical control system for preventing sink marks.

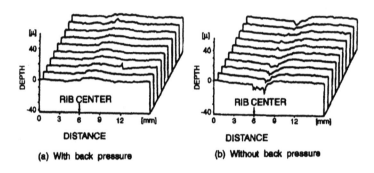

(a) With back pressure   (b) Without back pressure

Fig.14.   Surface roughness of a ribbed panel.

CONCLUSION

It should be arranged   a responsible moulding process by design   on surface quality or mechanical behaviour concerning transportation vehicles being considered the surroundings. The numerical results obtained here are in good agreement with the experiment values. It might be possible to improve sink marks and weld lines by suitably adjusting the configuration of the mould and SMC charging pattern, but a more effective method for preventing sink marks by applying back pressure in the rib part was developed.

REFERENCES

(1) T. Hirai, Rheology of Carbon Fiber Composite Prepreg Materials, Developments in Reinforced Plastics, 5,ed, by G. Pritchard, Elsevier Applied Science, London,(1978).
(2) V.P. Pervadchuk and V. Yankov, Heat Transfer-Soviet Research, 10. 1. 11 (1978)

Inverse-manufacturing Technology for Polymer Based Composite Materials

Kiyoshi Kemmochi*, Masayuki Sakurai*, Limin Bao* and Mikiya Ohara*
*Faculty of Textile Science and Technology, Shinshu University
3-15-1, Tokida, Ueda, Nagano 386-8567, Japan
kemm@giptc.shinshu-u.ac.jp

ABSTRACT

Advanced composite materials (ACM) such as carbon fiber reinforced plastics (CFRP) have a high specific strength and rigidity and advantage of being able to be designed to meet specific requirements. ACMs have been extensively developed and applied to a wide range of industrial fields including aeronautics and astronautics, ship, vehicles, civil engineering and architecture, sports and recreational articles.

Off-setting these various advantages, ACMs also have detrimental which effects may be harmful to the global ecological system after the end of product life. Some experts point out that, under present circumstances, this may produce more man-made garbage which is hard to incinerate or recycle and finally result in devastation of the ecological system.

A recent trend of research and development of industrial materials and products, the so-called Inverse Manufacturing, reuse, reduce and recycling technology for discarded products, consequently making promote saving resource and energy without consumption of natural resources.

This paper places special emphasis on original composites compatible with the requirements for recycling as an Inverse-manufacturing. A concept of environmentally consciousness, smart composites, complex materials and repairing technology for infrastructure using fiber reinforced laminates are able to satisfy various functional requirement and impose fewer hazards for the global environmental system is introduced.

INTRODUCTION

Global warming and desertification attributed to mass consumption and the waste of energy and products have become serious problems. The preservation of the earth's environment and natural resources is pertinent to the survival of all human beings. Recently, studies on environmentally conscious composite materials have received considerable attention. With the development of production technology and improved production methods, wood composites can be manufactured by simple processes. Various wood composites, such as those reinforced with fiber and plastics, are currently being studied(Kawai 97).

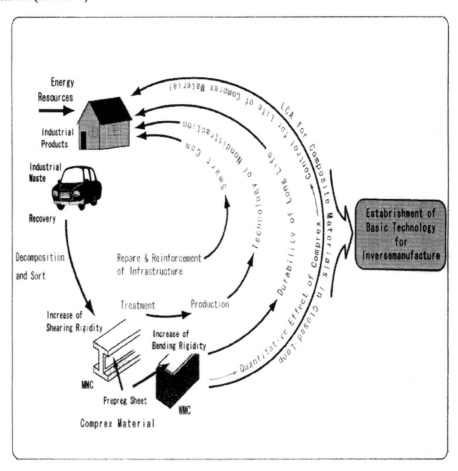

Figure 1. *Concept of Inverse-manufacturing System.*

As new materials and products are developed, it is very important to investigate and predict their mechanical properties. In this study, a composite structure composed of a small amount of unidirectional (UD) carbon fiber-reinforced plastic and wood was manufactured in order to improve the performance of wood and utilize the used wood, and mechanical properties were evaluated with the use of tensile and bending tests. The effect of the ply orientation and thickness ratio of a UD layer on the rigidity of wood composites were investigated with the use of the laminated plate theory and rule-of-mixtures.

EXPERIMENTAL PROSEDURE

Materials used

Vertically sawn western hemlock (*Tsuga heterophylla* Sarg., specific gravity   0.43 in dry air) is used for wood. Prepreg,P2053-15 (carbon fiber T800H, epoxy resin 2500, weight percent 30 of resin) produced by Toray Co., Ltd., Japan, was used. It adheres to a 31 cm ×31 cm wooden board at a temperature of 170℃, a pressure of 0.5 MPa, and a holding time of 90 min. Specimens were cut out from board by using a numerically controlled router. The tensile specimens shown in Fig.2 were processed according to the Japanese Industrial Standard Z2112 scaled down to 290mm from 390mm. Table 1 lists the specimen dimensions under the tensile and bending tests. Specimens of T and B, shown in Table 1, are the specimens without the carbon fiber-reinforced plastic. The number of wood composite specimens was five, whereas the numbers of Specimens T and B were 15 and 10, respectively. Tensile and three-point bending tests were carried out with an Instron testing machine. For the three-point bending test shown in Fig. 3, the distance between supporting noses was 240mm, and specimens were loaded at a loading rate of 10mm/min. The elastic constants of the UD layer and wood are shown in Table 2.

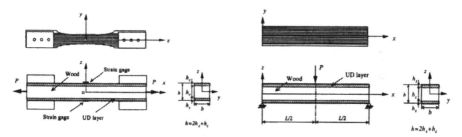

Figure 2. *A schematic diagram of a tensile test.*

Figure 3. *A schematic diagram of a three-point bending test.*

The shear modulus of the UD layer was calculated from Hayashi's equation for anisotropic plates. The bending modulus of wood was obtained from the deflection caused by the bending moment obtained by reducing shear deformation. Because the transverse Young's modulus and the shear modulus of wood were not dominant, they are assumed to be one-twentieth (Sawada 70) of the longitudinal Young's modulus. Poisson's ratio of wood was assumed to be 0.4 (Sawada 70).

Table 1. *Specimen dimensions of wood composites.*

Kinds of loading	Specimen No.	Number of plies	UD layer thickness (mm)	Thickness (mm)	Width (mm)
	T	0	-	15	5
Tensile	TP1	1	0.137		
	TP2	3	0.412	15	5
	TP3	5	0.686		
	B	0	-	17	17
Bending	BP1	1	0.134		
	BP2	3	0.424	17	17
	BP3	5	0.695		

Table 2. *Elastic constants of UD layer and wood.*

Properties	UD layer Tensile	Wood Tensile	Wood Bending
E_x (GPa)	162.0	11.4	12.5
E_y (GPa)	8.81	0.57	0.69
E_{xy} (GPa)	4.57	0.57	0.69
ν_x	0.332	0.40	0.40

RESULTS AND DISCUSSION

Calculation of tensile and bending rigidities of wood composites

Consider wood composites of a UD layer thickness h_S, wood thickness h_C, and width b, as illustrated in Figs.2 and 3. Wood composites were subjected to tensile and three-point bending load as shown in Figs.2 and 3, respectively. The coordinate system is also shown in Figs.2 and 3. The tensile and bending specimens were symmetrical with respect to the x-axis. In the analysis, the principal material directions of the wood coincided with the x- and y-axes and were fixed, whereas the principal material directions of UD layer varied. The effects of the ply orientation and thickness ratio of the UD layer on the rigidity of wood composites were investigated with the use of the laminated plate theory and the rule-of-mixtures.

Laminated plate theory (Tsai et al., 1980)

Based on Hooke's law, the force tensor $\{N\}$ and the moment tensor $\{M\}$ of a wood composite can be written as

$$\left\{ {\{N\} \atop \{M\}} \right\}^{(k)} = \left[{[A] \mid [B] \atop [B] \mid [D]} \right]^{(k)} \left\{ {\{\varepsilon^0\} \atop \{\kappa\}} \right\}^{(k)} \qquad [1]$$

For a symmetrical wood composite, the matrix elements of A_{ij}, B_{ij}, and D_{ij}, respectively, are expressed as follows

$$A_{ij} = 2h_s \left\{Q'_{ij}\right\}_s + h_c \left\{Q'_{ij}\right\}_c = h\left\{\bar{Q}_{ij}\right\}$$

$$B_{ij} = 0,$$

[2]

$$D_{ij} = I_s \left\{Q'_{ij}\right\}_s + I_c \left\{Q'_{ij}\right\}_c,$$

where I_s and I_c are the geometrical moments of inertia of the UD layers and wood, respectively. $\left\{Q'_{ij}\right\}_s$ and $\left\{Q'_{ij}\right\}_c$ are the off-axis modulus components for the UD layers and wood, respectively. $\left\{\bar{Q}_{ij}\right\}$ is the mean modulus component.

Considering that the wood composites are subjected to tensile load as shown in Fig.2, Young's modulus $E_x(\theta)$ in the x-axis direction can be obtained by $E_x(\theta) = \bar{\sigma}_x / \varepsilon_x^0$. We get

$$E_x(\theta) = \left(\bar{Q}_{11}\bar{Q}_{22}\bar{Q}_{66} + 2\bar{Q}_{12}\bar{Q}_{16}\bar{Q}_{26} - \bar{Q}_{11}\bar{Q}_{26}^2 - \bar{Q}_{22}\bar{Q}_{16}^2 - \bar{Q}_{66}\bar{Q}_{12}^2\right)$$

$$/\left(\bar{Q}_{22}\bar{Q}_{66} - \bar{Q}_{26}^2\right)$$

[3]

For a beam under three-point bending with a span of L as shown in Fig. 3, the deflection and strain at the center of a beam can be given by

$$\delta(\theta) = \frac{\beta_x PL^3}{48}, \quad \varepsilon(\theta) = \frac{\beta_x PLh}{8},$$

[4]

where

$$\beta_x = \left(D_{22}D_{66} - D_{26}^2\right)/\Delta,$$

$$\Delta = D_{11}D_{22}D_{66} + 2D_{12}D_{26}D_{16} - D_{12}^2 D_{66} - D_{26}^2 D_{11} - D_{16}^2 D_{22}.$$

[5]

Rule-of-mixtures.

The tensile modulus is given by

$$E_x(\theta) = \frac{2h_s E_s + h_c E_c}{2h_s + h_c},$$

[6]

where E_S is the modulus of UD layer, written as

$$\frac{1}{E_s} = \frac{\cos^4 \theta}{E_x} + \left(\frac{1}{G_{xy}} - \frac{2v_x}{E_x}\right)\sin^2\theta\cos^2\theta + \frac{\sin^4\theta}{E_y},$$

[7]

and E_C is the modulus of wood in the principal material direction.

For the wood composite beam, the deflection and strain at the center of the beam can be given by

$$\delta(\theta) = \frac{PL^3}{48EI}, \quad \varepsilon(\theta) = \frac{hPL}{8EI}$$

[8]

Relation between the observed deflection and strain and the calculated ones.

Table 3 shows the mean experimental Young's modulus and the calculated one based on the laminated plate theory under the tensile load. The coefficients of variance are also shown in Table 3. For the tensile test, Young's moduli of the wood composites increased when the thickness of the UD layers was increased. The moduli and strength of reliability of the wood composites could be largely improved by using only a little of the fiber-reinforced plastic. For the tensile property, Young's modulus increased by 56% when the wood was substituted for a UD layer of only 1.8 %.

Table 3. *Comparison of the experimental Young's modulus with the calculated one.*

Specimen No.	E_{exp}		E_{cal} (GPa)	E_{exp}/E_{cal}
	Mean (GPa)	C.V.* (%)		
T	11.4	19.5	-	-
TP1	17.8	18.1	14.2	1.25
TP2	23.3	12.6	21.7	1.07
TP3	26.6	9.6	25.2	1.06

* C.V.: Coefficient of variance.

Table 4. *Experimental deflections caused by the bending moment at a bending load of 1kN compared with the calculated ones.*

Specimen No.	δ_{exp}		δ_{cal} (mm)	$\dfrac{\delta_{exp}}{\delta_{cal}}$
	Mean (mm)	C.V.* (%)		
B	3.86	7.5	-	-
BP1	2.61	3.2	2.73	0.96
BP2	1.63	2.1	1.59	1.03
BP3	1.25	3.0	1.25	1.00

* C.V.: Coefficient of variance.

Table 5. *Experimental strains at a bending load of 1kN compared with calculated ones.*

Specimen No.	ε_{exp}		ε_{cal} (10^{-3})	$\dfrac{\varepsilon_{exp}}{\varepsilon_{cal}}$
	Mean (10^{-3})	C.V.* (%)		
B	5.80	9.7	-	-
BP1	3.64	3.4	4.07	0.90
BP2	2.31	4.5	2.22	1.04
BP3	1.61	4.9	1.64	0.98

* C.V.: Coefficient of variance.

Because the span depth ratio in the bending test was below 15, the shear deformation (Sawada *et al.*, 1968) had to be considered and was calculated based on the energy method. Shear deformation was reduced from the observed total deflection. Table 4 shows the mean experimental deflections caused by the bending moment and the calculated ones based on the laminated plate theory under the bending load. Table 5 shows the mean experimental strains and calculated ones based on the laminated plate theory under bending load. The coefficients of variance are also shown in Tables 4 and 5.

The bending rigidity and strength reliability could be largely increased when three-ply UD layers were used. It can be shown that the bending rigidity will slowly decrease as the UD layer thickness increases.

Figures 4 and 5 contain the analytical results for the bending test. The wood composites with various ply orientations and the thickness ratio ($2h/h$) of the UD layer were analyzed based on the laminated plate theory and the rule-of-mixtures. Figure 4 shows the properties of normalized deflection and strain with respect to ply orientation. The normalized deflection and strain increased when the ply orientation was increased. Normalized deflection and strain were lowest at $0°$ and highest at $90°$, remaining almost constant at an angle greater than $45°$. It can be shown from the stress analysis of the UD layers and wood that the stresses predicted by the lamination theory vary more slowly with the ply orientations than those predicted by the rule-of-mixtures. It has been shown that the analytical error between the laminated plate theory and the rule-of-mixtures was larger in Specimen BP1 than in Specimen BP3. The analytical error between the laminated plate theory and the rule-of-mixtures   was highest at a ply orientation of $15°$ to $18°$. The reason for this difference is that the laminated plate theory describes the off-axis stress-strain relation of the UD layer, whereas the rule-of-mixtures does not.

Figure 5 shows the variation of normalized deflection and strain with respect to thickness ratio. The calculated values of normalized deflection based on the laminated

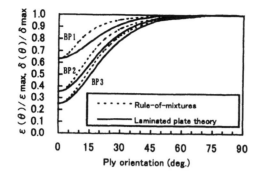

Figure 4. *Effect of ply orientation on normalized deflection or strain.*

plate theory coincided with the calculated values of normalized strain. The calculated values of normalized deflection based on the rule-of-mixtures, however, were larger than the calculated values of normalized strain.

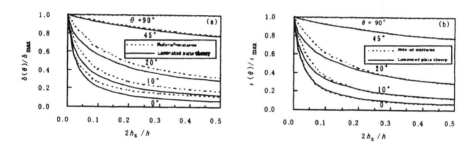

Figure 5. *Effect of thickness ratio on normalized (a) deflection and (b) strain.*

CONCLUSIONS

Tensile and bending rigidities and strength reliability can increase drastically when a small UD layer is used. Experimental results show that the tensile property can be significantly improved by using one-ply UD layer, and the bending property can be significantly improved with a three-ply UD layer.

Tensile and bending properties were calculated based on the laminated plate theory and rule-of-mixtures. The effect of the ply orientation and thickness ratio of a UD layer on the rigidity of the wood composites was investigated. In analyses of bending deflection and strain, the largest analytical errors were between the two procedures within a ply-orientation range of 15° to 18° and at a thickness ratio of 2.5%. The reason of these differences is that the laminated plate theory involves the off-axis stress-strain relation of the UD layer, whereas the rule-of-mixtures does not.

REFERENCES

Kawai S., Current trends in research and development on wood composite products, Mokuzai Gakkaishi, vol.43, 1997, p.617-622.

Sawada M., Strength properties of wooden sheet materils, Mokuzai Gakkaishi, vol.16, 1970, p.251-256.

Sawada M., Yamamoto H., Studies on wooden composite beams: Deflection characteristics within proportional limit of wooden composite beams, Research Bulletins of the College Experiment Forests, College of Agriculture, Hokkaido University, vol.26, 1968, p.11-44.

Tsai S.W., Hahn H.T., Introduction to Composite Materials, Technomic, Connecticut, 1980, p.217-276.

Particle shape distribution in Al-Al$_3$Ni functionally graded materials fabricated by a centrifugal *in-situ* method

Koichi Matsuda, Ryuho Sato and Yoshimi Watanabe

Department of Functional Machinery and Mechanics
Shinshu University
3-15-1 Tokida, Ueda 386-8567, Japan
zai5002@giptc.shinshu-u.ac.jp

ABSTRACT

It was shown that ring-shaped Al-Al$_3$Ni functionally graded material (FGM) could be successfully fabricated by a centrifugal *in-situ* method. It was found that the size of Al$_3$Ni particles within the FGM rings is distributed in a gradually graded manner as well as the volume fraction of Al$_3$Ni particles. However, although mechanical properties of particle-dispersed FGMs are depended on particle shape as well as volume fraction and particle size, there is no information on the particle shape. Therefore, in this article, particle shape distributions within the ring-shaped and cylindrical Al-Al$_3$Ni FGMs fabricated by the centrifugal *in-situ* method were studied.

INTRODUCTION

Functionally graded materials (FGMs) are a relatively new class of composite materials which have gradual compositional, mechanical and chemical properties. FGMs have been fabricated by many processing methods.[1] Centrifugal method, proposed by Fukui *et al.*, is one of the FGM fabrication methods.[2] The fabrication of the intermetallic compound-dispersed FGMs by centrifugal method can be classified into two categories according to the difference in processing temperature.[3] If the melting point of the intermetallic compound is notably higher than the processing temperature, the intermetallic compound remains solid in a liquid matrix. This situation is similar to ceramic-dispersed FGMs [4], and this method is referred to as a centrifugal solid-particle method. On the other hand, if the melting point of the intermetallic compound is lower than the processing temperature, centrifugal force is applied during the solidification to both the intermetallic compound and matrix. This solidification is similar to the production of *in-situ* composite using the crystallization phenomena, and this method is referred to as a centrifugal *in-situ* method.[5]

In the previous studies [5, 6], it was shown that ring-shaped Al-Al$_3$Ni functionally graded material (FGM) could be successfully fabricated by a centrifugal *in-situ* method. It was found that the size of Al$_3$Ni particles within the FGM rings is distributed in a gradually graded manner as well as the volume fraction of Al$_3$Ni particles. It is known that mechanical properties of particle-dispersed FGMs are depended on particle shape as well as volume fraction and particle size. Therefore, it is necessary to have a good knowledge of the particle shape of intermetallic compound-dispersed FGMs. Since no such data exist for the Al-Al$_3$Ni FGM fabricated by a centrifugal *in-situ* method, the present study was undertaken to investigate the particle shape distributions within the ring-shaped and cylindrical Al-Al$_3$Ni FGMs fabricated by the centrifugal *in-situ* method.

EVALUATION OF PARTICLE SHAPE

In order to describe the particle shape distributions in Al$_3$Ni FGMs quantitatively, the following two methods were adopted. One is box-counting method using fractal dimension, D, which is given by flowing equation:

$$N(r) = kr^{-D} \qquad \cdots(1)$$

where $N(r)$ is number of divided square, k is a constant, r is a size of square boxes and D in the scale range (r) is fractal dimension. Figure 1 shows a schematic illustration of the box-counting method. The second one is circularity, F, which is given by following equation:

$$F = R^2/4\pi S \qquad \cdots(2)$$

where R is perimeter length of the particle and S is the area of the cross-section of the particle. The circularity, F, becomes 1 for a perfect circle.

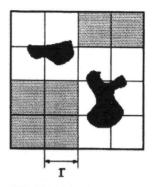

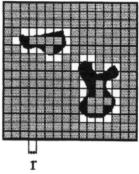

EXPERIMENTAL

Master ingot of Al-20mass%Ni alloy was used, and the applied G number (ratio of centrifugal force to gravitation) fixed to be 50, in this experiment. The melting point of the Al$_3$Ni intermetallic compound is

Fig.1 Schematic illustration of the box-counting method applied to the estimation of fractal dimension, D.

lower than the processing temperature of 730℃, the centrifugal force was applied during the solidification of the Al$_3$Ni intermetallic compound. Two different shaped FGMs are fabricated by the centrifugal *in-situ* method using different apparatuses, as shown in Fig. 2. One is a ring-shaped FGM with 90mm in outer diameter and roughly 25mm in wall thickness. The other is a cylindrical FGM with 30mm in length. After the casting, in the case of ring-shaped FGM, the mold-heating furnace is removed and the mold is cooled in air until complete solidification occurs. In contrast, furnace cooling was conducted for cylindrical FGM after the casting. Consequently, larger cooling rate was obtained with ring-shape than with cylindrical FGM. Therefore, cooling rates and temperature gradients are generated within ring-shape FGM. Specimen was divided into ten regions of equal width, along the direction the centrifugal force was applied, and the volume fraction, fractal dimension and circularity of Al$_3$Ni particles in each region are evaluated.

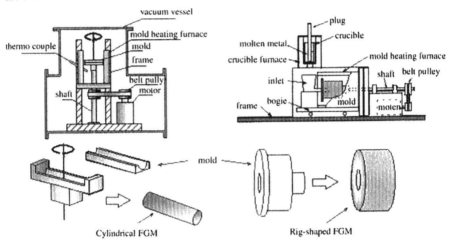

Fig. 2 The apparatuses for the centrifugal *in-situ* method performed in this study.
(a): cylindrical FGM and (b): ring-shaped FGM.

RESULTS and DISCUSSION

Microstructures of *in-situ* Al-Al$_3$Ni FGMs

Typical microstructures of the cylindrical FGM fabricated by the centrifugal *in-situ* method are shown in Fig.3. (a), (b) and (c) which were taken at the normalized distance of top, center and bottom, respectively.   Figure 4 shows microstructures of the ring-shaped FGM fabricated by the centrifugal *in-situ* method.   (a), (b) and (c) were taken at the normalized thickness of inner, interior and outer, respectively.   The black particles in Figs. 3 and 4 are stoichiometric Al$_3$Ni intermetallic compound and gray region is the Al matrix.   As can be seen, the Al$_3$Ni primary crystals are gradually distributed in the specimens.

The volume fraction of Al$_3$Ni particles in the FGMs was measured, and the obtained volume fraction distribution profiles of Al$_3$Ni particles in the FGMs are shown in Fig. 5. It is noteworthy that the graded distributions of Al$_3$Ni particles in the cylindrical FGM are steeper than those in the ring-shaped FGM, although the applied G number was thesame.   This is due to the fact that the cooling rate of the cylindrical FGM is much slower than that of the ring-shaped FGM.

The most remarkable point shown in Fig. 3 is that the particle shape of the cylindrical FGM is gradually distributed in each region.   This phenomenon will be discussed next section.

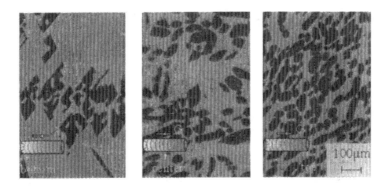

Fig. 3 Typical microstructures of the cylindrical FGM fabricated
by the centrifugal *in-situ* method ($G = 50$).
Normalized distance in (a), (b) and (c) is top, center and bottom, respectively.

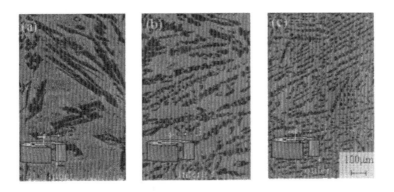

Fig. 4 Typical microstructures of the ring-shaped FGM fabricated
by the centrifugal *in-situ* method ($G = 50$).
Normalized thickness in (a), (b) and (c) is inner, interior and outer, respectively.

Evaluation of particle shape

Fig. 6 shows the fractal dimension distributions of the Al_3Ni particles in the FGMs.   Circularity distributions of the Al_3Ni particles in the FGMs are shown in Fig. 7. It can be seen that the cylindrical FGM has fractal dimension and circularity gradients, although fractal dimension and circularity are almost constant within the ring-shaped FGM.   It was shown that the fractal dimension of $Al-Al_3Ni$ FGM fabricated by a semisolid forming also shows graded distribution. [7]   The difference in the particle shape distributions may be caused by the difference in cooling rate and deformation of particles. Thus, we can obtain information on the formation mechanism of the composition gradient within the FGM fabricated by the centrifugal *in-situ* method by investigation of particle shape.   We are currently investigating the relationship between particle shape and processing conditions.   This may lead to obtain the important information for control of the composition gradient within the FGMs fabricated by the centrifugal *in-situ* method.

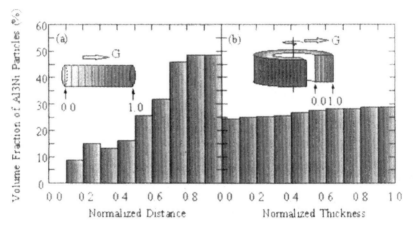

Fig. 5 The distributions of Al_3Ni particles in cylindrical
and ring-shaped FGMsfabricated under $G = 50$.

CONCLUSIONS

In this study, the volume fraction, fractal dimension and circularity distributions within the cylindrical and ring-shaped Al-Al$_3$Ni FGMs fabricated by a centrifugal *in-situ* method were investigated.   It was found that the cylindrical FGM has fractal dimension and circularity gradients, although fractal dimension and circularity are almost constant within the ring-shaped FGM.   We believe that, by detailed investigation of particle shape, we can obtain important information on the formation mechanism of composition gradient within the FGM fabricated by the centrifugal *in-situ* method.

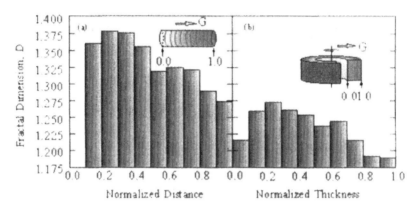

Fig. 6 Fractal dimension distributions of the Al$_3$Ni particles in the FGMs.
(a) is cylindrical FGM and (b) is ring-shaped FGM.

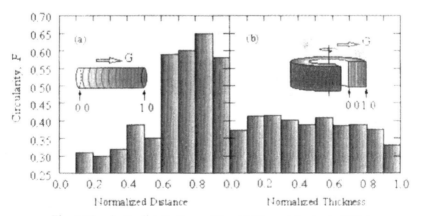

Fig. 7 Circularity distributions of the Al$_3$Ni particles in the FGMs.
(a) is cylindrical FGM and (b) is ring-shaped FGM.

ACKNOWLEGEMENT

This work was supported by Grant-in-Aid for COE Research (10CE2003) by the Ministry of Education, Culture, Sports, Science and Technology of Japan

REFERENCE

1. S. Surech and A. Mortensen, Fundamentals of Functionally Graded Materials, Processing and Thermomechanical Behavior of Graded Metals and Metal-Ceramic Composites, IOM Communications Ltd, London, 1988.

2. Y. Fukui "Fundamental Investigation of Functionally Graded Material Manufacturing System using Centrifugal Force" JSME Int. J. Series Ⅲ, 34 [1], 144-148, 1991.

3. Y. Watanabe and Y. Fukui "Microstructures and Mechanical Properties of Functionally Graded Materials Fabricated by a Centrifugal Method", "Rec. Res. Devel. Metall. Mater. Sci., 4, 51-93, 2000.

4. Y. Watanabe, A. Kawamoto and K. Matsuda "Particle Size Distribution in Functionally Graded Materials Fabricated by the Centrifugal Solid-Particle Method" Comp. Sci. Tech., 62, 881-888, 2002.

5. Y. Fukui, K. Takashima and C. B. Ponton, "Measurement of Young's Modulus and Internal Friction of an *in situ* Al-Al$_3$Ni Functionally Gradient Material" J. Mater. Sci. 29, 2281-2288, 1994.

6. K. Matsuda, Y. Watanabe and Y. Fukui "Particle Size Distribution in *in-site* Al-Al$_3$Ni FGMs Fabricated by Centrifugal Method" Ceramic Trans., 114, 491-498, 2000.

7. K. Matsuda, Y. Watanabe, M. Kamijo and Y. Fukui "Evaluation of Particle-Shape Graded Materials by Fractal Analysis" Functionally Graded Materials, 15, 125-130, 2002.

MATERIALS CHARACTERIZATION

Ultrasonic evaluation of thick composites for near surface flaws

A. Fahr, B. Roge, M. Brothers and D.G. Zimcik

Structures, Materials and Propulsion Laboratory
Institute for Aerospace Research
National Research Council Canada
Ottawa, Ontario, K1A 0R6
Correspondance to: abbas.fahr@nrc.ca

ABSTRACT

Composite components proposed for future structural applications include very thick sections that may be subjected to extremly high static and cyclic surface loads. There is a concern that minute material imperfections close to the outer surface region resulting from fabrication or surface finishing processes of these components may initiate delamination and eventual separation of the outer fibers during operation or cyclic testing. Conventional ultrasonic C-scan immersion techniques, commonly used for inspection of composite parts, are not sufficiently sensitive to identify small defects in thick composite structures which may be as large as 10 cm in thickness. Novel approaches based on ultrasonic Rayleigh waves have been developed and successfuly applied to the outer fiber surface of a thick composite structure after repeated cyclic testing. The aim is to identify minute manufacturing imperfections close to the surface region and to determine if such flaws have caused cracks or delamination during the cyclic tests.

In this paper, first the ultrasonic Rayleigh wave methods of flaw detection and ways of optimizing detection sensitivity and resolution will be described. Then, results of the optimization and sensitivity tests on carbon fibre composite specimens containing small simulated flaws 0.1 to 2 mm below the surface will be provided. Finally, the results of the application of the Rayleigh wave techniques on an actual thick composite section will be presented.

INTRODUCTION

Composite components have been used extensively in stuctural application in aerospace and other safety-concious industries.   Nondestructive evaluation (NDE) is often included in the original qualification/acceptance procedure and continuing inspection is conducted during service to ensure structural health and integrity. Inspection of composite parts less than 2 cm in thickness is easily done using ultrasonic pulse-echo or through-transmission techniques in an automated C-scan system. However, composite parts developed for some future structural applications include very thick sections (about 10 cm in thickness) that may be subjected to extremly high static and cyclic surface loads. There is a concern that minute material imperfections close to the outer surface region resulting from fabrication or surface finishing processes of these components may initiate delamination and eventual separation of the outer fibers during cyclic tests or in operation.

The sensitivity of conventional ultrasonic methods in such applications is inadequate due to increased attenuation, particularly at high frequencies (>5 MHz) which may be necessary for resolution of minute defects. In particular, detection of small flaws near the outer surface using the normal pulse-echo tests is a major problem since the surface echoes and the transducer ringing prevents identification of small signals resulting from flaws close to the surface. Also, in the normal through-transmission tests, the high attenuation of the thick composite material makes it impossible to detect small defects. Novel approaches based on ultrasonic Rayleigh waves have been developed and applied to the outer fiber surface of a thick composite structure.   Application and results of this approach are described in this paper.

ULTRASONIC RAYLEIGH WAVE TECHNIQUES

Rayleigh waves are generated when the ultrasonic incident beam strikes the test material at or above the critical angle of longitudinal and shear waves (1-2). When immersion mode is used (i.e. test component is immersed in water for acoustic coupling between the probe and the test piece), as Rayleigh waves propagate, their energy is quickly dissipated or "leaked" into the adjacent fluid. As a result, "leaky" Rayleigh waves propagate only a few wavelengths (λ) on the surface before their energy drops below the noise level. Also, their effective range is only about one wavelength below the surface.   Therefore, leaky surface waves are often used to characterize thin films and coatings. Leaky Rayleigh waves can be generated and detected using several approaches, two methods are described in this paper.

Method A:

Leaky Rayleigh waves are generated in a simple way by placing a flat or focused transmitting/receiving probe at the oblique Rayleigh wave angle as illustrated in Fig. 1.

The Rayleigh wave angle is slightly larger than the critical angle of the shear or transversal wave in the test material. Snell-Descartes' law defines the critical angle of a wave for a given material as:

$$\sin(\theta_c) = \frac{V_{inc}}{V} \tag{1}$$

where θ_c is the critical angle, V_{inc} is the velocity of the incident wave and V the velocity of the wave in the material. The critical angle exists only if $V_{inc} < V$. For an isotropic material, the Rayleigh wave velocity is a function of both longitudinal and transversal waves (3) as shown below:

$$V_R = A(V_T / V_L)V_T \tag{2}$$

where V_L and V_T correspond to the velocities of the longitudinal and transversal waves in the material, respectively. $A(V_T / V_L)$ is a function of both velocities and since it is always less than unity:

$$V_R < V_T < V_L \tag{3}$$

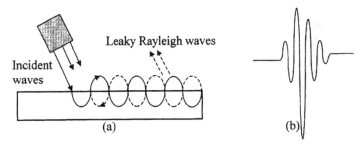

Figure 1 - (a) Generation and detection of ultrasonic Leaky Rayleigh waves using an ultrasonic probe at critical angle with respect to the test specimen. (b) A typical ultrasonic Leaky Rayleigh wave signal.

From equations 1 and 3, it is evident that the Rayleigh wave angle is larger than the critical angle of the transversal wave. When Rayleigh waves are used for inspection of surface and near-surface flaws, if there is no discontinuity in the beam path, the return energy is basically the scattering of the beam by the material surface texture. However, when a discontinuity is present in the beam path, a portion of the energy is reflected back. This energy is leaked into the coupling agent and picked up by the same probe resulting in an increase in the overall returned signal. Optimal echo amplitude is obtained when impinging surface waves are perpendicular to the plane of the flaw. Thus, flaw orientation would have an effect on the returned signal. Optimization of leaky Rayleigh waves is achieved by using a sharp edge of the specimen as a target and adjusting the probe angle and distance until the optimal signal is obtained. Surface and near-surface features are detected by mapping the returned Rayleigh wave echo as the function of the probe location. C-scan images generated in this way can indicate the location of flaws.

However, flaw images usually do not resemble the actual shape or size of the flaw due to the fact that the probe is at an angle with respect to the test material.

Method B:

Alternatively, an ultrasonic probe with a large aperture (larger than the critical angle) can be placed perpendicular to the surface of the test material in such a way that its focal point is below the surface, i.e. defocused. This approach is widely used in acoustic microscopy (4-5). In this way, in addition to the direct reflection of the normal incident beam (or specular waves), leaky Rayleigh waves resulting from the skimming of the beam striking the material at or above the critical angle and edge waves radiated from the transducer edge are generated and detected as illustrated in Fig. 2. Distinction and optimization of leaky Rayleigh waves from direct reflections and edge waves are achieved by defocusing the probe i.e. by moving the probe towards the sample surface until the Rayleigh echo is adequately separated from the other unwanted signals and an optimal amplitude is obtained. Fig. 2-b shows typical signals obtained at a defocused condition. Signal "A" corresponds to the specular reflection and "B" corresponds to the leaky Rayleigh waves. It must be noted that in this way Rayleigh waves are generated in a full circle, allowing flaw detection independent of the orientation of the flaw.

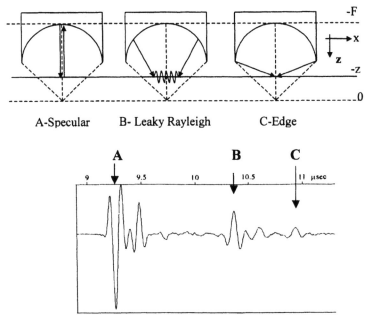

Figure 2 - Wave types and corresponding signals when a large aperture probe is defocused. Signal A is the specular reflection, signal B is leaky Rayleigh wave echo and signal C is the edge echo.

It is believed that two different mechanisms contribute to the flaw detection using leaky Rayleigh waves, as illustrated in Fig. 3. Crack-like discontinuities that are close to perpendicular to the surface interrupt the propagation of Rayleigh waves and therefore can be detected from the change in the corresponding echo amplitude.   On the other hand, planar defects, such as delamination, that are close to parallel to the surface may create a resonance effect, if an appropriate wave frequency is chosen, resulting in a change of amplitude and frequency of the returned echo. Both amplitude and frequency analysis were done to identify the different flaw types in this study.

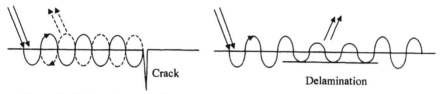

Crack   Delamination

Figure 3 - Mechanisms contributing to the Rayleigh wave detection of cracks and delamination.

EXPERIMENTS

Test Procedure

A commercial transmitter/receiver was used to excite a 10 MHz frequency, 12.5 mm focal length and 12.5 mm diameter probe. Scanning of the probe was done in a fully automated ultrasonic immersion system using water as the coupling media. The transducer index size was 0.2 mm. C-scan images were obtained by placing a gate on the echo corresponding to the Rayleigh waves and monitoring the signal amplitude as a function of the probe location.

Calibrations

In Rayleigh wave inspection, like many other NDE methods, it is necessary to establish the optimal test configuration and the sensitivity of the inspection system. The goal of the inspection was to detect small manufacturing or test-induced flaws close to the surface of a thick composite structure made of filament-wound carbon fiber epoxy. In order to check the sensitivity and resolution of the Raleigh wave techniques for detecting near-surface flaws, a calibration specimen was used. The specimen was made of a unidirectional carbon/epoxy system using the same material and fabrication method as the actual component. A series of 0.40 mm diameter flat-bottom holes were drilled in the calibration specimen such that the holes ended at different distances from the top surface as shown in Fig. 4.

RESULTS AND DISCUSSION

As previously mentioned, Rayleigh waves penetrate only about one wavelength below the surface of the material and thus, they are not effective beyond this range. At a frequency of 10 MHz, the Rayleigh wavelength in a unidirectional carbon fiber epoxy composite is approximately 0.2 mm. Therefore, it is expected that only holes # 1, 2 and 3 from the left, would be detected. The experimental result is shown in Fig. 5 that confirms this prediction. Reference lead tapes were used in order to identify the location of the holes on C-scan images since the holes were very small and their indications could be confused with other possible manufacturing flaws. It is evident that at 10 MHz, 0.4 mm discontinuities as near as 0.1 mm below the surface could be detected with good resolution using the leaky Rayleigh wave technique.

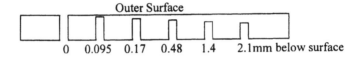

Figure 4 - The cross-section of a calibration specimen containing drilled holes below the surface

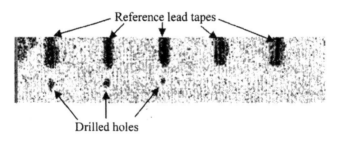

Figure 5 - Amplitude pulse-echo C-scan at Rayleigh wave angle. Probe characteristics are 10 MHz; focal length 12.5 mm; diameter 12.5 mm.

After calibration tests, the thick composite structure was inspected using the same 10 MHz frequency probe and the Method A described above. Note that at this frequency, only the detection of discontinuities located at approximately 0.2 mm below the surface is possible. Fig. 6-a shows the amplitude C-scan trace of a section of the thick composite structure. In this case, the probe was positioned such that the Rayleigh waves propagated in the direction of fibers. The Rayleigh angle set in this case was 49.2°. This is slightly larger that the critical angle for transversal waves. Fig. 6-b shows a similar C-scan where the probe was positioned such that the Rayleigh waves propagated in a direction perpendicular to the fiber orientation. The Rayleigh angle set in this case was 20°. This is slightly larger than the critical angle for the longitudinal waves in this direction. The critical angle for the transversal waves in the direction perpendicular to the fibers is too large and not very effective for flaw detection. The critical angle for a given wave type is different since the wave velocity is different in the fiber direction and normal to the fibers.

In the images shown in Fig. 6-a and 6-b, feature R represents a reference lead tape that was attached to the surface; feature D shows large defects; and feature F indicates fiber separation from the surface. As seen in Fig. 6-a, a large number of porosity-like defects were present close to the surface. Some of these, such as those identified as feature D, are probably large delamination-like cavities. In Fig. 6-b, the same large defects are also shown. In addition, two white indications are present on both sides of one of these defects that are identified as feature F. These are images of areas of the top surface where some fibers were separated during surface loading of the structure. It is clear that one of the detected near-surface defects exceeded the critical size and caused this undesirable failure during loading. The background noise in Fig. 6-b is due to scattering of the ultrasonic beam after striking the fibers at a normal angle and is expected.

In addition, Method B was used to look at the very end of the areas where fibers had separated in an attempt to identify possible cracks perpendicular to the outer surface. An example of the results is shown in Fig. 7-a indicating the presence of cracks along the fibers. These cracks were neither visible with the naked eye nor under a microscope. However, using solid rubber replication followed by microscopy of the replica, these cracks could easily be identified as seen in Fig. 7-b and-c that verify the ultrasonic results.

CONCLUSION

Novel approaches based on ultrasonic leaky Rayleigh waves have been developed to detect small flaws on and a few plies below the surface of thick composite structures. Near surface flaws cannot be detected using the conventional ultrasonic inspection techniques but are of great concern in some structural composites when subjected to extremely high surface loads. The detection sensitivity and resolution of the Rayleigh wave procedures have been established using reference specimens with known defects. These tests indicated that the Rayleigh wave methods have the ability to identify small defects as close as 0.17 mm below the surface with good sensitivity and confidence. The techniques have been applied to a real structure approximately 10 cm in thickness where delamination-like cavities have been detected just below the surface. During high surface loading of the structure, some of the detected cavities caused separation of fiber bundles from the surface. Further tests using Rayleigh waves also have identified surface cracks that could neither be seen by the naked eye nor under optical microscopes, but could be verified by optical examination of the rubber replica of the cracks.

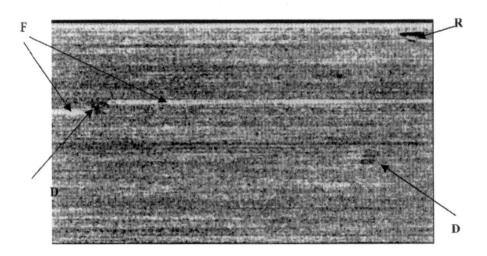

(a)

(b)

Figure 6 - Amplitude C-scans of a thick composite structure at the Rayleigh wave angle
(a) when the Rayleigh waves travel in the fibers direction and (b) when the Rayleigh
waves travel perpendicularly to the fibers direction.

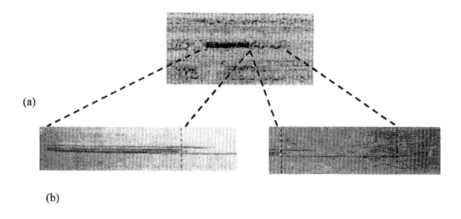

(a)

(b)

Figure 7 - (a) Amplitude C-scans of the tip of the surface damage using ultrasonic defocusing Rayleigh wave method and (b-c) the corresponding micrographs of the rubber replica.

REFERENCES

1. I.A. Viktorov, Rayleigh and Lamb waves, Plenum Press, New York, 1967.

2. R. Briers et al., "A liquid wedge as generating technique for Lamb and Rayleigh waves", The Journal of the Acoustical Society of America, Volume 102(4), 1997, 2117-2124.

3. E. Dieulesaint, D. Royer, Ondes élastiques dans les solides, Masson et Cie Editeurs, Paris, 1974.

4. J. Kushibiki, "Influence of reflected waves from the back surface of thin solid-plate specimen on velocity measurements by line-focus-beam acoustic microscopy", IEEE transactions on ultrasonics, ferroelectrics, and frequency control, Volume 47(1), 2000, 274-284.

5. J. Zhang et al., "PVDF large aperture spherical transducer in the transient mode", IEEE Ultrasonic symposium, 1994, 517-520.

Biaxial failure envelopes of angle-ply and multi-directional lay-ups of FRP tubulars

Fernand Ellyin
Advanced Composite Materials Engineering Group
Department of Mechanical Engineering
University of Alberta
Edmonton, Alberta, Canada T6G 2G8
fernand.ellyin@ualberta.ca

ABSTRACT

This paper discusses the concept of failure (functional) envelopes in the design of fibre reinforced polymeric (FRP) composites for pressure retaining structures. The envelopes are constructured for a given lay-up geometry by connecting the failure stress states in a biaxial stress coordinates. Experimental data are presented for two types of filament wound tubulars: angle-ply and multi-directional geometries. It is shown that for an angle-ply lay-up, the envelopes are highly asymmetric. In contrast, through judicious combination of fibre orientations, one can obtain a nearly symmetric failure envelopes. The corresponding damage mechanisms and failure modes for each configuration are also discussed.

INTRODUCTION

Fibre reinforced polymeric composites are seen as an attractive alternative to products made from conventional materials due to their corrosion resistance and high strength-to-weight ratio. Because of their anisotropic properties, the composites can be designed for minimum material under optimal performance conditions.

An area in which the progress has been rather slow, is in pressure retaining structures, particularly in high pressure applications. The reluctance in adopting composite materials in pressure retaining products is mainly due to limited understanding of the material behaviour and damage development under complex loading (multiaxial, creep, fatigue) and environmental conditions.

The focus of the *Advanced Composite Materials Engineering* (ACME) group at the University of Alberta has been to investigate and characterize the behaviour and damage development in composite materials at various length scales (micro-scale, macro-scale and structural). This is accomplished using a combination of unique experimental techniques (multiaxial, time-dependent and environmental)(1-3) and numerical modeling techniques (structural and constitutive) (4-6). The ultimate goal of this research program is to develop validated design methodologies for a wide range of loading and service condition for high pressure components.

It is the objective of this paper to report on the investigation of fibre reinforced polymer (FRP) composite tubulars under biaxial monotonic loading. Composite tubular products are most effectively manufactured using the filament winding method. Two families of glass-fibre/epoxy tubulars were tested: (a) angle-ply and (b) multi-directional geometries. The rationale for testing these type of geometries and results will be presented in the following sections.

Performance of FRP Composite Tubulars

Schematic of a fibre reinforced polymeric composite tubular is shown in Fig. 1. The winding angle is measured with respect to the tube axis (axis of revolution). Referring to Fig. 1, when the 0° fibres are absent, then the lay-up is termed angle-ply $[\pm\theta_n]$ otherwise multi-directional, e.g. $[0_m, \pm\theta_n]$ as in the figure. The subscripts n and m refer to the number of similar layers.

In those cases where principal stress directions, as well as their ratio are fixed, then an "optimum" fibre orientation can be determined from the "netting" analysis by the following relation:

$$\tan^2 \theta = \sigma_H / \sigma_A \tag{1}$$

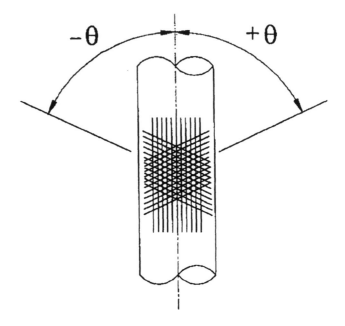

Figure 1 - Schematic diagram of a fibre reinforced polymeric composite tubular

In the above σ_H is the average global hoop stress and σ_A is the average global axial stress. For example, in a closed end pressure vessel $\sigma_H : \sigma_A = 2:1$ and from (1), $\theta \approx 55°$, and the fibres are then placed in an angle-ply pattern $[\pm 55_n]$. However, in most industrial applications the in-plane principal stress ratio may vary, particularly when the applied loads are decoupled. In these circumstances a designer would be interested to know the effect of such variations in a fixed lay-up geometry. This is achieved through the construction of a failure envelope described below.

Generation of Biaxial Failure Envelopes

For a fixed angle-ply geometry $[\pm \theta_n]$, a failure envelope in a biaxial stress state (σ_H, σ_A) is constructed by applying a limited number of proportional path loadings with fixed ratio of σ_H / σ_A, until failure. Through connecting these failure points, a failure envelope is obtained. Figure 2 is a schematic illustration of three failure envelopes for three lay-up geometries. One is optimized with respect to a "pure" pressure loading resulting in $[\pm 70]_s$ lay-up, the other for a pressure vessel type loading $\sigma_H : \sigma_A = 2:1$, and the third one for an equibiaxial loading $\sigma_H : \sigma_A = 1:1$. Note that in each case the

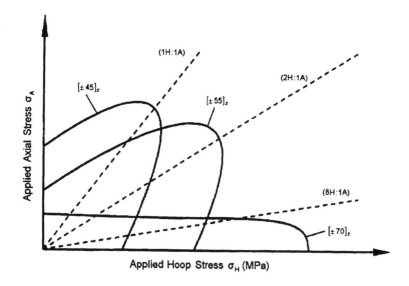

Figure 2 - Relationship between functional (leakage) failure envelopes
for three angle-ply fibre orientations

maximum strength is obtained for the angle given by eq. (1). A combination of lay-up angles can be constructed to obtain pressurized components which have high resistance for both pure pressure and pure axial and their combination, e.g. $[0_m, \pm 70_n]$ to be discussed later on.

In the case of pressurized components, a failure envelope is generally defined with respect to the functional failure (leakage) rather than structural failure. The leakage failure occurs when the intra-laminar matrix cracks coalesce and propagate through the wall thickness. In the following experimental results are presented for the angle-ply and multi-directional geometries.

EXPERIMENTAL INVESTIGATION

Results of an experimental investigation will be presented for two lay-up geometries, an angle-ply optimized with respect to pressure vessel loading, 2H:1A, and a multi-directional lay-up with high strength for both pressure and axial loadings. (For the sake of simplicity letters H and A are used for the hoop stress and axial stress, respectively.) It was mentioned earlier that filament winding process is the most economical manufacturing method for the pressure retaining structures; hence, the specimens used in this investigation were made by this technique.

Failure Envelopes for [± 55₇], Angle-Ply Tubulars

It was mentioned earlier that winding angles of ± 55° are common in the FRP pipes and pressure vessels, because this angle is near optimum design for 2H:1A loading ratio. The resulting maximum principal stress is approximately 55° off axis and, correspondingly, the fibres are oriented in the direction of the maximum stress. Furthermore to simplify the results, the axial load and pressure data are converted into an average stress according to the ASTM D2992-91(7):

$$\sigma_A = \frac{load + pressure\,(\pi r_i^2)}{\pi(r_o^2 - r_i^2)} \qquad\qquad \sigma_H = pressure\,(\frac{2r_o - h}{2h}) \qquad\qquad (2)$$

where r_i and r_o are the internal and external radii, respectively, and h is the wall thicknes.

For each specified stress ratio a von Mises type equivalent stress was determined.

$$\sigma_{eq} = [\,\sigma_A^2 - \sigma_A\sigma_H + \sigma_H^2\,]^{1/2} \qquad\qquad (3)$$

and the applied stress rate, δσ/δt, was calculated from

$$\delta\sigma/\delta t = \delta\sigma_{eq}/\delta t \qquad\qquad (4)$$

In these tests three stress rates were applied: (a) ≈ 500 kPa/s, (b) ≈50 kPa/s and (c) ≈ 5 kPa/s. Note that each loading rate varied by an order of magnitude from each other. This was chosen so as to investigate the rate-dependency of the fibrereinforced epoxy resin composite tubulars.

Figure 3 depicts the biaxial failure envelopes for the above three loading rates. It is seen that filament wound tubulars exhibit time-dependent behaviour. It is also observed that the 2H:1A load ratio does not indicate any noticeable viscous effect within the range of the applied loading rates. These results therefore indicate that the failure envelopes not only change size with different loading rates, but they also change shape.

The observed failure modes are discussed by Carroll et al. (8), and an interested reader is encouraged to consult this reference. Here only a brief summary of the observed modes will be given. In general, the failure modes can be related to the strain response. For the positive axial strain, matrix cracking parallel to the fibres occurred which led to leakage failure. For hoop dominated loading with large hoop strains, a burst failure was observed involving fibre breakage and delamination. For compression tests with compressive axial strains, the tubes failed in a matrix dominated shear mode.

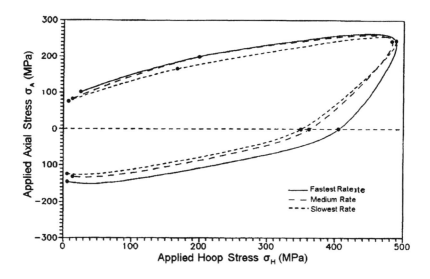

Figure 3- Experimentally determined failure envelopes of [± 55₇]ₛ angle-ply tubulars for three loading rates, after ref. (8)

FAILURE ENVELOPES FOR MULTI-DIRECTIONAL TUBULARS

A multi-directional lay-up geometry of filament wound tubular was tested. This tubular had a rather complex geometry, from inside to outside consisting of [± 45₄, + 45, 0, -70, ±45₂ , ±70], 17 layers in total. This tube was designed for a down hole tubular in oil industry where there is a variation in loading from pure pressure to pure axial and in-between ratios. The ± 70°lay-up is to withstand the pure pressure whereas the 0° layer is for the axial loading, and ± 45°is for the in-between ratios. Two loading rates were applied, the fast one two orders of magnitude higher than the slow rate (5kPa/s). The biaxial envelopes for both the fast and slow loading rates are depicted in Fig. 4. Note that the stress values used are the averages through the tube wall calculated according to eq. (2). It is seen that there is a definite loading rate effect. The shape of the envelope, in contrast to that of the [± 55₇]ₛ angle-ply tubulars, is less asymmetric.

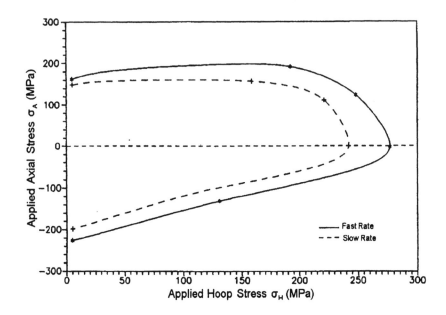

Figure 4 - Experimentally determined failure envelopes of multidirectional [0/± 45₇/±70₂] tubulars for two loading rates, after ref. (9)

The multi directional tubes produced some complex failure modes under the applied loading conditions. These tubulars were comprised of a complicated winding scheme that incorporated three different fibre angle orientations, i.e. 0°, ±45° and ± 70°. The differing laminae properties and interaction of each angle-ply contributed to the failure mode and damage accumulation. Under pure axial tension (0H:1A) matrix cracking first occurred parallel to the ± 70° fibres. Final failure took place when a critical axial strain was reached and the 0° fibres fractured. This caused extensive delamination between the dissimilar fibre angle layers. Matrix cracking was also observed in the 0° layers and in the ± 45° layers on the inside wall. Leakage preceded just before sudden structural failure (burst).

In the case of pure hoop loading (1H:0A), the specimens failed when critical hoop strain caused the ± 70° fibres to break. The ± 70° layers then delaminated from the rest of the tube, some matrix cracking parallel to the ± 45° fibres was observed on the inside wall.

For both 1H:1A and 2H:1A loadings matrix cracking parallel to ± 70° fibres were observed as the test progressed. This matrix cracking led to failure by leakage when cracks coalesced and a path through thickness was formed.

Shear failure with a ± 45°angle through the wall thickness was observed when the tubes were subjected to an axial compression. In the case of 1H:-1A, loading the ± 45°fibres were loaded in either tension or compression which resulted in a fibre breakage in the ± 45° layers and delamination between 45° and 70° layers, resulting in a burst failure. In essence the failure mode for this loading was a combination of that of 1H:0A and 0H:-1A which occurs at a localized region.

THEORETICAL PREDICTION OF FAILURE ENVELOPES

The design of FRP pressure retaining structures is usually based on experimentally determined allowable average stress values as described in the previous section. Currently, no universally accepted methodology exists to predict accurately the failure of FRP pressure retaining structures (10). The difficulty with predicting these functional (leakage) failure envelopes stems from the complexity of damage development process. There are multiple interacting and competing failure mechanisms which operate during initiation and accumulation of damage with the increasing load intensity. Furthermore, depending on the lay-up geometry and applied loads, the stress-strain response can be highly nonlinear. It is for this reason that ref. (10) found variation over 300% among predicted failure values for simple cross-ply laminates. Nonlinear viscoelastic constitutive models and proper failure theories validated by experiments, such as that presented in ref. (6), will be required for the accurate prediction of failure envelopes. This, however, is beyond the scope of the present paper.

CONCLUSIONS

From the results presented here, the following conclusions are drawn:

- The functional failure (leakage) stresses and failure modes were found to depend on the lay-up geometry, biaxial loading ratio and rate of loading.
- Time is required for a fluid to permeate through a crack network (which is a function of applied pressure and wall thickness). Leakage is not only governed by the formation of matrix cracks within each ply, but also by their coalescence and inter-connectivity through the wall thickness, i.e. formation of delaminations.
- In the tested multi-directional tubulars, matrix cracking was found to be governed by the axial tension while delamination in between dissimilar laminae was governed by the combined internal pressure and axial load.

Acknowlegement

The research investigation reported here is part of a larger research program of the TCPL/NSERC Senior Industrial Research Chair. The Chair holder (F. E.) wishes to thank the sponsors, as well as, those of the ACME group who contributed to this investigation. In particular he wishes to thank former graduate students, M. Carroll, M. Martin, J. Wolodko who conducted part of the experiments reported herein.

REFERENCES

1. F. Ellyin. and J. Wolodko, "Testing Facilities for Multiaxial Loading of Tubular Specimens". In Multiaxial Fatigue and Deformation Testing Techniques, ASTM STP 1280, S. Kalluri and P.J. Bonacuse eds. Amer. Soc. Testing Mater. Philadelphia, PA, 1997, 7-24.

2. J.W. Hoover, D. Kujawski and F. Ellyin, "Transverse Cracking of Symmetric and symmetric Glass-Fibre/Epoxy-Resin Laminates", *Composite Science and Technology*, Vol. 57, 1997, 1513-1526.

3. F. Ellyin and C. Rohrbacher, "Effect of Aqueous Environment and Temperature on Glass-Fibre Epoxy Resin Composites", *J. of Reinforced Plastics and Composites*, Vol. 19, 2000, 1405-1427.

4. Y. Hu, Z. Xia and F. Ellyin,"Mechanical Behaviour of an Epoxy Resin Under Multiaxial Loading, Part I: Experimental Investigation and Part II: Comparison of Viscoelastic Constitutive Model Predictions", *Polymer & Polymer Composites*, Vol. 8, 2000, 11-18, and 157-166.

5. Y. Chen, Z. Xia and F. Ellyin, "Evolution of Residual Stresses Induced During Curing Processing Using a Viscoelastic Micromechanical Model", *J. Composite Materials*, Vol. 35, 2001, 522-542.

6. F. Ellyin, Z. Xia and Y. Chen, "Viscoelastic Micromechanical Modeling of Free Edge and Time Effects in Glass Fiber/Epoxy Cross-Ply Laminates", *Composites: Part A,* Vol. 33A, 2002, 399-409.

7. ASTM Designation: D2992-91, "Standard Practice for Obtaining Hydrostatic or Pressure Design Basis for "Fiberglas" (glass-fiber reinforced thermosetting-resin) Pipe and Ffittings. Amer. Soc. Testing Mater. Philadelphia, PA.

8. M. Carroll , F. Ellyin, D. Kujawski and A.S. Chiu, "The Rate-dependent Behaviour of ± 55° Filament Wound Glass-fibre Epoxy Tubes Under Biaxial Loading", *Composite Science and Technology*, Vol.55, 1995, 391-403.

9. F. Ellyin, M. Carroll, D. Kujawski and A.S. Chiu, "The Behavior of Multidirectional Filament Wound Fibreglass/Epoxy Tubulars Under Biaxial Loading" *Composites Part A*, Vol. 28A, 1997, 781-790.

10. P.D. Soden, M.J. Hinton and A.S. Kaddour, "A Comparison of the Predictive Capabilities of Current Failure Theories for Composite Laminates". *Composite Science and Technology*, Vol. 58, 1998, 1225-1254.

Investigation of the Thermal Creep properties of Al/SiC Composites made by the Powder Metallurgy Technique

R.N. Saraf and Prof. R.J.Klassen

Department of Mechanical and Materials Engineering

The University of Western Ontario

London, Ontario, Canada N6A 5B9

rklassen@eng.uwo.ca

ABSTRACT

We report on the effect of Cu additions on the high-temperature hardness and creep resistance of non-reinforced and reinforced Al-alloys made by a powder metallurgy technique consisting of mixing powders of elementally pure metal and SiC followed by sintering and hot-pressing. Indentation hardness tests were performed at room temperature, 100, 200, 300, 400, and $500^{\circ}C$ while creep tests were performed at $550^{\circ}C$. The hardness and the creep resistance of the non-reinforced material decreased significantly with increasing temperature and Cu content while the hardness of the reinforced (20 vol% SiC) material decreased by a much smaller amount. The creep resistance of the reinforced material also decreased significantly with increasing Cu content. we suggest that the fabrication technique used in this study results in non-homogeneous Cu distribution through the material. This causes decreased hardness and creep resistance with increasing Cu content in the non-reinforced material. The addition of SiC as a reinforcement produces an MMC with hardness that is not affected by the Cu content but with creep resistance that decreases with increasing Cu content. The creep resistance of the reinforced material is, therefore, more dependent on the properties of the matrix than is the indentation hardness.

Introduction

Solid-state, Powder Metallurgy (PM), techniques are frequently used to fabricate Metal Matrix Composites (MMCs) that contain uniform distributions of reinforcing particles [1-5]. Alloyed metals can be produced by several PM techniques including the mixing of elementally pure metal powders followed by sintering to allow for the inter-diffusion of the metal elements. Aluminum-alloy matrix, SiC particulate reinforced composites made by this technique have been shown to have excellent strength and creep resistance to temperature up to about 400°C [4]. Although considerable effort has been directed to understanding the role of reinforcements on the creep deformation of MMCs, few investigations have studied the role of matrix alloy additions. This is particularly of interest in MMCs made by PM techniques since incomplete inter-diffusion, and the resulting non-homogeneous distribution of alloy elements, may influence the high temperature strength and creep resistance.

We report here an investigation to determine the effect of Cu alloy additions on the high-temperature strength and creep resistance of non-reinforced and reinforced Al-alloys made by a PM technique that involves the inter-diffusion of elementally pure metal powders.

Procedure

Three Al-alloy compositions, containing 1, 2, and 4 wt%Cu, were fabricated in the non-reinforced and in the reinforced, with 20 vol% SiC, condition (Table 1). The alloys, referred to as Materials 1 to 6, were fabricated as follows. Pure Al, Cu, Mg, and Mn powders (average diameter of 45 μm) were mixed with SiC powder (average diameter of 25 μm) in an acetone medium. The materials all contained 1.4 wt% Mg and 0.4 wt% Mn but had various amounts of Cu (1, 2, and 4 wt%). Materials 2,4,and 6 were reinforced with 20 vol% SiC while Materials 1,3, and 5 were kept in the non-reinforced condition.

After mixing, the powders were cold-pressed into cylindrical pellets and then sintered at 600°C for 2 hours in vacuum followed by hot-pressing in air at about 550°C and at 35 MPa stress. Cylindrical samples of 10 mm diameter and 10 mm length were machined from the hot-pressed pellets.

Material 7 (Table 1) is a commercially available MMC (Goodfellow Cambridge Ltd, Huntingdon England) that has approximately the same chemical composition and SiC particle distribution as Material 6. Material 7 is included in this investigation as a reference MMC with which to compare the properties of the other materials.

The microstructure of the fabricated material was evaluated using Scanning Electron Microscopy (SEM) to confirm that the material was free of porosity and had a uniform distribution of reinforcing particles. Analytical SEM was also used to determine the chemical composition of particles distributed within the material and to observe the deformation around micro-hardness indentations that were made in the materials at 500°C.

Micro-indentation hardness tests were performed on samples, of about 2 mm thickness, that were cut from Materials 1 to 7. The tests were performed at room temperature, 100, 200, 300, 400, and 500 °C using a high-temperature micro-indentation tester (Micro Materials Ltd, Wrexham UK) with a diamond Berkovich indenter. Five indentations were made, with a 1N indentation load, on each material at every temperature.

Constant uniaxial compressive load creep tests were performed at 550°C in air on Materials 1 to 6. This temperature was chosen to ensure that all the Cu in the materials should be dissolved in the α-Al phase [6]. Creep tests were performed at three or more stress levels, between 3 to 15 MPa, for each material by applying a constant compressive load to the sample which was located in the centre of a vertical tube furnace. The displacement of the sample was measured with an LVDT and displacement measurements were made at approximately 1 minute intervals over the two hour duration of each test. This time duration was sufficient for considerable creep deformation to occur because of the high temperature of the tests ($T/T_m \approx 0.88$).

Results

Figures 1 (a, b) show the microstructure of the non-reinforced Material 5 and the reinforced Material 6. The figures indicate the microstructure around indentations that were made at 500°C. The SiC particles are uniformily distributed throughout the reinforced material. The non-reinforced material has large cracks in the region around the indentation while the reinforced material has no cracks.

Figures 2 (a, b) indicate the average Berkovich indentation hardness versus temperature for the reinforced and the non-reinforced materials. The reinforced material has a higher indentation hardness than the non-reinforced material at all temperatures. The hardness of the non-reinforced material decreased with increasing Cu content and temperature while the hardness of the reinforced material did not decrease in a systematic way with increasing Cu and only decreased to a small extent with increasing temperature. The error bars in Figure 2 indicate that there was considerable scatter in the individual hardness data. It should be noted that while the Berkovich indentation hardness is a function of the flow stress of the material, it does not indicate the numerical value of the flow stress because of the complex triaxial stress imposed during the indentation process.

Figure 3 indicates the creep strain versus time plot of a sample made from Material 4 tested at 12 MPa and 550°C. Materials 1 – 6 all showed creep curves of similar shapes, depending upon the applied stress, to that in Figure 3.

The steady-state creep strain rate of Materials 1 to 6 at 550°C is shown in Figure 4(a, b) where the steady-state creep strain rate is plotted versus stress, on a logarithmic scale. The steady-state creep rate at any stress level, is lower for the reinforced than for the non-reinforced material however the power-law stress dependence increases from n $\approx$ 5 for the non-reinforced material to n $\approx$ 6 –10 for the reinforced material. The steady-state creep rate at any level of stress increases dramatically with increasing Cu content in both the non-reinforced and the reinforced material.

Discussion

The data presented here allow for an assessment to be made of the effect of alloy additions on the high-temperature strength and creep resistance of MMCs made by a PM technique where an alloyed metal matrix is made by mixing and sintering elementally pure metal powders.

Figure 5 shows the hardness of Materials 6 and 7 versus temperature. Both materials have essentially the same chemical composition and SiC particle distributions; Material 6 was fabricated by the PM technique described above while Material 7 is a commercially available MMC. The hardness of the two materials is relatively the same with a similar temperature dependence. This indicates that the technique used in this investigation results in MMCs with comparable strength to what is commercially available.

Figure 2 (a) indicates that the hardness of the non-reinforced materials decreases at all temperatures with increasing additions of Cu. This observation is unexpected since Cu is known to strengthen Al. The SEM images of the non-reinforced material, Figure 1(a), indicate clearly that considerable cracking occurred during indentation at 500°C. This suggests that there are weak regions in the non-reinforced alloy that crack first when the material is stressed at high temperature. These observations suggest that non-homogeneous distribution of Cu throughout the material contributes to the reduction in hardness.

Figure 2(b) indicates that the hardness of the reinforced material does not decrease in a systematic way with increasing additions of Cu. The hardness also decreases only slightly with increasing temperature. Figure 1(b) indicates that no large cracking occurred during indentation of the reinforced material at 500°C. Since the same fabrication techniques was used for all the material, one can conclude that the presence of 20 vo% SiC either: enhances the inter-diffusion of the alloy additions in the Al matrix or, reduces the stress on the matrix by carrying the bulk of the imposed load, thereby preventing pre-mature cracking of weak regions in the matrix.

The stress dependence of the steady-state creep rate of the reinforced material was much higher than that of the non-reinforced material, Figures 4(a, b). This is in agreement with the findings of others regarding the effect of SiC reinforcements on the creep resistance of Al-based MMCs [7].

Conclusion

We report here an investigation of the effect of Cu additions on the high-temperature hardness and creep resistance of non-reinforced and reinforced Al-alloys made by a powder metallurgy technique. Aluminum based-alloys of compositions 1.4 wt% Mg and 0.4 wt% Mn with various amounts of Cu (1, 2, and 4 wt%) were fabricated in the non-reinforced (0 % SiC) and in the reinforced (20 vol% SiC) condition by mixing powders of elementally pure metals and sintering at 600°C followed by hot-pressing at about 550°C.

The indentation hardness of the non-reinforced and the reinforced material was measured at room temperature, 100, 200, 300, 400, and 500°C while the creep resistance was measured at 550°C.

The hardness of the non-reinforced material decreased significantly while the hardness of the reinforced material remained relatively constant with increasing temperature and Cu content. Although the creep resistance of all the reinforced material was significantly greater than that of the non-reinforced material, it also decreased significantly with increasing Cu content for both materials.

Based upon these observations we suggest that the creation of non-reinforced Al alloys, containing Cu, by using a PM technique involving the inter-diffusion of elementally pure metals results in incomplete inter-diffusion of the Cu and, hence, a microstructurally inhomogeneous material. This causes a decrease in hardness and creep resistance with increasing Cu content in the non-reinforced material. The addition of 20 vol% SiC as a particulate reinforcement produces an MMC with hardness that is not decreased, but creep resistance that is decreased, by the addition of Cu. This results from the SiC particulates carrying the bulk of the load imposed upon the sample, thus preventing premature cracking of weak regions in the matrix material, and indicates that the creep resistance is more dependent on the properties of the matrix than is the indentation hardness.

References

1. K. Park, E.J. Lavernia, and F.A. Mohamed, *Acta Metall. Mater.*, 38 (1990) 2149 – 2159.
2. V.V. Bhanuprasad, R.B.V. Bhat, A.K. Kuruvilla, K.S. Prasad, A.B. Pandey, and Y.R. Mahajan, *Int. J. of Powder Metall.*, 27(1991) 227 – 235.
3. A.B. Pandey, R.S. Mishra, and Y.R. Mahajan, *Acta Metall. Mater.*, 40 (1992) 2045 - 2052.
4. A.B. Pandey, R.S. Mishra, and Y.R. Mahajan, *Metall. and Mat. Trans.*, 27A (1996) 305 – 316.
5. Y. Li and F.A. Mohamed, *Acta Mater.* 45(1997) 4775 – 4785.
6. Hansen, <u>Constitution of Binary Alloys</u> 2nd Ed., McGraw Hill, (1958) 84.
7. Y. Li and T.G. Langdon, *Metall. and Mat. Trans.*, 30A (1999) 315 – 324.

Acknowledgements

The authors would like to thank the Natural Science and Engineering Research Council for supporting this investigation through a research grant provided to RJK.

Table 1:   Chemical composition of the seven materials used in this investigation.

Material	Matrix Composition				Particulate Reinforcement
	Cu (wt%)	Mg (wt%)	Mn (wt%)	Al	(vol%)
1	1	1.4	0.4	bal.	0
2	1	1.4	0.4	bal.	20
3	2	1.4	0.4	bal.	0
4	2	1.4	0.4	bal.	20
5	4	1.4	0.4	bal.	0
6	4	1.4	0.4	bal.	20
7	3.3	1.2	0.4	bal.	26

(a)   (b)

Figure 1: Scanning electron micrographs of Berkovich indentations made at 500°C on: (a) the non-reinforced material (Material 5) and (b) the reinforced material (Material 6).

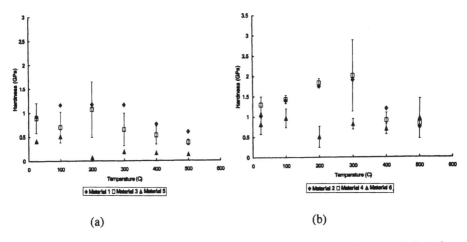

(a)   (b)

Figure 2: Berkovich indentation hardness versus temperature of: (a) the non-reinforced material (Materials 1,3,5) and (b) the reinforced material (Materials 2,4,6).

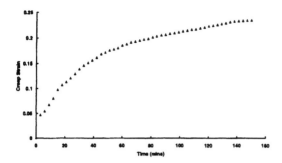

Figure 3: Creep strain versus time for the reinforced Material 4 tested at 550°C.

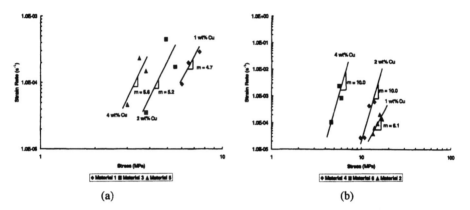

Figure 4: Steady-state creep strain rate versus stress at 550°C for: (a) the non-reinforced material (Materials 1,3,5) and (b) the reinforced material (Materials 2,4,6).

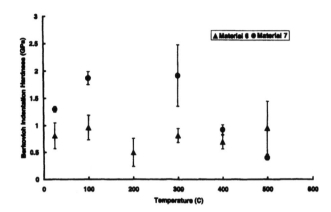

Figure 5:   Comparison of the Berkovich indentation hardness versus temperature of Materials 6 and 7 (Table 1).

Development of compression bending test method for advanced composites – A review –

Hiroshi Fukuda
Department of Materials Science and Technology, Tokyo University of Science
2641 Yamazaki, Noda, Chiba 278-8510, Japan

ABSTRACT

This paper reviews a series of our research works to have dealt with a compression bending test method during these years. Starting from the basic concept, some refinements of data reduction methodology and some advancements of test fixtures are described in conjunction with the application to flat coupons as well as pipe configurations. The most recent work of an eccentric compression bending is also reviewed.

INTRODUCTION

It is well known that the bending strength of CFRP coupon is strongly affected by the stress concentration due to a hard loading device. The works of Whitney [1] and Cui and Wisnom [2] are some examples to have dealt with this stress concentration. The work of Hojo, *et. al.* [3] focused on reducing the stress concentration by inserting soft film between the test specimen and the loading devices.

To compensate for this undesirable effect, we took another way. That is, the authors have hitherto demonstrated and proposed a compression bending test method [4-7] which is based on Euler buckling of a column. Applying "the elastica [8]" of Fig.1, a methodology to calculate the bending modulus and strength of flat coupons has already been almost established. That is, if we measure only the applied load and the crosshead movement during the buckling process of the column, we can calculate both the bending strength and the bending modulus with the aid of the elastica.

The undesirable effect of the loading device used for 3- or 4-point bending will be more serious for pipe configuration ; in most cases the thin-wall CFRP

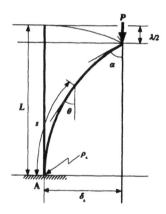

Fig. 1 Compression bending.

pipes will be crushed by the loading nose rather than a bending failure. Then, we tried to apply the above compression bending test method to CFRP pipes [9]; parametric study of various configurations of CFRP pipes has been conducted.

During a series of test, another difficulty took place. That is, if the diameter of the pipe is large, the Euler buckling has least likely occurred. Then, as a successive work, we tried a bending by means of eccentric compression as shown in Fig.2.

In the present paper, the results of both the compression bending and the eccentric compression bending will be shown.

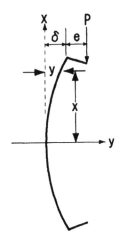

Fig. 2 Eccentric compression bending.

THE ELASTICA

Since details of compression bending test method have already been reported elsewhere several times [4-7], only a brief summary is described here.

Figure 1 shows the half length of the present specimen the experimental methodology of which is based on the Euler buckling. The bending moment at the midspan A is

$$M_A = P\delta_A \qquad (1)$$

where P is the applied load and δ_A is the midspan deflection. The bending stress (skin stress) at point A is, under linear elasticity condition,

$$\sigma_A = \frac{M_A t}{2I} \qquad (2)$$

where I is the moment of inertia of section and t is the thickness of the specimen. The bending strength is obtained by substituting the bending moment at failure in eq.(2).

The bending modulus is

$$E = \frac{M_A \rho_A}{I} \qquad (3)$$

where ρ_A is the radius of curvature at point A.

According to the elastica [8] and succeeding formulation [5], the following equations hold:

$$\frac{\delta_A}{L} = \frac{2p}{K(p)} \qquad (4)$$

$$\frac{\rho_A}{L} = \frac{1}{2pK(p)} \qquad (5)$$

where λ is the crosshead movement, $p=\sin\alpha/2$, and $K(p)$ is the perfect elliptic integral of the first kind. Because δ_A and ρ_A are mutually related through a parameter p, ρ_A can be calculated by measuring δ_A.

There is another sophisticated way of calculating ρ_A or even δ_A. By further calculation [6] of the elastica,

$$\frac{\lambda}{L} = 4 - \frac{E(p)}{K(p)} \qquad (6)$$

holds where $E(p)$ is the perfect elliptic integral of the second kind. From equations (4) - (6) we can calculate δ_A and ρ_A by measuring λ. That is, by measuring the applied load, P, and the crosshead movement, λ, both the bending strength and the bending modulus can be calculated.

RESULTS OF FLAT COUPONS

In the first paper [4], the bending strength was evaluated, that is, by measuring δ_A of Fig.1 at failure, the bending strength was calculated by eq.(2). Figure 3 compares the bending strength by means of the compression bending with that of 4-point bending of CFRP coupons where the alphabets U and Q represent unidirectional and quasi-isotropic, respectively, and the numerals 8 and 3 stand for T800 and T300 carbon fibers. The results show that for T300 composites, the bending strengths by both methods are almost the same whereas for T800 composites, the strength obtained by the compression bending was much higher than that of 4-point bending. Generally speaking, if we can get higher strength by a new test method, the test method is superior. In that sense, the compression bending test is suitable especially for high-performance composites such as T800 composites.

In the first paper [4], however, the bending modulus could not taken into account. Then we next tried to establish the methodology to get the bending modulus from the compression bending test. In Reference [5], equations (4) and (5) were used to calculate E whereas eq.(6)

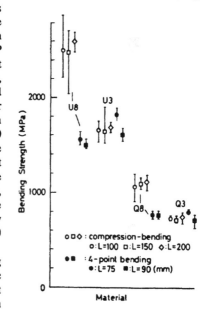

Fig. 3 Comparison of bending strength.

was adopted in Reference [6]. Figure 4 is the result of the bending modulus [5]. In the figure, the elastica method means the present method of using eq.(5) to get ρ_A which is necessary to calculate the bending modulus by eq.(3). On the other hand, in the straingage method, a pair of strain gages were glued on both surfaces at midspan of each test coupon. In this case, the radius of curvature, ρ_A, can be obtained by

$$\rho_A = \frac{t}{\varepsilon_t - \varepsilon_c} \qquad (7)$$

where ε_t and ε_c are the strains of tension-side surface and compression-side surface, respectively, and t is the specimen thickness.

Comparing with the 4-point bending, the present method may be acceptable to obtain the bending modulus.

APPLICATION TO PIPES

The above compression bending test was next applied to CFRP pipes [9]. Figure 5 [9] is the case for the pipes of the diameter of 5mm and the wall thickness of 0.3mm. Again the superiority of the present method was demonstrated because the measured strength was much higher than that of 3-point bending, although there exhibited some scatter in the test results.

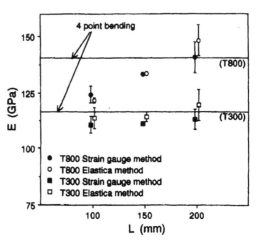

Fig. 4 Comparison of bending modulus.

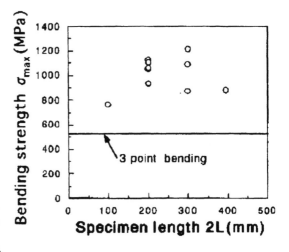

Figure 5. Comparison of bending strength (pipe).

ECCENTRIC COMPRESSION BENDING
– ANOTHER NEW METHOD –

In the case of eccentric compression bending of Fig.2, the bending moment of an arbitrary point x is [10]

$$M_x = P(e + \delta - y) \qquad (8)$$

and the maximum bending moment at the center of the pipe is

$$M_{max} = P(e + \delta) \qquad (9)$$

instead of eq.(1) for compression bending. Then the maximum bending stress is

$$\sigma_{max} = \frac{M_{max}}{Z} \qquad (10)$$

where Z is the section modulus of the pipe. If the axial compressive stress should be taken into amount,

$$\sigma_{max} = \frac{M_{max}}{Z} + \frac{P}{A} \qquad (11)$$

is the maximum stress, where A is the cross sectional area of the pipe.

To realize above idea, a set of test fixture shown in Fig.6 was designed and machined. Using this device, three kinds of CFRP specimens of 15mm diameters were tested [11], although details of specimen configuration are not shown here.

Figure 7 is a major result for Type B specimens the wall thickness of which is about 0.5mm. The failure stress was calculated by eq.(11). According to our experiment, the failure stress was almost constant against the amount of

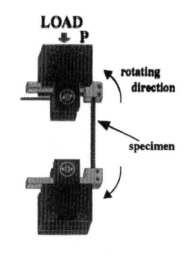

Fig.6 Schematic view for eccentric compression bending.

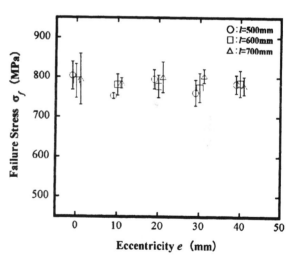

Fig. 7 Failure stress vs. eccentricity (Type B).

eccentricity and specimen length. Thus, it may be concluded that the present eccentric compression bending is reasonable test method for large diameter pipes. In the case of 15mm-diameter pipes, 3- or 4-point bending test was not conducted because it was no more a bending test to get reliable bending strength.

The compression bending, even the eccentric compression bending, inevitably includes compressive stress in addition to bending stress. If the ratio of the compressive stress to the bending stress is large, it is no more called a bending test. Figure 8 shows this ratio where

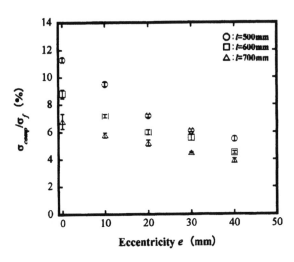

Fig. 8 Ratio of compressive stress to failure stress (Type B)

$$\sigma_{comp} = \frac{P}{A} \qquad (12)$$

With increasing the eccentricity and the specimen length, the ratio of compressive stress to bending stress decreased, as was expected. Thus, it may be concluded that the present eccentric compression bending test is acceptable from practical view point.

CONCLUSIONS

In this review paper, the idea of the compression bending test and data reduction scheme were first described together with step-by-step development of the idea.

This idea was first applied to flat coupons and then pipe-shape specimens. The applicability of the compression bending test was demonstrated. Comparing to the 3-point bending, the bending strength by means of the compression bending was much larger which leads the superiority of the present method. As for the bending modulus, measured values were comparable to those of 3-point bending. Thus, the method proposed here may become one candidate to get reliable bending properties of pipe-shaped components.

It may also be concluded that the eccentric compression bending is superior to the compression bending especially for large diameter pipes although the fixture becomes somewhat complicated.

ACKNOWLEDGEMENTS

The author thanks Toray Industries for providing CFRP flat coupons and Daiwa Seiko Co. for preparing tremendous number of CFRP pipes. Dr. Masaaki Itabashi and Dr. Atsushi Wada are acknowledged for designing and machining many types of fixtures, conducting tests, and preparing some parts of manuscripts. Thanks should also be addressed to many students including Mr. Asao Koike, Mr. Hiroshi Nakane, Mr. Hisataka Katoh, Mr. Hiromu Uesugi, Mr. Tetsuya Watanabe, Mr. Makoto Yamazaki, Mr. Osamu Watanabe and Mr. Isao Watanabe for conducting tests.

REFERENCES

1. Whitney, J. M., "Elasticity analysis of orthotropic beams under concentrated loads," *Composites Sci. Tech.*, Vol.22, 1985, pp.167-184.
2. Cui, W.-C. and M. R. Wisnom, "Contact finite element analysis of three- and four- point short-beam bending of unidirectional composites," *Composites Sci. Tech*, Vol.45, 1992, pp.323-354.
3. Hojo, M., Y. Noguchi, H. Furue and J. Matsui, "On the bending test method of advanced composite materials – Effect of cushion materials," *Proc. 28th JSASS/JSME Strength Conf.*, 1986, pp.370-373 (in Japanese).
4. Fukuda, H., "A new bending test method of advanced composites," *Experimental Mechanics*, Vol.29, 1989, pp.330-335.
5. Fukuda, H., "Compression bending test method for advanced composites," *J. Japan Soc. Aero. Space Sci.*, Vol.41, 1993, pp.482-487 (in Japanese).
6. Fukuda, H., H. Katoh and H. Uesugi, "A modified procedure to measure bending strength and modulus of advanced composites by means of compression bending," *J. Composite Mater.*, Vol.29, 1995, pp.195-207.
7. Fukuda, H. and M. Itabashi, "Simplified compression bending test method for advanced composites," *Composites, Part A*, Vol.30, 1999, pp.249-256.
8. Timoshenko, S. P. and J. M. Gere, "Theory of Elastic Stability," 2nd ed., McGraw-Hill, 1961, pp.76-82.
9. Fukuda, H., T. Watanabe and M. Itabashi, "Compression bending test method for CFRP pipe," *Proc. ACCM-2000*, Kyongju, Korea, 2000, pp.1187-1192.
10. Timoshenko, S., "Strength of Materials, Part I," 3rd ed., Van Nostrand Reinhold, 1955, p.259.
11. Fukuda, H., O. Watanabe, M. Itabashi and A. Wada, "Eccentric compression bending test for CFRP pipe," *Proc. JISSE-7*, Tokyo Big Sight, 2001, pp.967-970.

Mechanical Properties of Silk Fiber Reinforced Thermoplastic Composites

T.KIMURA, S.KATORI and K.HANADA
Advanced Fibro-Science
Kyoto Institute of Technology
Matsugasaki, Sakyo-ku, Kyoto
606-8585, Japan
tkimura@ipc.kit.ac.jp

S.HATTA
Kyoto Municipal Textile Research Institute
Kamidacyuri-agaru, Karasuma-dori
Kamigyo-ku, Kyoto
602-0869, Japan
hatta@city.kyoto.jp

ABSTRACT

The present paper discussed the applicability of silk fiber as a reinforcement for fiber reinforced thermoplastic composites. Nowadays, much attention has been focused upon the development of fiber reinforced composites which have superior mechanical properties. Recently, attention has been shifting from glass and carbon fibers to natural fibers because of their renewable nature, low cost, low density and low energy consumption. In this study, the silk textile fabrics was used as a reinforcement, and the compression molding method was performed by using PP and PBS films as matrix materials. The tensile and bending tests were performed and their properties were discussed. The strength of silk composites was increased with increasing volume fraction of silk fiber for both tensile and bending tests. The adhesion between silk fiber and matrix materials affects little on the mechanical properties because of the continuous fiber mode of textile fabrics used here. However, the bending strength of Silk/PP composites treated by maleic anhydride polypropylene(MAPP) was larger than that of Silk/PP composites without MAPP. It is an indication from the results that the silk fabrics acts effectively as a reinforcement of thermoplastic composites.

INTRODUCTION

The increased emphasis has been placed on developing the eco-materials with the goals of protecting the environment. Recently, attention has been shifting from glass fiber and carbon fiber to natural fibers as reinforcement for the composite materials because of their renewable nature, low cost, low density and low energy consumption. Generally, the natural fiber can be classified into two types; one is cellurosic fiber and the other is protein fiber. Application of the cellulosic fibers such as jute, flax, hemp and kenaf to the reinforcement of composite materials has been discussed in many papers [1-4]. However, the applicability of a protein fiber to a reinforcement of the composite materials has been rarely discussed until now. Therefore, in this study, the silk fiber reinforced composite materials were tried to mold by a compression molding method.

In generally, for the matrix materials, polypropylene (PP) has been used for the wide range as a usual thermoplastic resin. Meanwhile, biodegradable polymers have been paid much attention in recent years, because their use should bring decrease in environmental problems caused by the wastes of synthetic non-biodegradable polymers. Aliphatic polyesters such as poly L-lactic acid (PLLA), poly 3-hydroxybutyrate (PHB) and polybutylenes succinate (PBS) are well known to the representative biodegradable polymers. The biodegradable composites are expected to use for the geo-materials and constructions.

Under these circumstances, the present paper discusses the mechanical properties of silk fiber reinforced PP and also PBS composites.

EXPERIMENTAL PROSEDURE

Materials used

Raw silk fiber was used as a reinforcement. Raw silk fiber consists of two types of protein such as fibroin and sericin. The sericin in the raw silk used here was not remove at the degumming process. Figure 1 shows the silk textile with plain fabric construction used here. The properties of silk fabrics and silk fibers pulled out from the fabrics are shown in Tables 1 and 2, respectively. The properties of raw silk fiber used for the fabrication of silk textile were also shown in Table 2. It is clearly seen in Table 2, the tensile strength of silk fibers pulled out from the fabrics is slightly smaller than that of silk fiber before fabrication. However, the value of strength is higher by one digit than those of usual thermoplastics resin plates. Therefore, the silk textile used here can be expected to be a reinforcement of the composite materials.

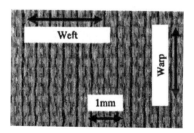

Fig.1   Aspect of silk fabric used

Table1   Properties of silk fiber

treatment (min)	Tensile strength (MPa)	Tensile modulus (GPa)	Breaking elongation (%)	Fineness (denier)
0	522	14	25	27
30	468	10	16	22
90	499	11	17	22

Table2   Properties of silk fabric

	Density (yarns/cm)	Tensile strength (kN) (JIS L 1096)
warp direction	55	8.74
weft direction	32	4.11

Polypropylene (PP) and Polybutylene succinate (PBS) were used as matrix materials in this study. To determine the molding condition, the thermogravimetry (TG) measurement of silk fiber and the differential scanning calorimetry (DSC) measurement of PP and PBS were performed and the results were shown in Figs. 2(a) and (b), respectively. It should be noted here that the melting points of PP and PBS are 165℃ and 114℃, respectively and are smaller than the deterioration temperature 250℃ of silk fiber shown in Fig.2(a). PP and PBS resins were used in film form in order to make it easy to mold together with silk fabrics in textile form. The thickness of PP and PBS films were 100 μ m and 20 μ m, respectively.

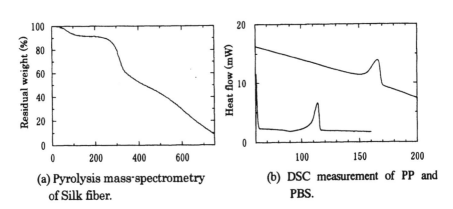

(a) Pyrolysis mass-spectrometry of Silk fiber.

(b) DSC measurement of PP and PBS.

Fig.2   Property of materials used

Molding method of composite

The silk fiber reinforced thermoplastic composites were molded by a compression method. Figure 3 shows the aspect of compression molding method used here. The silk textile fabric and film type matrix resin were laminated mutually. The volume fraction of silk fabrics was varied by adjusting the laminating number of silk fabric and resin films. The heating conditions were 180°C/3min. and 130°C/3min. for PP and PBS, respectively,   in consideration of their melting points. Preheating of 60°C/120min. for PP and 40°C/120min. for PBS was

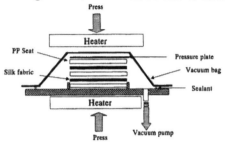

Fig.3 Schematic illustration of hot press

performed for both composites to avoid the deterioration of resin during the compression molding. The dimensions of the molded composites were 150mm x 220mm x 2mm(thickness).

Mechanical tests

The tensile and the three points bending tests were performed at an ambient temperature of 20°C by using Instron Universal Testing Machine. According to JIS K7118 and JIS 7055, the crosshead speeds were 5 mm/min. and 2 mm/min. for tensile and bending tests, respectively.

RESULTS AND DISCUSSION

Aspect of cross section and interface

Figures 4(a) and (b) show the cross section of molded Silk/PP and Silk/PBS composites, respectively. The resin layer without silk fiber can be clearly seen between neighboring two fabrics because of the laminating molding used here. It is cleared here that the melting resin impregnates well between fiber bundles for both composites. However, the lack of adhesion between silk fiber and PP resin can be seen in Fig.5(a).

The adhesion of interface between fiber and matrix is very important to achieve a good mechanical   properties   for   the   composites.   Therefore,   the   maleic   anhydride polypropylene(MAPP) was added as the compatibilizer to improve the adhesion of Silk/PP composite. In generally, MAPP was blended with PP before molding. However, in this study, MAPP powder was sprinkled only on the adhesive surface of the silk fabrics before compression molding in order to simplify the molding method. Figure 6 shows the MAPP powder used here and the value of addition was $0.01g/cm^2$.   As a result, the good adhesion between silk fiber and PP matrix was obtained as shown in Fig.7.

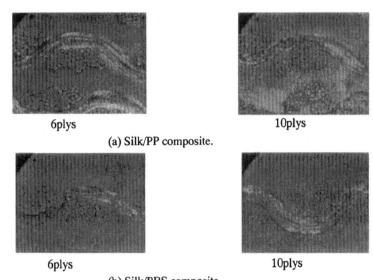

6plys   10plys
(a) Silk/PP composite.

6plys   10plys
(b) Silk/PBS composite.
Fig.4 Observance of cross section for silk composites

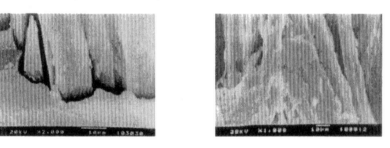

(a) Silk/PP composite   (b) Silk/PBS composite
Fig.5 SEM microphotograph of cross section for Silk/PP composites and Silk/PBS composites

Fig. 6 SEM microphotograph of MAPP powder

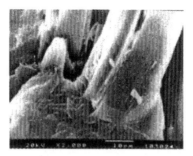

Fig.7 Silk/PP composites with MAPP

Tensile properties

Tensile stress-strain curves of Silk/PP and Silk/PBS are shown in Figs.8(a) and (b), respectively. It is noted from these figures, the initial slopes of the curves become more steeper

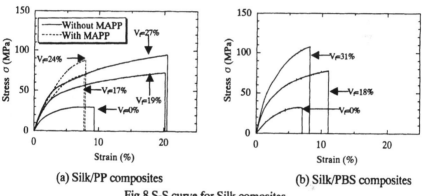

(a) Silk/PP composites

(b) Silk/PBS composites

Fig.8 S-S curve for Silk comosites.

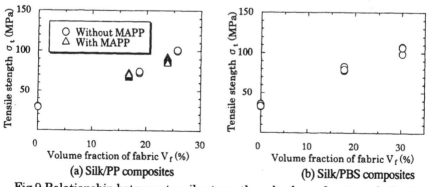

(a) Silk/PP composites

(b) Silk/PBS composites

Fig.9 Relationship between tensile strength and volume fraction of silk fabrics.

than those of matrix materials and the maximum stresses for the composites are fairly larger than those of matrix materials. It should be noted here that the breaking strain for Silk/PP composite without MAPP is considerably larger than those of PP and Silk/PP composite with MAPP. This fact indicates that the good adhesion between silk fiber and PP matrix leads the brittle property of composite.

The tensile strength of Silk/PP composite was shown in Fig.9(a). It is cleared here that the silk fiber becomes a good reinforcement for the composite and the strength increases largely with increasing volume fraction of silk fiber. However, the containing of MAPP scarcely affects on the strength of composites reinforced by the continuous silk fiber. Figure 9(b) shows the strength of Silk/PBS composites. The strength increases largely with increasing volume fraction of silk fiber as same as the results of said Silk/PP composites.

Bending properties

Figures 10(a) and (b) show the typical stress-deflection curves for each composite obtained from the bending tests. The large deference of curves can be seen between Silk/PP composites with and without MAPP. Namely, Silk/PP composite with MAPP has a brittle property differ from the ductile property of matrix material and Silk/PP composite without MAPP. The ductile property also can be seen for Silk/PBS composites.

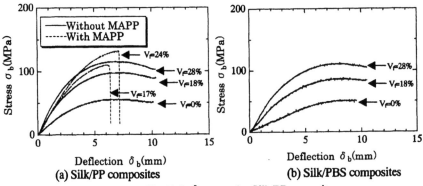

(a) Silk/PP composites

(b) Silk/PBS composites

Fig.10 S- δ curve for Silk/PP comosites.

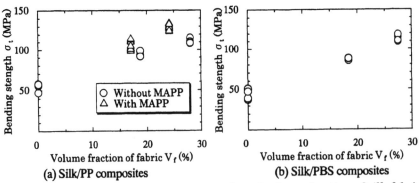

(a) Silk/PP composites

(b) Silk/PBS composites

Fig.11 Relationship between bending strength and volume fraction of silk fabrics.

Figure 11(a) shows the bending strength of Silk/PP composites. It is cleared here that the bending strength increases largely with increasing volume fraction of silk fiber. It should be noted in the figure that the strength of Silk/PP composite without MAPP is slightly smaller than that of Silk/PP composite with MAPP. This may be caused by the separation of the interface between silk fiber and PP matrix in the compression side of the composite without MAPP because of the lack of adhesion. The bending strength of Silk/PBS composites increases largely with increasing volume fraction of silk fiber as same as Silk/PP composites.

CONCLUSION

Composite materials based on protein fiber such as silk fiber were prepared by compression molding method. The silk textile fabrics were used for the reinforcement, and PP and PBS resins were selected as the thermoplastic matrix. It was found that the tensile and bending strength increased with increasing volume fraction of silk fiber. Consequently, the silk fiber is regarded as a good reinforcement for the eco-composites.

REFERENCE

1 X.Chen, Q.Guo, Y.Mi,"Bamboo Fiber-Reinforced Polypropylene Composites: A Study of the Mechnical Propertys", J. Applied Polymer Science, Vol.69, 1891-1899 (1998).

2 M.Avella, L.Casale, R.Dell'erba, B.Focher, E.Martuscelli, A.Marzetti, "Broom Fibers as Reinforceing Materials for Polypropylene-Based Composites", J. Applied Polymer Science, Vol.68, 1077-1089 (1998).

3 Norma E.Marcovich, Maria M.Reboredo, M.I.Aranguren,"Mechanical Properties of Woodflour Unsaturated Polyester Composites", J. Applied Polymer Science, Vol.70, 2121-2131 (1998).

4 Michael P.Wolcott, S.Yin, Timothy G.Rials,"Using dynamic mechanical spectroscopy to monitor the crystallization of PP/MAPP blends in the presence of wood", Composite Interfaces, Vol.7, No.1, pp.3-12(2000)

Cyclic Mechanical Properties of Composites Based on Shape Memory Polymer

Takeru Ohki, Qing-Qing Ni and Masaharu Iwamoto

Division of Advanced Fibro-Science, Kyoto Institute of Technology,
Matsugasaki, Sakyo-ku, Kyoto 606-8585, Japan
E-mail: nqq@ipc.kit.ac.jp

Abstract

Recently, shape memory polymer (SMP) as one of functional materials has received much attention and its mechanical properties have been investigated. Shape memory polymer of polyurethane series has the glass transition temperature (Tg) around the room temperature. According to the large change in modulus of elasticity above and below Tg, the SMP material has excellent shape memory effect. In this study, the glass fiber reinforced shape memory polymer was developed for the improvement of the mechanical weakness of SMP bulk and for wider applications in the fields of industry, medical treatment, welfare and daily life. The specimens with different fiber weight fractions were fabricated and their mechanical behavior and shape memory effect were investigated experimentally. It was confirmed that the cyclic mechanical properties of the shape memory polymer was improved by the reinforcement fibers and at the same time the shape memory effect was measurably kept in the developed composites.

INTRODUCTION

Shape memory materials have received much attention in industries and the other fields, particularly for Shape Memory Alloys (SMAs) that are a group of metallic alloy and exhibit a shape memory effect. On the other hand, shape memory polymers (SMPs) are another group of shape memory materials and they may also be used as a temperature sensor or an actuator. Compared with SMAs, SMPs have the advantages, such as lightweight, large recovery ability, superior processability and lower cost. Most of SMPs has the glass transition temperature (Tg) around the room temperature. Although SMPs indicated that the deformed shape returned to the original shape by heating, the mechanism of shape memory effect and the change of mechanical properties in the SMPs were different from those in SMAs. It is well known that the mechanism of the shape memory polymer was based on thermo-elastic phase transformation and its reversal at the temperatures above and/or below Tg. When the SMPs were heated at fluxing temperature over Tg and formed into a specified shape by compression, extrusion or injection moldings they memorized the formed shape (Process 1). The formed component was easily deformed to form an arbitrary shape by heating at glass transition temperature and external force (Process 2). Then, the deformed shape was fixed by cooling below Tg (Process 3). This was called the shape fixity in shape memory behavior. After the use of the deformed shape obtained in Process 3, SMPs could recover their memorized shape (Process 1) by heating over glass transition temperature with free load condition (Process 4). In the SMPs, the polyurethane series has following advantages: the forming processes for other thermoplastic polymer could be used; the shape recovery temperature could be set at any value within ±50K around the room temperature; there existed the large differences of the mechanical properties [1-3], the optical property and the water vapor permeability at the temperatures above and/or below Tg. Based on these advantages, the SMPs of polyurethane series were expected to have wider applications in the field of industry, medical treatment, welfare and daily life. However, the use of these materials was quite limited due to their low strength.

In this study, fiber reinforced composites based on SMP were developed in order to overcome the mechanical weakness in SMP bulk. The materials developed were the short glass fiber reinforced SMP of the polyurethane series with different fiber weight fractions [4]. For the practical use of the developed composites, it is important to clarify the fundamental mechanical and shape memory behavior to meet the reliability requirement. Additionally, the thermo-mechanical behaviors, such as stress-strain-temperature relations, are also important due to the influence of thermal processes. Thus, the tensile properties of the developed composites at the testing temperatures of Tg-20K, Tg and Tg+20K were evaluated. The mechanical cyclic test under constant strain condition was performed at room temperature (Tg-20K) for the evaluation of the mechanical properties for practical use. Then, thermo-mechanical cycle tests with consideration of both mechanical and thermal factors were carried out and the influence of fiber weight fraction and the thermal condition on shape memory effects was investigated.

EXPERIMENTAL PROCEDURE

Fabrication of specimens

The shape memory polymer (DIARY, MM4510: DIAPLEX Co., Ltd.) with Tg of 318K was used as the matrix and the chopped strand glass fibers with fiber length of

3mm (03MA411J,ASAHI FIBER GLASS Co., Ltd.) were used as the reinforcement in the developed composites.

The matrix and reinforcements were compounded by a twin screw extruder (LABOTEX-300, produced by JAPAN STEEL WORKS Co. Ltd) at the cylinder temperature of 483K and the screw rotation of 200rpm. The fiber weight fractions were specified as SMP bulk, 10wt%, 20wt% and 30wt%, respectively. Dumbbell type specimens (JIS K7113 Type1) were fabricated by an inline screw type of injection molding machine (Plaster Ti-30F6, produced by TOYO MACHINERY and METAL Co., Ltd.) after enough drying at 353K for the compounded materials. The fabricated specimens were non-weld. The cylinder temperature, mold temperature and injection speed were 483K, 303K and 27.4 cm³/sec, respectively. Figure 1 illustrates the geometry and size of a specimen.

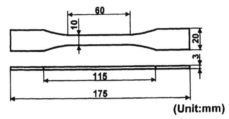

(Unit:mm)

Figure 1 — Geometric shape and size of a specimen

Experimental equipment

The experimental equipment used in this study was an Instron Universal Testing Instrument (Type 4466) with a temperature-controlled chamber. Heating or cooling for specimens was controlled by compressed and heated or cooled air at the atmosphere and the temperature was measured by a thermocouple near the specimen. The tip of the thermocouple was put between two 1.5mm thickness plates with the same material as that in the specimen to make the same temperature condition.

Static tensile test

The static tensile test was performed at the crosshead speed of 5mm/min. within the temperature-controlled chamber. The temperature in the chamber was specified at 298K(Tg-20K), 318K(Tg) and 338K(Tg+20K), respectively.

The strain was calculated by the ratios of the elongation obtained by the crosshead displacement to the span length (60mm) with a maximum of 300% due to the limit of the chamber.

Mechanical cycle test

In order to investigate the cyclic behavior of the developed composites, the constant strain cycle test was performed at room temperature (298K=Tg-20K), where the upper limit strain value was set to be 50% of the strain at the maximum stress obtained in the static tensile test. The test was performed at the crosshead speed of 5 mm/min. until the cycle numbers of 20, 40 and 60, respectively. The influences of cycle number and fiber weight fraction on the mechanical behavior of the developed composites were discussed.

Thermo-mechanical cycle test

Thermo-mechanical cycle tests were performed to investigate the shape memory effect of the developed composite. Figure 2 shows the schematic of the thermo-mechanical cycle.

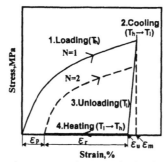

Figure 2 — The schematic stress-strain curve in the thermo-mechanical cycle test

The specimen was loaded to the strain ε_m at the constant crosshead speed of 5 mm/min. at the temperature T_h (Process 1). Then, it was cooled to the temperature T_l by keeping the same strain ε_m (Process 2). After five minutes at the temperature T_l, the load on the specimen was taken off (Process 3), and then the specimen was heated from T_l to T_h in ten minutes under free-loading (Process 4). One thermo-mechanical cycle consisted of these four processes and then the cycle test was repeated to N cycles. The conditions in the thermo-mechanical cycle test were as follows: ε_m=100%, T_h=338K, T_l=298K, the crosshead speed of 5mm/min. and N=5. The strain was measured as done in the static tensile test.

RESULTS AND DISCUSSIONS

Static tensile property

Firstly, the results of static tensile test in previous research are cited. Figure 3 shows the maximum stresses within the strain range of 300%. The maximum stress increased greatly at any temperature with the increment of fiber weight fraction. At T=298K(Tg-20K), the maximum stresses of each specimen with reinforcement fiber of 10wt%, 20wt%, and 30wt% were 9%, 23% and 44% higher than that of bulk specimen, respectively. At the higher temperature of T=318K and 338K (Tg and Tg+20K), they also got the increment from 8% to 15% in comparison with bulk specimen. This indicates that the fiber weight fraction plays an important role on the maximum stress. Figure 4 shows the relationship between temperature and Young's modulus of all materials. It is known that the elastic modulus of polymer material at the temperatures above and/or below Tg is considerably different. The higher elastic modulus at the temperature below Tg is due to the energy elasticity in crystalline phase and glass state amorphous phase, while the lower elastic modulus at the temperature above Tg may be considered as the entropy elasticity caused by the micro-Brownian movement of amorphous phase. This variation of elastic modulus around Tg is a key point to utilize and control the shape memory effects of the materials based on SMP. The large change of Young's modulus above and below Tg was observed for all of specimens as shown in Fig.4. In other words, the result in Fig.4 means that the developed materials may have shape memory effect.

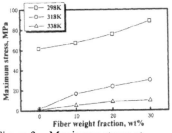

Figure 3 — Maximum stress at each testing temperature

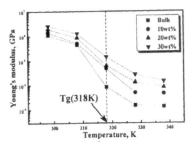

Figure 4 — The Relationship between temperature and Young's modulus

The results in the static tensile tests can be remarked briefly as follows: a obvious increment of the tensile strength was obtained in the developed materials when fiber weight fraction increased; a large change in Young's modulus above and below Tg for all materials were observed and its shape memory effect was partially kept.

Mechanical cycle property

Figure 5 shows the stress-strain curves in constant strain cycle tests. For the specimens with different fiber weight fractions, a large hysteresis loop was observed at the first cycle and there was no obvious difference in the loop shape except the slope of the loop, which corresponded to the Young's modulus. It is considered that the large hysteresis loop at the first cycle is mainly contributed by matrix deformation and failures around fibers. However, the loops following the first cycle showed almost no hysteresis due to the characteristics of SMP with a training effect.

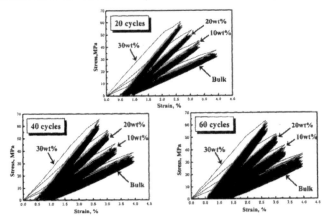

Figure 5 — The stress-strain curves in a mechanical cycle test.

Figure 6 shows the stress decrement per each cycle, which may be a parameter for measuring relaxation due to constant strain in this test. The stress decrement was defined as the ratio of stress value at each cycle to that at the first cycle. In the bulk specimen, the stress decreased about 30% at 60 cycles and the decrement slope was almost invariant. On the other hand, the stress of specimens with reinforcement fiber indicated the decrement of from 10% to 20% at 60 cycles and the slope of stress decrement curve tended to be smooth with the increment of cycle number. Figure 7 shows the total residual strain after the prescribed cycle numbers of 20, 40, and 60. The total residual strain in the bulk and 10wt% specimens increased between 20 cycles and 40 cycles, and tended to be constant after 40 cycles. But it seems to be unchanged in 20wt% and 30wt% specimens even though the cycle number increased. Figure 6 and 7 indicated that reinforcement fibers mixed in the SMP caused stress decrement and stabilized the cyclic behavior of the developed materials.

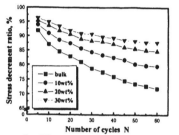

Figure 6 — The stress decrement ratio for
each fiber weight fraction

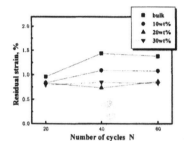

Figure 7 — The Residual strain
after cyclic loading

Thermo-mechanical cycle property

Figure 8 shows the stress-strain curves in a thermo-mechanical cycle test. The maximum stresses and Young's modulus in each cycle increased with the increment of the cycle number N. This may be caused by the strain hardening of the materials.

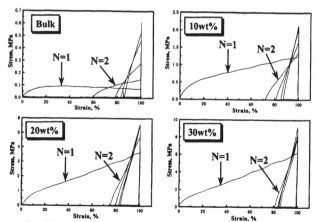

Figure 8 — The stress-strain curves in thermo-mechanical cycle tests

Here, let us look at the recovery strain ε_r (see Fig.2), i.e., the strain recovered when the materials were heated from T_l to T_h without loading (Process4). The relationship between the strain recovery ratio and the number of cycle are shown in Fig.9. The strain recovery ratio was defined by the value of $\varepsilon_r / \varepsilon_m$. The recovery ratio at first cycle appeared to be quite different for the specimens with different fiber weight fractions. It is clear that the recovery strain ε_r in the specimens with fiber weight fractions of 10, 20 and 30wt% were much smaller than that in the bulk specimen. However, the recovery strain ε_r after the second cycle was almost of the same value.

In order to investigate the shape memory effect after the cyclic loading, the thermo-mechanical cycle test was performed to the specimen after mechanical cycle test (20 cycles). Figure 10 shows the strain recovery ratios at the first thermo-mechanical cycle with and/or without mechanical cyclic loading history. Strain recovery ratios in the bulk and 10wt% specimens with cyclic loading history were about 10% and 5% lower than those without cyclic loading history, respectively. On the other hand, the influence of cyclic loading history hardly appeared in the 20 and 30wt% specimen. These differences caused by fiber weight fraction agree with the result of the degrading mechanical properties by cyclic loading as shown in Fig.6 and 7. Consequently, it is considered that the improvement of tolerance under cyclic loading may prevent the degradation of shape recovery effect in the practical use even under the complex conditions of both mechanical cyclic loading and thermal action.

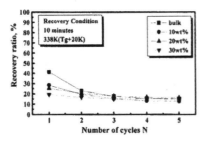

Figure 9 — The relationship between recovery ratios and the number of cycle

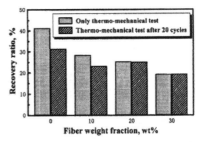

Figure 10 — The effects of cyclic loading on recovery ratios

CONCLUSION

In this study, the composites based on the shape memory polymer were developed and their cyclic behavior and shape memory effects were investigated by the experimental approach. The results obtained are remarked as follows.
1. The tensile strength of the developed materials became higher with the increment of fiber weight fraction under each temperature condition.
2. The resistance to cycle loading of the developed composites with SMP was clearly improved due to reinforcement fiber.
3. It was confirmed that the developed smart composites measurably kept the shape memory effect.

REFERENCES

Journal References

1. H. Tobushi, S. Hayashi and S. Kojima, "Mechanical Properties of Shape Memory Polymer of Polyurethane Series," Transactions of the Japan Society of Mechanical Engineers, A, 57 (1991) 146 (in Japanese).
2. H. Tobushi, S. Hayashi and S. Kojima, "Cycle Deformation Properties of Shape Memory Polymer of Polyurethane Series," Transactions of the Japan Society of Mechanical Engineers, A, 58 (1992) 139 (in Japanese).
3. H. Tobushi, S. Hayashi and S. Kojima, "Constitutive Modeling for Thermo-mechanical Properties in Shape Memory Polymer of Polyurethane Series," Transactions of the Japan Society of Mechanical Engineers, A, 64 (1998) 186 (in Japanese).

Edited Conference Proceeding

4. Q-Q. Ni, N. Ohsako, M. Sakaguchi, K. Kurashiki and M. Iwamoto, "Mechanical Properties of Smart Composites Based on Shape Memory Polymer," JCOM: JSMS COMPOSITES-29 of the society of Material Science, Japan, (2000-3) 293 (in Japanese).

Characterization of nonlinear stress-strain behavior of carbon/epoxy unidirectional and angle-ply laminates

Shinji Ogihara and Yusuke Hirakawa
Department of Mechanical Engineering
Tokyo University of Science
2641 Yamasaki, Noda, Chiba 278-8510, Japan
ogihara@rs.noda.sut.ac.jp

Satoshi Kobayashi
Department of Mechanical Engineering
Graduate School of Engineering
Tokyo Metropolitan University
1-1 Minami-Ohsawa, Hachioji, Tokyo 192-0347, Japan

Nobuo Takeda
The University of Tokyo
4-6-1 Komaba, Meguro-ku, Tokyo 153-8904, Japan

ABSTRACT

Nonlinear mechanical behavior of carbon/epoxy unidirectional and angle-ply laminates under uniaxial tensile loading is investigated experimentally. The one-parameter plasticity model is used to characterize the nonlinear behavior of the unidirectional laminates. The mechanical behavior of the angle-ply laminates is predicted by using the combination of the classical lamination theory and the one-parameter plasticity model of the unidirectional laminates.

INTRODUCTION

It is well known that fiber reinforced plastics exhibit nonlinear stress-strain response under off-axis loading (1-15). Nonlinear behavior of composites has been modeled by two approaches, i.e. macroscopic and microscopic. In the macroscopic approach, composites are considered as a homogeneous nonlinear elastic or plastic body. In the microscopic approach, attempts are made to describe the effective composite response using the properties of the fiber and matrix.

In the present study, nonlinear mechanical behavior of carbon/epoxy unidirectional and angle-ply laminates under uniaxial tensile loading is investigated experimentally. The validity of the generalized method of cells for predicting the nonlinear stress-strain relation of the unidirectional laminates under off-axis tensile loading is discussed. The one-parameter plasticity model is applied to characterize the nonlinear behavior of the unidirectional laminates and to predict the stress-strain relation of angle-ply laminates.

EXPERIMENTAL PROCEDURE

The material system used is a T700S/2500 carbon/epoxy composite. The 2500 epoxy system is a 120°C cure conventional epoxy resin. Unidirectional (4 plies) and angle-ply ($[\pm\theta]_s$) panels were fabricated by following the manufacturer's recommended cure cycle.

The unidirectional specimens were cut from the unidirectional panels. 0°, 15°, 30°, 45°, 60° and 90° specimens were prepared. The angle-ply panels, $[\pm\theta]_s$, where $\theta=15°$, 30°, 45° and 60°, were also fabricated, from which the angle-ply specimens were cut. The specimen size for both unidirectional and angle-ply laminates was 150mm long, 10mm wide and 0.52mm thick. 25mm long GFRP tabs were glued on both ends of the specimens which results in the specimen gage length of 100mm. The fiber volume fraction for the composite was about 0.55. Resin specimens were also prepared. The size of the resin specimens was 100mm long, 10mm wide and 2.0mm thick. No tabs were used for resin specimens and the grip length was 20mm for each end so that the specimen gage length was 60mm.

Tensile tests were performed on the unidirectional, angle-ply laminates and resin specimens at room temperature. The crosshead speed was 0.5mm/min. The strain data was obtained by averaging the data from the strain gages attached on both sides of the specimens. At least three specimens were tested for each condition.

ANALYSIS

One-Parameter Plasticity Model

The one-parameter plasticity model was developed by Sun and Chen (2) to model the nonlinear mechanical behavior of unidirectionally reinforced composite materials. Starting with a general 3-dimensional yield function, simplification is made by assuming that there is no plastic deformation in the fiber direction, which results in the yield function as

$$2f = \sigma_{22}^2 + 2a_{66}\sigma_{12}^2 \tag{1}$$

where a_{66} is a plastic parameter which should be determined from experiments. The parameter is chosen so that the effective stress-effective plastic strain curves from various off-axis tension data reduce to a single master curve. In this study, the one-parameter plasticity model is used to characterize the nonlinear behavior of the T700S/2500 carbon/epoxy unidirectional laminate.   An attempt is made to predict the stress-strain behavior of the angle-ply laminates based on the one-parameter plasticity model of the unidirectional laminate.

RESULTS AND DISCUSSION

Figure 1 shows the typical stress-strain curves for (a) unidirectional (0°, 15° and 30° specimens), (b) unidirectional (45°, 60° and 90° specimens), (c) angle-ply laminates (±15° and ±30°) and (d) angle-ply laminates (±45° and ±60° specimens).   In the unidirectional laminates, the stress-strain relation is almost linear for 0° specimens.   All the off-axis specimens and 90° specimens exhibit nonlinear stress-strain behavior.   In the angle-ply laminates, only ±15° specimens show an almost linear relation between stress and strain. The ±45° specimens exhibit very high nonlinearity and large failure strain.

In the present study, the one-parameter plasticity model is also used to characterize the unidirectional stress-strain behavior.   It is assumed that the total strain at a certain stress level is decomposed into a linear elastic part and nonlinear part and that the nonlinear part is plastic strain.   By subtracting the linear elastic part from the total strain, the relation between the stress and the plastic strain is obtained as shown in Fig.2 (a).   In the one-parameter plasticity analysis, the curves are converted to effective stress-effective plastic strain curves using an assumed value of the parameter a_{66}.   The following relations   are used.

$$\sigma = h(\theta)\sigma_x$$
$$\varepsilon = \varepsilon_x^p / h(\theta)$$
$$h(\theta) = \left\{ \frac{3}{2}\left(\sin^4\theta + 2a_{66}\sin^2\theta \cos^4\theta \right) \right\}^{\frac{1}{2}}$$

$$(2)$$

where σ_x is the uniaxial stress in the tensile direction, ε_x^p is the plastic strain in the tensile direction and θ is the off-axis fiber angle.   Our task is to find the value of a_{66} which makes the curves of various off-axis data gather on a master curve.   We obtain a value of $a_{66}=2.0$ and resulting master curve is shown in Fig.2 (b).   The master curve is fitted in the form

$$\varepsilon^p = \alpha\sigma^n$$

$$(3)$$

where α and n are constants.   The values we obtained are $\alpha=7.5\times10^{-14}$ (MPa^{-n}) and $n=5.1$.   The experimentally-obtained mechanical properties of the T700S/2500 carbon/epoxy unidirectional laminate   are summarized in Table 1.   By using these data we can predict the stress-strain curve at any off-axis angle.

In this paper, an attempt is made to predict the nonlinear behavior of angle-ply laminates under an assumption that the nonlinear behavior of the unidirectional composite is known, that is, the nonlinear behavior of unidirectional laminate can be described by the Sun and Chen one-parameter plasticity model.   We perform a classical lamination analysis using the one-parameter plasticity model.

Figure 3 shows the comparison between the experimental results and the analytical predictions for stress-strain curves of the angle-ply laminates. A good agreement is obtained which implies the validity of the combination of the one-parameter plasticity model and the classical lamination theory.   It may be possible to predict the stress-strain behavior of general multidirectional laminates by using the one-parameter plasticity model.

CONCLUSION

Nonlinear mechanical behavior of carbon/epoxy unidirectional and angle-ply laminates under uniaxial tensile loading is investigated experimentally.   The one-parameter plasticity model is also used to characterize the nonlinear behavior of the unidirectional laminates.   By combining the one-parameter plasticity model with the classical lamination theory, the mechanical behavior of the angle-ply laminates is predicted and good agreement with the experimental results is obtained.   As a lamina-level macroscopic approach, the one-parameter plasticity model will be useful to predict the mechanical behavior of multidirectional laminates.

REFERENCE

1.   Hahn, H.T. and S.W.Tsai. 1973. *J. Composite Materials*, 7:102-118.
2.   Sun, C.T. and J.L.Chen. 1989. *J. Composite Materials*, 23: 1009-1020.
3. Kenaga, D., J.F.Doyle and C.T.Sun. 1987. *J. Composite Materials*, 21: 516-531.
4. Ogi, K. and N.Takeda. 1997. *J. Composite Materials*, 31: 530-551.
5.   Tamuzs, V. J.Andersons, K.Aniskevich, J.Jansons and J.Korsgaad. 1998. *Mechanics of Composite Materials*. 34: 321-330.
6. Aboudi, J. 1987. *Int.J.Engng Sci.* 25:1229-1240.
7. Aboudi, J. 1990. *J. Reinforced Plastics and Composites*, 9: 13-32.
8. Paley, M. and J.Aboudi. 1992. *Mechanics of Materials*, 14: 127-139.
9. Bednarcyk, B.A. and M.-J.Pindera. 1997. *NASA Contractor Report* 204153, Cleveland, OH: NASA Lewis Research Center.
10. Pindera, M.-J. and B.A.Bednarcyk. 2000. *Composites: Part B*, 30: 87-105.
11. Dvorak, G.J. and Y.A.Bahei-El-Din. 1982. *Transactions of the ASME*, 49: 327-335.
12. Bahei-El-Din, Y.A. and G.J.Dvorak. 1982. *Transaction of the ASME*, 49:740-746.
13. Sun, C.T. and J.L.Chen. 1991. *Composites Science and Technology*, 40:115-129.
14. Sun, C.T., J.L.Chen, G.T.Sha and W.E.Koop. 1993. *Composites Science and Technology*, 49:183-190.
15. Wang, C. and C.T.Sun. 1997. *J. Composite Materials*, 31: 1480-1506.

Table 1.   Elastic properties and one-parameter plasticity parameters for T700S/2500 unidirectional composite.

E_1 (GPa)	121
E_2 (GPa)	10.4
ν_{12}	0.31
G_{12} (GPa)	4.80
a_{66}	2.0
α (MPa^{-n})	7.5×10^{-14}
n	5.1

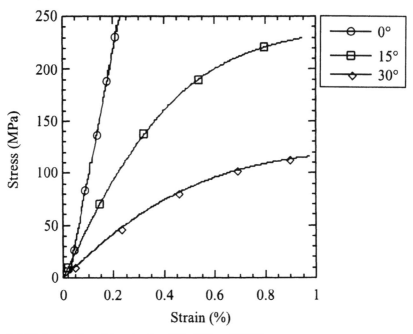

(a) Unidirectional laminates (0°, 15° and 30°)

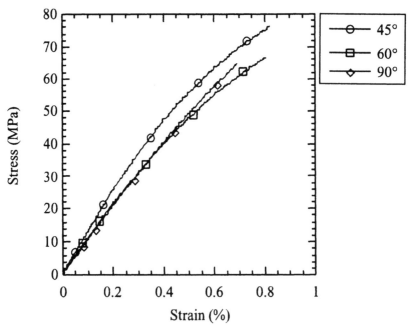

(b) Unidirectional laminates (45°, 60° and 90°)

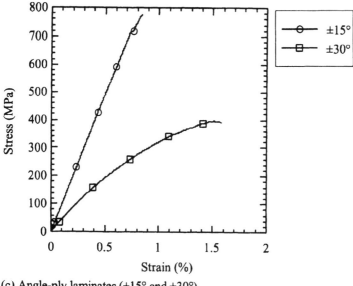

(c) Angle-ply laminates (±15° and ±30°)

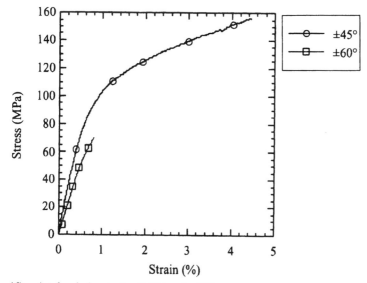

(d)   Angle-ply laminates (±45° and ±60°)

Figure 1.   Typical stress-strain curves for (a) unidirectional laminates (0°, 15° and 30°), (b) unidirectional laminates (45°, 60° and 90°), (c) angle-ply laminates (±15° and ±30°), and (d) angle-ply laminates (±45° and ±60°).

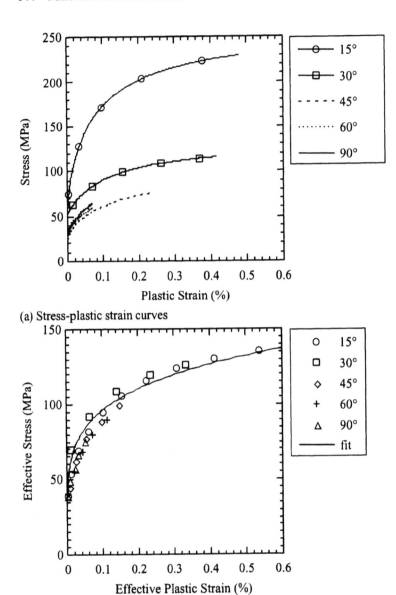

(a) Stress-plastic strain curves

(b) Effective stress-effective plastic strain curves

Figure 2.   Relations between (a) stress and plastic strain and between (b) effective stress and effective plastic strain using $a_{66}=2.0$ with fitting curve, for T700S/2500 unidirectional laminates.

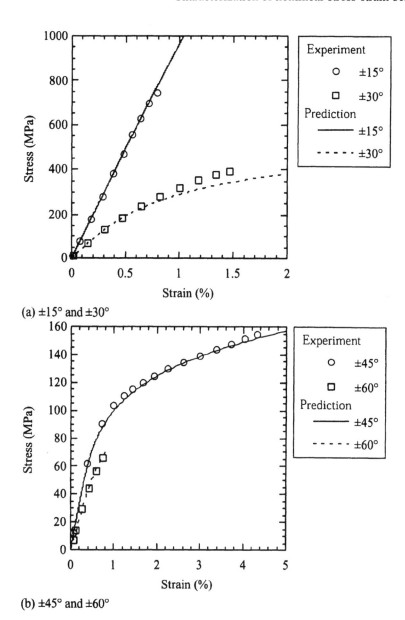

(a) ±15° and ±30°

(b) ±45° and ±60°

Figure 3.   Comparison between the experimental results and the analytical prediction using the classical lamination theory and the one-parameter plasticity model for the stress-strain curves of T700S/2500 angle-ply laminates.

Identification of equivalent elastic parameters of triaxial woven fabric composites for golf club shafts

M. Zako and Y. Nakanishi
Graduate School of Osaka University
2-1 Yamadaoka, Suita, Osaka
565-0871, Japan
zako@mapse.eng.osaka-u.ac.jp, y-naka@mapse.eng.osaka-u.ac.jp

K. Matsumoto
Faculty of Education
Mie University
1515 Kamihamcho, Tsu, Mie
514-8507, Japan
matumoto@edu.mie-u.ac.jp

N. Matsumoto
Fujikura Enterprise Ltd.
Haramachi, Fukushima
975-0027, Japan
mailto:matsu-no@rub.fujikura.co.jp

ABSTRACT

Triaxial woven fabric (TWF) composites are used for space structures, golf club shafts and so on. Though many researchers have reported the static characteristics of TWF composites, the dynamic behaviors have not have been discussed. The dynamic properties of TWF are very important for the structural analysis. The purpose of this study is to make clear the dynamic properties of TWF composites by an experiment, and to propose the identification method by which the elastic properties can be estimated. The excitation tests of a basic weave and a bi-plain weave TWF composites plates are carried out. From the experimental results, the equivalent elastic parameters of each material are identified. As one of the applications of the proposed method, the natural frequencies of the golf club shafts made of bi-plain TWF composites and unidirectional composites are analyzed. Finally, measured natural frequencies of golf club shafts by excitation tests are compared with numerical ones by using identified equivalent elastic parameters. As the identified result has a good agreement with the experimental one, it is recognized that the proposed method is very useful for the evaluation of mechanical properties.

INTRODUCTION

Golf club shafts are one of the popular goods made by composite materials in Japan. The unidirectional carbon fiber reinforced plastics (CFRP) are the main materials for the shafts. CFRP introduces the reduction of weight and high stiffness of the shafts. However, the low shearing stiffness causes the torsional deflection of golf shaft because CFRP shows the anisotropic properties strongly. In order to make high specific stiffness shafts without torsion, we have applied TWF composites for golf club shafts because of their high shearing modulus.

To analyze the properties of golf club shafts, it is necessary to identify the elastic parameters exactly. However, it is difficult to determine the elastic parameters of TWF composites by either theoretical or experimental approaches. We have proposed an inverse analysis method, which identifies the elastic parameters of laminated composite materials using the finite element eigenvalue analysis [1].

Deobald has determined two Young's moduli, the in-plane shear modulus and a poisson ratio for aluminum plate and graphite epoxy plate by using the natural frequencies measured by an impulse technique [2]. Qian et al. have presented a method for identifying elastic and damping properties of composite laminates by using measured complex modal parameters [3]. On the other hand, we have proposed an inverse analysis method to identify the elastic parameters for laminated composite materials using the FEM eigenvalue analysis and the nonlinear optimization method.

In this paper, we have two purposes. One is to obtain the dynamic properties of TWF composites by excitation tests, the other is to identify the equivalent elastic parameters by the proposed method. The identification of equivalent parameters can be treated a nonlinear optimization problem. The quasi-Newton method is employed for the nonlinear problem. A shell element with the first-order shear deformation theory is used for the modeling of TWF composites. The excitation tests of a lamina with basic weave (Fig.1a) and a lamina with bi-plain weave (Fig.1b) are carried out. From the experimental results, the equivalent elastic parameters of each material are identified. And then, the natural frequencies of the golf club shafts made of bi-plain TWF and unidirectional composites are analyzed. Finally, measured natural frequencies of golf club shafts obtained by excitation tests are compared with numerical ones obtained by identified equivalent elastic parameters.

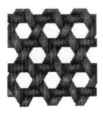

(a) Basic weave   (b) Bi-plain weave

Fig.1 Typical triaxial woven fabric composites

EXPERIMENT AND IDENTIFICATION OF TWF COMPOSITES

Identification method

In order to identify the equivalent elastic parameters along the material principle direction, the quasi-Newton method is employed. We define the error function as shown in Eq. (1).

$$g_n(x) = f_{En} - f_n(x) \tag{1}$$

where, f_{En} is the natural frequency measured by the excitation test, and $f_n(x)$ is the ones obtained by the eigenvalue analysis. The identification is considered as a nonlinear optimization problem to find a solution of x that minimizes the error norm $\phi(x)$.

$$\phi(x) = \frac{1}{2} \sum_{n=1}^{N_{TM}} g_n^{2}(x) \tag{2}$$

where N_{TM} is the total number of modes.

Quasi-Newton method takes an initial value x_0, and the improved value x_k by the iteration formula using a search direction vector d and a step size parameter λ chosen by the line searcher algorithm can obtain as follows,

$$x_{k+1} = x_k + \lambda_k d_k \tag{3}$$

A search direction vector d_k can be given as a solution of Eq. (4)

$$H_k d_k = -J^T(x_k) f_n(x_k) \tag{4}$$

where H and J are the Hessian and Jacobian matrices of error function $g(x_k)$ respectively.

Identification of equivalent elastic parameters for TWF composite plates

Two kinds of TWF composites plates have been treated; the one is made of basic weave, the other is bi-plain weave. The dimensions of the plates are of 300 mm long, 210 mm wide and 0.89 mm thick. The density of the basic weave and the bi-plain plates are 1273 kg/m³ and 1395 kg/m³, respectively.

In order to measure the vibration characteristics of TWF composites plates, the excitation tests are carried out. The experimental device is illustrated in Fig.2. The plate is suspended by the fine threads and excited by a loudspeaker system. The displacement of plate is measured by a laser sensor. The results of the excitation test are shown in Table1 and 2.

Equivalent elastic parameters of TWF composite plates are identified by the obtained natural frequencies. We have considered that TWF composite plates consist of a single layer with orthotropic body. The equivalent elastic parameters of each material are identified. The identified elastic parameters of the lamina are shown in Table3. Poisson's ratios ν_{LT} are 0.3. The identified natural frequencies in Table1 and 2 mean the results of the eigenvalue analysis of an equivalent shell model of the TWF composite plates using the identified equivalent elastic parameters. It is recognized that the experimental results have a good agreement with the numerical results. From these assumptions, it is obtained that the shearing modulus G_{LT} of TWF is higher than ones of general unidirectional composite materials.

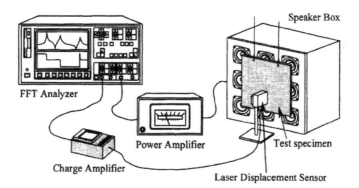

Fig.2 Experimental setup

Table1 Comparisons between experimental and numerical natural frequencies with basic weave

Mode shape	1st	2nd	3rd	4th
Experiment, Hz	22.59	27.09	55.81	58.44
Analysis, Hz	22.54	26.87	55.78	58.90
Error, %	−0.2	−0.8	−0.1	−0.8

Table2 Comparisons between experimental and numerical natural frequencies with bi-plain weave

Mode shape	1st	2nd	3rd	4th
Experiment, Hz	44.06	45.38	94.45	100.9
Analysis, Hz	43.27	43.28	98.13	100.7
Error, %	−1.8	−4.6	−3.9	−0.2

Table3 Identified equivalent elastic parameters of lamina, GPa

	E_L	E_T	G_{LT}
Basic weave	20.4	14.4	10.0
Bi-plain weave	31.6	25.7	12.2

Experiment and analysis of golf club shafts

The natural frequencies of CFRP golf shafts are analyzed using beam models and are compared with the experimental results of golf shaft made of TWF and unidirectional composites are $(UD[(\pm45)_2 / 0_3] + TWF[90])$ (See Fig.3a) and the unidirectional composites $(UD[(\pm45)_2 / 0_3] + UD[0_3])$ (See Fig.3b). The golf club grip is secured in a vise grip. An impact hammer is then used to excite the club on the face of the club-head. The mechanical properties of a UD lamina are E_L=120 [GPa], E_L=10 [GPa], G_{LT}=5 [GPa], v_{LT} =0.3, respectively.

The natural frequencies of golf shafts are shown in Fig.4. Though the numerical results are different from experimental ones, the natural frequency of TWF composite club is bigger than one of the unidirectional composite clubs can be recognized. From these results, it is considered that TWF composite is effective because of the shearing rigidity of shaft.

CONCLUSION

We have proposed the identification method for mechanical properties by a vibration test. The numerical results of natural frequencies by using identified properties have a good agreement with ones obtained by experiment. Though the mechanical properties in-plane of plate can be obtained easily by tensile test, it is very difficult to get the mechanical properties along thickness of plate. However, both of the mechanical properties for in-plane and direction of thickness can be estimated by the proposed method.

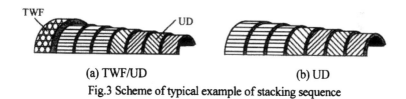

(a) TWF/UD   (b) UD

Fig.3 Scheme of typical example of stacking sequence

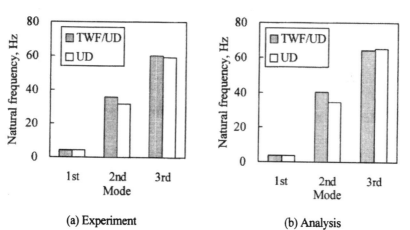

(a) Experiment   (b) Analysis

Fig.4 Comparison with experimental and numerical natural frequencies

References

1. K. Matsumoto, M. Zako, M. Furuno, Identification of Anisotropic Parameters for Laminated Composites by using FEM Eigenvalue Analysis, Third International Conference on Composeite Engineering, Vol.3, pp.567-568, 1996.

2. L.R.Deobald and R.F.Gibson, Determination of Elastic Constants of Orthotropic Plates by a Modal Analysis/ Rayleigh-Ritz Technique, 1988, *Journal of Sound and Vibration*, **Vol.124**, No.2, pp.269-283.

3. G.L.Qian, S.V.Hoa and X.Xiao, Identification of Elastic and Damping Properties of Composite Laminates from Vibration Test Data, 1995, *Proceedings of International Conference of Composite Materials*, **Vol.10**, No.4, pp.IV25-32.

Identification of damping parameters for laminated shallow cylindrical shells

Dr. K. Hosokawa
Faculty of Engineering
Chubu University
1200 Matsumotocho, Kasugai
Aichi, 487-8501 Japan
hosokawa@isc.chubu.ac.jp

Dr. K. Matsumoto
Faculty of Education
Mie University
1515 Kamihamacho, Tsu
Mie, 514-8507 Japan
matumoto@edu.mie-u.ac.jp

ABSTRACT

An inverse analysis method has already been proposed by the authors to identify damping parameters of laminated composite materials using the FEM eigenvalue analysis. The purpose of this study is to apply the proposed method to a cross-ply laminated shallow cylindrical shell. First, by applying the experimental modal analysis technique to a cross-ply laminated shallow cylindrical shell with free boundary conditions, natural frequencies, mode shapes, and modal damping ratios are obtained. Next, by considering the obtained ones of the cross-ply laminated shallow cylindrical shell, damping parameters for the lamina of the shell are identified.

INTRODUCTION

Since composite materials such as fiber reinforced plastics (FRP) have high specific strength and high specific modulus, they have been used in many structural applications and aerospace structures. It is therefore very important to make clear the dynamical properties of the laminated composites for the design and the structural analysis. Especially, damping parameters are essential for the vibration analysis.

The authors have already proposed an inverse analysis method to identify the damping parameters for laminated composite materials using the FEM eigenvalue analysis [1]. Also, the proposed inverse analysis method was applied to a symmetrically laminated square plate. From the comparison of the identified damping parameters of the lamina and experimental ones, one can see the good agreements [2]. However, excepting those aforementioned, one can find few reports of the identified damping parameters for the laminated composites [3].

For the above reason, in this paper, the proposed method is applied to a laminated shallow cylindrical shell. The proposed method mainly consists of the FEM eigenvalue analysis and the nonlinear optimization method considering the relation between modal damping ratios of the laminated shallow shell and damping parameters of the lamina as a nonlinear system. First, by applying the experimental modal analysis technique to the cross-ply laminated shallow cylindrical shell with free boundary conditions, natural frequencies, mode shapes, and modal damping ratios are obtained. Next, considering the obtained vibration characteristics of the cross-ply laminated shallow cylindrical shell, the damping parameters for the lamina of the cross-ply laminated shallow cylindrical shell are identified numerically.

IDENTIFICATION METHOD

Model of Laminated Composites

The stiffness matrix in the stress-strain relations for the rth orthotropic lamina in the material principle direction LTV (See Figure 1) is given by

$$Q_r = \begin{bmatrix} Q_{11} & Q_{12} & 0 & 0 & 0 \\ Q_{12} & Q_{22} & 0 & 0 & 0 \\ 0 & 0 & Q_{44} & 0 & 0 \\ 0 & 0 & 0 & Q_{55} & 0 \\ 0 & 0 & 0 & 0 & Q_{66} \end{bmatrix}_r \tag{1}$$

Also, the stiffness matrix in the direction of the elemental coordinate axes xyz is given by transforming the stiffness matrix Q_r with the transfer matrix T_r calculated by the orientation angle.

$$\overline{Q_r} = T_r^T Q_r T_r \tag{2}$$

Using first-order shear deformation theory, for a laminated composite, the stiffness matrix in the generalized stress-strain relations for a finite shell element model is

$$D = \begin{bmatrix} D_p & D_c & 0 \\ D_c & D_b & 0 \\ 0 & 0 & D_s \end{bmatrix} \tag{3}$$

where the components of matrices (D_p, D_c, D_b, and D_s) in Eq.(3) are

$$D_{pij} = \sum_{r=1}^{NTL} \int_{h_{r-1}}^{h} \overline{Q_{ijr}} \, dz \qquad (i,j=1,2,6) \tag{4}$$

$$D_{cij} = \sum_{r=1}^{NTL} \int_{h_{r-1}}^{h} z\overline{Q_{ijr}} \, dz \qquad (i,j=1,2,6) \tag{5}$$

$$D_{bij} = \sum_{r=1}^{NTL} \int_{h_{r-1}}^{h} z^2 \overline{Q_{ijr}} \, dz \qquad (i,j=1,2,6) \tag{6}$$

$$D_{sij} = \sum_{r=1}^{NTL} k \int_{h_{r-1}}^{h} \overline{Q_{ijr}} \, dz \quad (i,j=4,5) \quad (7)$$

Here z is the distance from the neutral surface to each lamina, k is the shear correction coefficient (5/6), and NTL is the total number of laminae.

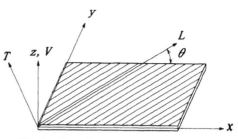

Fig.1   Coordinate System of Shell Element for Laminated Composites

Adams *et al.* [4] reported that the specific damping capacity (SDC) of the *r*th lamina of unidirectional composite materials could be expressed as a diagonal matrix in the form with

$$\psi_r = \begin{bmatrix} \psi_L & 0 & 0 & 0 & 0 \\ 0 & \psi_T & 0 & 0 & 0 \\ 0 & 0 & \psi_{TV} & 0 & 0 \\ 0 & 0 & 0 & \psi_{VL} & 0 \\ 0 & 0 & 0 & 0 & \psi_{LT} \end{bmatrix}_r \quad (8)$$

where the components ψ_{ij} are the specific damping capacity in each strain direction. They also suggested that components of the SDC matrix could be measured by vibration tests for some kinds of unidirectional specimens with various fiber orientations.

The energy dissipated during a cycle of the *n*th modal vibration for a laminated composite shell is calculated as follows :

$$\Delta U_n = \frac{1}{2} \int_S \varepsilon_n^T \psi \varepsilon_n \, ds \quad (9)$$

Here ψ is the SDC matrix of a shell element written by

$$\psi = \begin{bmatrix} \psi_p & \psi_c & 0 \\ \psi_c & \psi_b & 0 \\ 0 & 0 & \psi_s \end{bmatrix} \quad (10)$$

where the components of matrices (ψ_p, ψ_c, ψ_b, and ψ_s) in Eq.(10) are

$$\psi_{pij} = \sum_{r=1}^{NTL} \int_{h_{r-1}}^{h} \tilde{\psi}_{ijr} \, dz \quad (i,j=1,2,6) \quad (11)$$

$$\psi_{cij} = \sum_{r=1}^{NTL} \int_{h_{r-1}}^{h} z \tilde{\psi}_{ijr} \, dz \quad (i,j=1,2,6) \quad (12)$$

$$\psi_{bij} = \sum_{r=1}^{NTL} \int_{h_{r-1}}^{h} z^2 \tilde{\psi}_{ijr} \, dz \quad (i,j=1,2,6) \quad (13)$$

$$\psi_{sij} = \sum_{r=1}^{NTL} \int_{h_{r-1}}^{h} \tilde{\psi}_{ijr} dz \quad (i,j=4,5) \tag{14}$$

Since $\tilde{\psi}_{ijr}$ are the components of the energy-dissipated matrix $\tilde{\psi}_r$, which is calculated by the SDC matrix in Eq.(8) transforming into the direction of the elemental coordinate axes.

$$\tilde{\psi}_r = T_r{}^T \psi_r Q_r T_r \tag{15}$$

The maximum strain energy and the energy dissipated during a cycle of the nth modal vibration are given by

$$U_n = \frac{1}{2} \delta_n{}^T L^T \int_S B^T D B ds L \delta_n \tag{16}$$

$$\Delta U_n = \frac{1}{2} \delta_n{}^T L^T \int_S B^T \psi B ds L \delta_n \tag{17}$$

where δ_n is the nth modal vector, B is the displacement-strain relation matrix and L is the transfer matrix of elemental coordinates. Finally, the nth modal damping ratio is calculated as the ratio of the dissipated energy and the maximum strain energy.

$$\zeta_n = \frac{1}{4\pi} \cdot \frac{\Delta U_n}{U_n} \tag{18}$$

Nonlinear Optimization Method

Considering the relationship between the SDC matrix and the modal damping ratios as a nonlinear system, the quasi-Newton method can be used for identifying the damping parameters in the material principle direction. We define the error function as the difference between the nth modal damping ratio ζ_{En} measured by the excitation test and calculated one $\zeta_n(x)$ by the analysis, as follows:

$$g_n(x) = \zeta_{En} - \zeta_n(x) \tag{19}$$

Then, the identification is considered as a nonlinear optimization problem to find a solution x that minimizes the error norm $\Phi(x)$.

$$\Phi(x) = \frac{1}{2} \sum_{n=1}^{NTM} g_n(x)^2 \tag{20}$$

where NTM is the total number of referring modes.

The quasi-Newton method takes an initial approximation x_0, and attempts to improve x_0 by the iteration formula using a search direction vector d and a step size parameter λ.

$$x_{k+1} = x_k + \lambda_k d_k \tag{21}$$

A step size parameter λ is chosen by the line searcher algorithm, and a search direction vector d can be given as a solution of the equation followed by

$$H_k d_k = -\nabla\Phi(x_k) = -J^T(x_k) g(x_k) \tag{22}$$

where H and J are the Hessian and Jacobian matrices of error function $g(x_k)$, respectively.

Figure 2 shows the flow chart of the identification program. First, the data about the geometrical configuration and material properties of the specimen are given. Secondly, to calculate the natural frequencies and mode shapes, the eigenvalue analysis is carried out with the initial parameters. Subsequently, considering the obtained mode shapes, the modal damping ratios of the specimen are estimated. Thirdly, the error function $g(x)$ is estimated by the difference of the modal damping ratios between the analysis and experiment. After that, damping parameters of the lamina are identified by the quasi-Newton method. Finally, if the calculated modal damping ratios are converged into the experimental ones, the identification program is terminated.

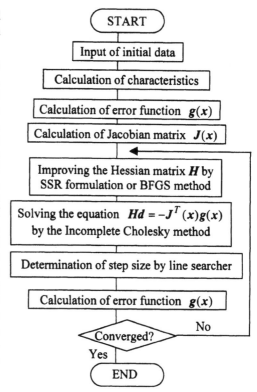

Fig.2   Flow Chart of Identification Program

IDENTIFICATION

Initial Data

To satisfy the free boundary conditions, the cross-ply laminated shallow cylindrical shell was hung from the ceiling by a fine string (See Fig. 3). As shown in Fig. 4, the configuration of the shallow cylindrical shell is a square planform ($a=b=0.2$[m], $h=1.60 \times 10^{-3}$[m], $R=0.4$[m], where h and R are the shell's thickness and the inside radius, respectively). The stacking sequence of the cross-ply laminated shallow cylindrical shell is $[0°_2/90°_2/90°_2/0°_2]$. Each layer material, that is lamina, is a carbon fiber reinforced plastics (CFRP). The estimated material properties of the lamina are: E_L =130[GPa], E_T =7.43[GPa], G_{TV} =2.00[GPa], $G_{VL} = G_{LT}$ =4.16[GPa]. Poison's ratios are assumed to be $v_{LT}=v_{TV}=v_{VL}$ =0.32. The density of the shell is 1557 [kg/m^3]. The shell was divided into many reference points. To measure the transfer function (accelerance), an accelerometer was attached to one reference point and then all reference points were impacted by an impulse force hammer (See Fig. 3). The mass of

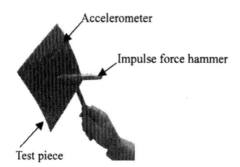

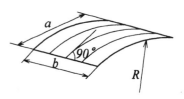

Fig.3 Vibration Test of Laminated Shallow Cylindrical Shell

Fig.4 Cross-ply Laminated Shallow Cylindrical Shell

the accelerometer is 0.48 [g]. From the obtained transfer function, the natural frequencies, mode shapes, and modal damping ratios of the cross-ply laminated shallow cylindrical shell were estimated by applying the experimental modal analysis technique. Table 1 shows the experimentally obtained natural frequencies, mode shapes, and modal damping ratios of the cross-ply laminated shallow cylindrical shell. For the mode shapes, upper and lower edges are the curved edges.

Identified Damping Parameters

From the experimental natural frequencies, mode shapes, and modal damping ratios shown in Table 1, the damping parameters for the lamina of the cross-ply laminated shallow cylindrical shell were estimated by the proposed inverse analysis method. The computations were carried out using the FEM eigenvalue program with triangular shell elements. The shell is discretized with 800 elements and 441 nodal points. In the numerical calculations, the mass of the accelerometer was not considered because the mass of the accelerometer is very small. The identified damping parameters of the lamina are shown in Table 2. In Table 2, the damping parameters ψ_L, ψ_T in the direction of the parallel and normal to the fiber, shear damping parameters ψ_{TV}, ψ_{VL}, and ψ_{LT} are shown (See Fig. 1). In order to confirm the identified damping parameters, the modal damping ratios of the laminated shallow cylindrical shell were estimated by using these identified damping parameters. From Table 1, one can see that the difference between the experimental and the numerically calculated modal damping ratios is about 3.1% at the most. And, one can see that the difference between the experimental and the numerically calculated natural frequencies is about 0.3% at the most. Also, in this paper, numerical mode shapes are not reported because one can find the excellent agreements between these mode shapes of the cross-ply laminated shallow cylindrical shell.

CONCLSIONS

The inverse analysis method to identify damping parameters of the laminated composite materials was applied to the cross-ply laminated shallow cylindrical shell with free boundary conditions. To confirm the identified damping parameters, the numerical calculations were carried out. From the results, one can see the good agreements with respect to the natural frequencies, mode shapes, and damping ratios. Accordingly, it follows that one can accurately estimate damping parameters for the lamina of the laminated shallow cylindrical shells by using the inverse analysis method proposed by the authors.

REFERENCES

1. K. Matsumoto, K. Hosokawa, M. Zako, and H. Kobayashi, "Identification Method of Vibrational Damping Parameters for Laminated Composite Materials", The Transactions of Japan Society of Mechanical Engineers (in Japanese), Vol.65, No.636, 1999, 3129-3134.
2. K. Matsumoto, K. Hosokawa, and M. Zako, "Identification of Vibrational Damping Parameters for Laminated Composites Material by Using FEM Eigenvalue Analysis", ASME PVP, Vol.374, 1998, 319-324.
3. G. L. Qian, S. V. Hoa, and X. Xiao, "Identification of Elastic and Damping Properties of Composite Laminates from Vibration Test Data", Proceedings of International Conference of Composite Materials, Vol. 10, No. 4, 1995, IV25-32.
4. D. X., Lin, R. G. Ni, and R. D. Adams, "Prediction and Measurement of the Vibrational Damping Parameters of Carbon and Glass Fiber-Reinforced Plastic Plates", J. Composite Materials, Vol. 18, 1984, 133-152.

Table 1   Comparison between Experiment and Analysis with Identified Parameters of Cross-ply Laminated Shallow Cylindrical Shell

Modal order		1st	2nd	3rd	4th	5th
Experimental mode shape						
Natural frequency	Experiment [Hz]	68.39	151.8	205.8	420.7	469.7
	Analysis [Hz]	68.18	151.9	206.4	420.4	469.2
	Error [%]	−0.3	0.1	0.3	−0.1	−0.1
Modal damping ratio	Experiment [%]	0.295	0.101	0.191	0.096	0.141
	Analysis [%]	0.298	0.099	0.191	0.099	0.138
	Error [%]	1.0	−2.0	0.0	3.1	−2.1

Table 2   Identified Damping Parameters for Lamina of Cross-ply Laminated Shallow Cylindrical Shell

ψ_L [%]	ψ_T [%]	ψ_{TV} [%]	$\psi_{VL} = \psi_{LT}$ [%]
0.630	2.77	0.765	3.88

Characterization of off-axis ply cracking behavior

in CFRP laminates

Satoshi KOBAYASHI[#], Akihisa TANAKA[#], Shuichi WAKAYAMA[#],

#: Department of Mechanical Engineering, Tokyo Metropolitan Univ.
1-1 Minami-osawa, Hachioji-city, Tokyo 192-0397, JAPAN
E-mail:koba@ecomp.metro-u.ac.jp

Abstract

To clarify the basic characteristics of off-axis ply cracking, unidirectional and angle-ply laminates were tested. Laminate configurations were $[\theta]_8$ and $[\pm\theta]_s$, where θ = 15, 30, 45, 60 and 75. For unidirectional composite, pre-cracks were introduced at the center of either side of the free edge using razor blade. Static tensile tests were conducted at room temperature. Microscopic damage progress was observed using video microscopy and soft X-ray radiography. For unidirectional laminates, mode I (G_I) and II (G_{II}) energy release rate associated with crack progress were calculate using FE method. Mode I and II energy release rate became larger as the crack progress. G_{II} ratio to G_I became larger as the ply angle θ became smaller. That is, mixed mode fracture criterion were necessary to predict the crack progress in small angle ply. For angle-ply laminates, damage mechanics analysis was used to predict the modulus reduction caused by matrix cracking. Analysis underestimated modulus reduction. Nonlinear analysis which include plastic deformation is necessary for accurate prediction.

INTRODUCTION

Carbon fiber reinforced plastic (CFRP) laminates are used in the aerospace field because of their high specific modulus and strength.   However, it is well known that various microscopic damages such as matrix crack and delamination initiate and grow before final fracture in the failure process of CFRP laminates.   These structures have been designed based on the safe-life concept, which do not allow any initial damages.   However, it is necessary that design criterion is shifted to risky side to reduce the weight of structures.   To establish such damage tolerance design, it is necessary to clarify the mechanism of damages nucleation and the effects of damages on mechanical properties.

There are many studies about the damages in the transverse plies, example for the shear lag theory [1] and Hashin's model [2].   Actually, it has also been confirmed that the matrix cracks initiate in off-axis plies, which are ±45° plies included in quasi-isotropic laminates, however, there are little research about mechanism of behavior of these damages.   In the present study, the damage behavior in off-axis plies were concerned. Matrix crack progress in unidirectional and angle-ply laminates were observed. For unidirectional laminates, mode I and II energy release rate were calculated using FE method based on experimental results.   Damage mechanics analysis was applied to predict the modulus reduction in angle-ply laminates.

EXPERIMENTAL PROCEDURE

Material system used in this study was carbon-epoxy composites, T700S/2500.   The mechanical properties of unidirectional composites are tabulated in Table 1.   For each material system, $[\theta]_8$ and $[\pm\theta]_s$ angle-ply laminates were fabricated.   The range of θ was from 15° to 75° to evaluate the effect of the fiber orientation on microscopic damage progress.   Specimen length was 150mm. Width and thickness were 20mm and 1.1mm for unidirectional laminates, and 10mm and 0.55mm for angle-ply laminates.   To avoid the stress concentration at the grip, GFRP tabs, which have 25mm length, were glued on both ends of specimens. To define the crack position, pre-crack along the fiber direction were introduced to the unidirectional laminates using razor blade.

Quasi-static tensile tests were carried out at room temperature. Cross-head speed were 0.01 mm/min for unidirectional laminates and 0.5 mm/min for angle-ply laminates. During tests on unidirectional laminates, crack progress were observed using a video microscope placed in front of the specimen.   In order to observe the microscopic damage progress in angle-ply laminates, the specimen was removed from testing machine to observe free edge by optical microscopy. Internal observation using soft X-ray radiography were also conducted. Observation area was 25mm length at the center. Matrix crack density was defined as the number of matrix crack per unit observed length.

EXPERIMENTAL RESULTS

Figure 1 shows the relation between crack extension and stress in unidirectional laminates with pre-crack. Critical crack extension, at this point unstable crack progress occurred, became smaller as the ply angle θ became larger. Stress saturates with increase in crack extension other than θ=75.. That is, fracture becomes unstable mode with increase in ply angle.

The results of tensile test on angle-ply laminates are summarized in Tables 2. Stiffness and strength decrease with the increase in lamination angle.

Figure 2 shows the result of free edge observation for [±30]$_s$ laminates. For the [±15]$_s$ and [±30]$_s$ laminates, similar damage behaviors were observed. A matrix crack in outer ply and a delamination at a crack tip initiated at first. Then a matrix crack in inner ply initiated from a tip of the delamination.

Figure 3 shows the results of free edge observation for [±45]$_s$ laminates of both materials. First, a matrix crack in outer ply and a little delamination at a crack tip was observed on both materials. Then a crack from a tip of the delamination is initiated in the inner ply. As load increased, additional cracks are initiated along the edges of specimen. Figure 4 shows matrix crack density as a function of the laminate strain in [±45]$_s$ laminate. Matrix crack densities in each ply increase almost equivalently.

The result of internal observation for [±15]$_s$ laminate is shown in Figure 5. For the [±15]$_s$ and [±30]$_s$ laminates, similar damage behavior is also observed in internal observation. It can be seen in the figure that triangular delaminations surrounded by the matrix cracks in outer and inner ply were initiated. As load increased, the delamination areas grew and new delaminations were initiated.

Figure 6 shows the result of internal observation for [±45]$_s$ laminates. Similarly to the free edge observation, matrix cracks were observed along the free edge. The delamination hardly grew along the width direction. Matrix cracks on each ply did not go through the width.

For the [±60]$_s$ and [±75]$_s$ laminates, no microscopic damages were observed before the final failure.   That is, fracture of these laminates was catastrophic.

DISCUSSION

Energy Release Rate

In the present study, FE method was used to obtain stress distribution around the crack tip.   Energy release rates were calculated using virtual crack closure technique.   A commercially available finite element program, MARC 7.1 was used as a solver.   6 node triangular plane stress shell element were used to model laminates with cracks. FE mesh model were constructed based on the experimental measurements of the crack length.   The energy release rates were calculated for the applied loads which were used in the experiments.   Constant displacement was applied to the upper side of the specimen in axial direction as the loading. Lower side was perfectly fixed in the axial direction. Midpoint at lower side of specimen was also fixed in transverse direction.

Figure 1 shows the relation between energy release rate and crack extension. Both mode I and II energy release rates increase with increase in crack length.   Fracture surface are shown in Figure 7.   Many fiber bridging was observed at fracture surface. Large increase in crack extension resistance was attributed to fiber bridging.

In all laminates, mode I energy release rates were larger than mode II, however, the ratio mode II to mode I increased with decrease in ply angles. That is, mixed mode fracture criteria are necessary to clarify the matrix cracking especially in small angle plies.

Damage Mechanics Analysis

Modulus reduction was calculated based on damage mechanics analysis proposed by Gudmundson and Zang [3].   The compliance matrix of laminate as functions of matrix crack densities in each plies is expressed as

$$S(\rho^k) = \left((S_0)^{-1} - \sum_{k=1}^{N} v^k \rho^k (A^k)^T \sum_{i=1}^{N} \beta^{ki} A^i \right)^{-1} \quad (1)$$

where v^k, ρ^k and β^{ki} denotes the volume fraction of ply k, the matrix crack density of ply k and the matrix related to the average crack opening displacements in ply k, respectively.

$\mathbf{A}^k$ is the matrix represented with the compliance matrix of ply k and the unit normal vector on the crack surfaces in ply k. The laminate stiffness $\mathbf{E}(\rho^k)$ is calculated from $\mathbf{S}(\rho^k)$. Figure 8 shows the comparisons of the experimental result of $[\pm45]_s$ laminate with the result calculated from the above equation. The experimental date just after damage initiation is lower than the calculation for each material. It is suggested that these results reflect the effect of the nonlinear shear behavior of unidirectional composites.

CONCLUSION

Microscopic damage behavior of angle-ply laminates under tensile load were studied. Following conclusions were obtained.
1. The effect of ply angles on the microscopic damage behavior was clarified experimentally.
2. Large increase in crack extension resistance caused by fiber bridging was observed in all laminates.
3. It was confirmed that there was the effect of nonlinear shear behavior of unidirectional composites on modulus reduction.

REFERENCE
1.   J. -W. Lee and I. M. Daniel, J. Comp. Materials, Volume 24, 1990, 1225-1243
2.   Z. Hashin, Engng Fract. Mech., Volume 25, 1986, 771-778
3.   P. Gudmundson and W. Zang, Int J. Solids Structures, Volume 30 1993, 3211-3231.

Table 1   Mechanical properties of T700S/2500 unidirectional composites.

Longitudinal Young's Modulus, GPa	121
Transverse Young's Modulus, GPa	10.4
In-Plane Shear Modulus, GPa	4.80
In-Plane Poisson's Ratio	0.312
Out-of-Plane Poisson's Ratio	0.49

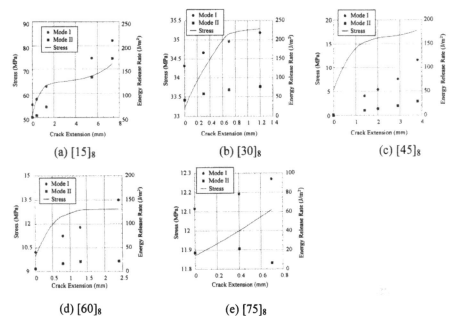

(a) $[15]_8$ (b) $[30]_8$ (c) $[45]_8$

(d) $[60]_8$ (e) $[75]_8$

Figure 1 Relationship between stress, energy release rate and crack extension in unidirectional laminates.

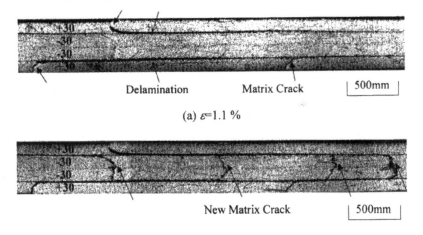

(a) ε=1.1 %

(b) ε=1.3 %

Figure 2 Damage progress in $[\pm30]_s$ laminate observed by optical microscopy. (ε: laminate strain).

Table 2   Results of tensile tests of T700S/2500 angle-ply laminates.

Lamination Angle	15°	30°	45°	60°	75°
Young's Modulus, GPa	101	42.1	16.1	11.1	10.6
Ultimate Strength, MPa	814	392	161	67.8	61.0
Failure Strain, %	0.96	1.64	7.78	0.78	0.63

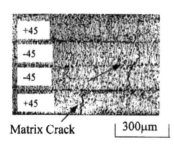
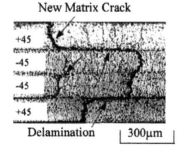

(a) ε=2.3 %   (b) ε=3.5 %
Figure 3   Damage progress in [±45]$_s$ laminate observed by optical microscopy. (ε: laminate strain).

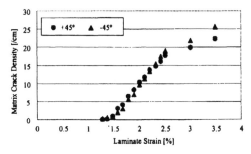

Figure 4 Matrix crack density as a function of laminate strain in [±45]$_s$ laminate.

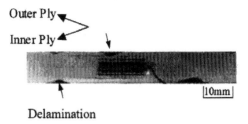

Figure 5 Damage progress in T700S/2500 [±15]$_s$ laminate observed by soft X-ray radiography (ε=0.8%, ε: laminate strain).

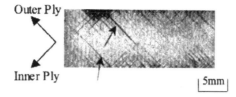

Outer Ply

Inner Ply

5mm

Figure 6 Damage progress in [±45]ₛ laminate observed by soft X-ray radiography (ε= 3.5%,ε: laminate strain).

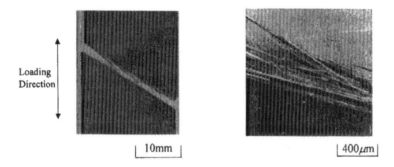

Loading
Direction

10mm

400μm

Figure 7   Fracture surface of [60]₈ laminates.

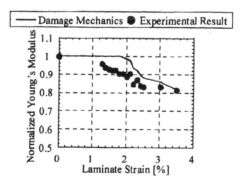

Figure 8 Modulus reduction in [±45]ₛ laminate.

A Method to Evaluate Interfacial Behavior of Composites

Qing-Qing NI

Division of Advanced Fibro-Science
Kyoto Institute of Technology
Matsugasaki, Sakyo-ku, Kyoto
606-8585, Japan
nqq@ipc.kit.ac.jp

ABSTRACT

A double cleavage drilled compression (DCDC) specimen was analyzed using the boundary element method (BEM). It was confirmed that the DCDC specimen had a mode I stress distribution for the hole offsetting displacement of b=0, which is of a symmetrical hole, whereas it had the mixed mode (mode I+ mode II) stress distribution for b≠0. Furthermore, the DCDC specimen of FRP composites was proposed and used to evaluate the interfacial behavior and fracture energy of interlaminar and/or intralaminar interface. As a result, the approach to calculation of interfacial fracture energy was established for both isotropic and anisotropic materials. It was shown that the DCDC test was a useful method to quantitatively evaluate the interfacial behavior of fiber/matrix in composites.

INTRODUCTION

For fiber reinforced composite materials, the interlaminar and/or intralaminar interface will have a great influence on their fracture mechanisms. The interface may be the weakest part in a fracture process. So, the improvement of interfacial behavior becomes increasingly important. This also forces us to develop testing methods to evaluate interfacial behavior quantitatively [1]-[4]. For the evaluation of interfacial behavior of composite materials, DCB (Double Cantilever Beam), ENF (End-notched Flexure), WLPT (Wedge Loaded Peel Test), MMF (Mixed Mode Flexure) *et al.*, have been proposed [5]-[7]. These methods have been widely used in the interlaminar behavior evaluation. The range of practical application of these methods may depend on the ratio of the stress intensity factor of mode I to mode II. However, for the materials with high interface fracture energy, the crack will propagate in a kinking way and may be hard to keep along the interface. Also it may be hard to use them to evaluate the intralaminar interface behavior.

On the other hand, the double cleavage drilled compression specimen (DCDC) has been developed for fracture mechanics studies on monolithic brittle materials and/or bimaterials [8]. The DCDC geometry, which is illustrated in Fig.1, offers experimental advantages such as easy fabrication of specimen, compressive loading, midplane crack stability, and auto-precracking [9]-[12]. Despite the experimental advantages of the DCDC fracture mechanics technique, it has seen only limited use, primarily because of the lack of a well established procedure for evaluating the stress intensity factors of mode I and mode II, where the mode I is defined in the fracture mechanics as the crack lips tend to separate symmetrically relative to the crack plane, and the mode II has anti-symmetry about the crack plane. Here, we have proposed a special DCDC specimen to evaluate the interfacial behavior between fiber and matrix for composite materials. The stress distribution and the stress intensity factors in the DCDC specimen were analyzed using the boundary element analysis, which may have a higher precision than the FEM due to the existed interface in the DCDC specimen, and then a procedure to evaluate the interfacial fracture energy of fiber/matrix was established.

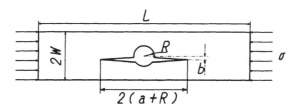

Fig.1. A schematic of the DCDC specimen.

DCDC SPECIMEN FOR INTRALAMINAR INTERFACE

The DCDC specimen has been shown in Fig. 1, in which R, a and b are the radius of the hole, crack length and the hole offsetting displacement, respectively. Under applied compressive load two cracks initiate from the hole and will extend along the long axis of the specimen.

To evaluate interfacial behavior on the intralaminar interface such as fiber/matrix, the DCDC specimen should contain the interface to be evaluated. Here, our focus is on the interface between fiber and matrix. The proposed DCDC specimen having the interface of fiber/matrix has been shown in Fig. 2. An epoxy resin (KR675) with the elastic modulus of 3.3 GPa and Poisson's ratio of 0.39 was used as the matrix. The thin carbon fiber layer was inserted in the specimen center along the long axis and its area was represented by the area ratio of the fiber/matrix interface, Fs ($=W_f L/TL$). The specimens with different Fs, hole diameters and hole offsetting displacements were fabricated, and their loads and crack lengths were measured in the experiments.

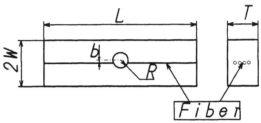

Fig. 2.   Geometry of the DCDC specimen for a fiber/matrix interface behavior.
(L=150 mm, T=8.5 mm, 2W=10 mm)

Analysis

For the DCDC specimen with a central hole of $b=0$, the crack will propagate along the loading direction from two sides of the hole, and the fracture mode I will be obtained due to the symmetry of the specimen configuration. However, for $b \neq 0$ the specimen is non-symmetrical and this will result in a mixed mode of I and II. So, the stress intensity factors, K_I and K_{II}, could be written in the following general forms:

$$K_I = g_I \sigma \sqrt{\pi R} , \tag{1}$$

$$K_{II} = g_{II} \sigma \sqrt{\pi R} , \tag{2}$$

where, g_I, g_{II} are dimensionless geometric coefficients for mode I and mode II. Since g_I and g_{II} only depend on the specimen size, such as a/R, W/R and b/R, they can be represented in the forms of $g_I(a/R, W/R, b/R)$ and $g_{II}(a/R, W/R, b/R)$. Here the boundary element analysis was used to calculate these dimensionless geometric coefficients as there is no theoretical solution for them.

Mode I Deformation

For the case of $b=0$, Fig. 3 has shown the relationship between $\sigma\sqrt{\pi R}/K_I$ and the relative crack length with different values of W/R. Almost linear relations were obtained. Based on these relations, the dimensionless geometric coefficient, g_I, was obtained according to Equation (3), and then K_I at an applied load could be calculated.

$$\frac{1}{g_I\left(a/R, W/R\right)} = \frac{\sigma\sqrt{\pi R}}{K_I} \tag{3}$$

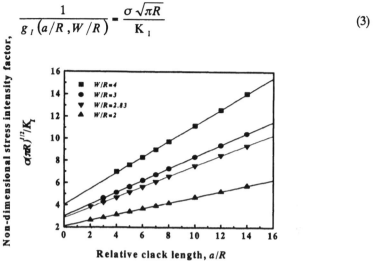

Fig. 3. Dimensionless $\sigma\sqrt{\pi R}/K_I$ vs. the relative crack length for different W/R ($b=0$).

Mixed Mode Deformation

For the case of $b\neq0$, the stress intensity factor of mode I and mode II could be calculated, and then the g_I and g_{II} were obtained with a similar formula to equation (5), while b/R should be put into the geometrical function as an additional parameter. According to the relations of $\sigma\sqrt{\pi R}/K_I$, $\sigma\sqrt{\pi R}/K_{II}$ and a/R, W/R, b/R, the loading phase angle defined by

$$\psi = \tan^{-1}\left(K_I/K_{II}\right), \tag{4}$$

could be represented in Fig. 4. When the crack propagates the loading phase angle reduces slowly and the contribution of the mode II will increase.

Using the results of the dimensionless geometric coefficients obtained in the above, the interfacial fracture energy, J, on the fiber/matrix interface was calculated by ignoring the influence of the thin carbon fiber layer on the elastic modulus of the specimen.

$$J = \frac{K_I^2 + K_{II}^2}{E}\left(1 - v^2\right), \tag{5}$$

$$\begin{cases} K_I = g_I(a/R, W/R, b/R)\sigma_{Fs}\sqrt{\pi R} \\ K_{II} = g_{II}(a/R, W/R, b/R)\sigma_{Fs}\sqrt{\pi R} \end{cases} \qquad (6)$$

where σ_{Fs} is the nominal applied stress which were measured in the experiments.

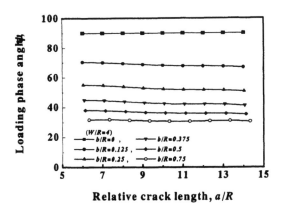

Fig. 4. Loading phase angle vs. the relative crack length for different hole offsetting displacements, b/R.

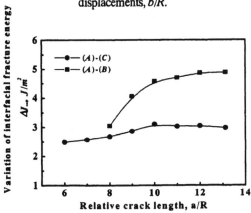

Fig. 5.   Relative variation of interfacial fracture energy in fracture mode I.

Interfacial Fracture Energy of Fibre/Matrix

Combining the dimensionless geometric coefficients obtained by the boundary element analysis with the experimental data: the crack length and the applied stress, the stress intensity factors, K_I and K_{II}, could be calculated, and then the interfacial fracture energy was obtained. For different fiber/matrix area ratio of Fs values, the interfacial fracture energy was quite different. In the case of $b=0$, the relative variation of the interfacial fracture energy between two specimens with different Fs was obtained as shown in Fig. 5. The reduction of the interfacial fracture energy from the specimen A with $Fs=18$ % to the specimen B with $Fs=36$ % was about 3 J/m^2, which corresponded to the change of the interfacial fracture energy for 18 percent interfacial area of matrix replaced by the fiber/matrix. The reduction of the interfacial fracture energy tended to a constant when the relative crack length increased.

DCDC SPECIMEN FOR INTERLAMIAR INTERFACE

For the interlaminar interface such as the interface in the laminated composites, the orthotropic anisotropy of the materials must be considered in the BEM analysis. Here, the program for the orthtropic body was developed and then was confirmed by the exact solution of the plate with a central crack.

Interfacial Fracture Energy of Laminates

Using the Rizzo's solution as the core function in the boundary element analysis, the dimensionless geometric coefficients could be obtained for the orthotropic body. Combining the dimensionless geometric coefficients with the experimental data of the crack length and the applied stress, the stress intensity factors, K_I and K_{II} in laminated composites could be calculated, and then the interfacial fracture energy, G, was obtained as follows.

$$G_i = H_i K_i^2 \quad (i = \mathrm{I}, \ \mathrm{II}) \tag{7}$$

And the relationship between the total interfacial energy and the energy component in each mode was represented by Eq. (8).

$$G = G_{\mathrm{I}} + G_{\mathrm{II}} \tag{8}$$

where, H_i was represented by the compliance of anisotropy, S_{ij}. For the case of the plane stress condition,

$$H_{\mathrm{I}} = \sqrt{\frac{S_{11} + S_{22}}{2}} \left(\sqrt{\frac{S_{22}}{S_{11}}} + \frac{2S_{12} + S_{66}}{2S_{11}} \right)^{1/2} \tag{9}$$

$$H_{\mathrm{II}} = \frac{S_{11}}{2} \left(\sqrt{\frac{S_{22}}{S_{11}}} + \frac{2S_{12} + S_{66}}{2S_{11}} \right)^{1/2} \tag{10}$$

For the case of the plane strain condition, it is only necessary to replace S_{ij} with T_{ij}, which is the function represented only by S_{11}, S_{22}, S_{12}.

$$T_{ij} = S_{ij} - \frac{S_{i3} - S_{3j}}{S_{33}} \tag{11}$$

Interfacial Fracture Energy of laminated Interface

The laminated composite DCDC specimens were fabricated by prepregs where the carbon fiber T700S reinforced an epoxy resin (No. 2500). The size and names are shown in Table 1 with a symmetrical central hole. Figure 6 showed the interfacial fracture energy in the laminated specimens of A2 and A3 and its value was about 3 kJ/m2.

Table 1.   Name and size of specimens.

Specimen	L(mm)	2W(mm)	T(mm)	R(mm)
A-1				4.0
A-2	140	9.0	9.0	4.5
A-3				5.0

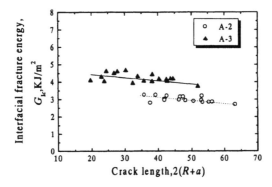

Fig.6. Interfacial fracture energy versus crack length in laminated composites.

CONCLUSIONS

A double cleavage drilled compression (DCDC) specimen has been analyzed by the boundary element method (BEM). At the same time, the DCDC specimen of FRP composites was proposed and used to evaluate the interfacial behavior and fracture energy in CFRP model and laminated composites. The results as follows could be remarked.

It was confirmed that the DCDC specimen generated a mode I and/or a mixed mode stress distribution with the results of the boundary element analysis. Based on the combination of both experimental and analytical works an approach to calculate the interfacial facture energy on the intralaminar and interlaminar interface has been proposed.

It was clear that the proposed composite DCDC specimen would be useful in evaluation of various interfaces for composite materials and this should also be acceptable for other interfaces of bi-materials.

Acknowledgment

I would like to thank the Ministry of Education, Culture, Sports, Science and Technology for the financial support by Grant-in-Aid for Scientific Research (C).

REFERENCES

Edited Conference Proceedings

1. C. Janssen, *Proc. 10th. Cong. on Glass*, 10.23 (1974).

Journal References

2. Q-Q. Ni, and K. Baba, Transactions of the Japan Society of Mech. Engineers, A, 63, 1069 (1997), In Japanese.

3. Q-Q. Ni, and K. Baba, Transactions of the Japan Society of Mech. Engineers, A, 64, 529 (1998), In Japanese.

4. M. R. Piggott, Polym. Composites, 8, 291 (1987).

5. T. Ohsawa et al., J. Appl. Polym. Sci., 22, 3203 (1978).

6. M. Hojo et al., Composites, 26, 243 (1995).

7. K.Tanaka et al., Composites, 26, 257 (1995).

8. A. G. Evans and B. J. Dalgleish, Acta Metall. Mater., 40, S, 295 (1992).

9. M. Y. He, B. Wu, Z. Suo, Acta Metall. Mater., 43, 3453 (1995).

10. M. R. Turner et al., Acta Metall. Mater., 43, 3459 (1995).

11. T. A. Michalske, E. R. Fuller, J. Am. Ceram. Soc., 68, 586 (1985).

12. T. A. Michalske et al., Engng. Fract. Mech., 45, 637 (1993).

Study on in-situ fiber reinforced composites: Polyolefin/Polycaprolactone blend

T. Semba and K. Kitagawa
Department of Applied Chemistry
Kyoto Municipal Institute for Industrial Research
17 Chudoji Minami-machi, Shimogyo-ku, Kyoto 600-8813, Japan

H. Hamada
Division of Advanced Fibro-science
Kyoto Institute of Technology
Goshokaido-cho, Matsugasaki, Sakyo-ku, Kyoto 606-8585, Japan

ABSTRACT

In-situ fiber reinforced composites were manufactured in polymer processing process with shear and elongational flows. In this study, the in-situ fiber reinforced composite was prepared by blending polypropylene (PP) and polycaprolactone (PCL). This work proved that the in-situ fiber formation in PP/PCL blends were due to a shear and elongational stress. The PP/PCL sheets were manufactured by use of single screw extruder equipped with slit type die which could be applied higher shear and elongational stresses and heat press machine with lower shear and elongational stresses. The mechanical properties of the extrusion sheets were considerably improved in contrast with compressive sheets attributed to the in-situ dispersed long fiber phase in extrusion processing. The DSC curves made clear the difference of crystallization between the fiber and spherical shape dispersive components.

INTRODUCTION

Morphology control of the dispersive phase is important technology in polymer composites materials. Many works concerning this control have been carried out and applied in industrial products [1]. Our study has been focusing that the morphology control would result in showing a reinforcable ability for composites.

Kitagawa et al.[2] found peculiar morphlogy in LDPE/PBT blend systems that a PBT dispersive phase was continuous long fiber rather than a fibril. The generation of in-situ long continuous fibers of PBT in LDPE matrix was observed. The drawing on extrusion affected on mechanical properties and morphology of the extrusion strands. In the injection molding of pellet obtained from the drawing strands, the most suitable processing temperature was existed. The injection moldings possessed the in-situ PBT reinforcement in LDPE matrix. Michel [3)-6)] et al. also researched peculiar morphology of PBT component. They mentioned the fibrillar morphology and the processing conditions. Furthermore, a LDPE/PCL (polycaprolactone) blend system in which melting point of PCL is lower than that of LDPE, also, showed a in-situ LDPE long fiber reinforcement formed in PCL matrix [7]. The tensile moduli of this blends exceeded the theoretical line calculated by the low of mixture, but were not sufficient to composites. Almost in-situ fibril or long fiber reinforced composites in polymer blends except the LCP system have not sufficient mechanical properties. In several years, we found that the mechanical properties of in-situ fiber reinforced PP (polypropylene)/PCL blend showed multiple effects without compatibilizer [8]. At present, the material of PCL is well generally known as biodegradable polymer, but the first use of the PCL was an additive for improving plastic materials, for example, improvements of impact strength and so on.

In this study, mechanical properties and the morphology of the PP/PCL blend system which is one of in-situ fiber reinforced composites were investigated.

EXPERIMENTAL

2-1. Materials

Materials used in this study were a PP (IDEMITSU PP J900GP, MI=13,Mw=200,000: Idemitsu petrochemical Co.,ltd) and a PCL (Celgreen PH7,MI=2.4: Daicel Chemical Industries,Ltd).

2-2. Fablication of PP/PCL blend sheets

Compressive molding method as the processing with low shear and elongational flow was applied to fablication of the PP/PCL blend sheets. First, the PP and PCL pellets were gradually added to a batch type mixer (LABO PLASTOMILL 100C100 : TOYOSEIKISEISAKU-SHO LTD.) at the temperature of 200 ℃ and the screw revolution of 30rpm for 90 seconds. The compositions of the PP/PCL blend were 0/10, 2/8, 5/5, 8/2 and 10/0 in weight. The compounds were held between two aluminum plates, with a polished teflon surface, and compressed by heat press machine(NF-50 : Shinto Metal Industries Ltd., max load=50tf). Compressive processing conditions were as follows; compressive pressure was 0.3MPa, the processing time was 60second. The prepared sheets, immediately, were quenched by water and the thickness of the sheets were about 1mm.

Furthermore, PP/PCL blend sheets were manufactured using a single screw extruder with a screw length by screw diameter ratio (L/D) of 32 (PSV30 : PULA ENGE CO., LTD.) equipped with slit type die(t=1mm), and drawing machine.   They generated high shear flow, and high elongational flow.   Processing conditions were as follows; the barrel temperature was between 170 and 200℃ between hopper and die, the screw revolution was fixed to keep the thickness 1mm of the sheets, the drawing rate was 715cm/min.   The extruded sheets was water-cooling.   The compounding ratio of the PP/PCL blend were also 0/10, 2/8, 5/5, 8/2 and 10/0.

2-3. Mechanical properties

The tensile tests of the blended sheets along flow direction were performed by universal testing machine (Auto graph AG-5000E: SHIMADZU CORPORATION) at cross head speed 5mm/min and at room temperature.   The tensile gage length was 50mm and 65mm of the compressive sheets and the extruded sheets, respectively.

2-4. Morphology

Owing to an effective SEM observation, the PCL component of the blend sheets was extracted using chloroform by a soxhlet extractor.   The surfaces of the residual PP were observed by SEM(JSM 5900-LV: JEOL Ltd.).   The skin area where the depth from surface is 100μ m and core area was observed in extruded sheets.

2-5. DSC analysis

Thermal analysis was carried out by differential scanning calorie meter (DSC7: Perkin-Elmer Co.,Ltd,. heating rate: 10℃/min.).

RESULTS and DISCUSSION
3-1. Tensile properties

Typical stress-strain curves of the PP/PCL blended sheets manufactured by use of the heat press machine(a) or the extruder(b) are shown in Figure1.   The 5/5 blend of the heat pressed sheets fractured at only 4.1% strain, however the other samples showed ductile behavior similar to the PP and PCL neat resin with the ultimate nominal strain of over 50%.

Figure2 shows the relationship between PP content and tensile modulus of the blended sheets in both processing.   The tensile moduli of the extrusive sheets with various compositions were considerably larger than theoretical line which was calculated by the law of mixture, whereas those of the heat pressed sheets nearly followed the line. The increasing of the modulus toward the theoretical line of extruded sheets were 6.6%, 26.2% and 18.7% in 2/8, 5/5 and 8/2 blended sheets, respectively.   Furthermore, the tensile modulus of the 8/2 blended sheets exceeded 6.7% over that of the neat PP sheets. Figure3 shows the relationship between tensile strength and PP content of the both blend sheets.   The tensile strengths of the pressed sheets showed the typical tendency of immiscible polymer blends that the values were quite below the theoretical line.   The tensile strengths of pressed sheets decreased by blending 20wt% PP in the PCL matrix, and were kept constantly from 2/8 to 5/5 compositions, finally, increased from 5/5 to 10/0 compositions.   However the tensile strength of the extrusive sheets had excellent property.   The strengths were higher than the theoretical line; 6.5%, 2.1% and 7.2% in 2/8, 5/5 and 8/2 blended sheets, respectively.   The above results confirmed that the shear and elongational flow in the extrusion process affected the tensile properties of the blended sheets.   It was suggested that these differences were resulting from the differences in the internal structures of both blended sheets.

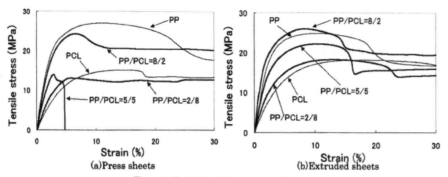

Figure 1 Typical tensile stress-strain curves
of the press and extruded PP/PCL blended sheets.

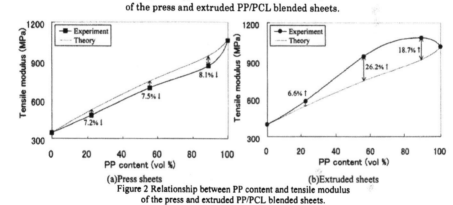

(a)Press sheets   (b)Extruded sheets
Figure 2 Relationship between PP content and tensile modulus
of the press and extruded PP/PCL blended sheets.

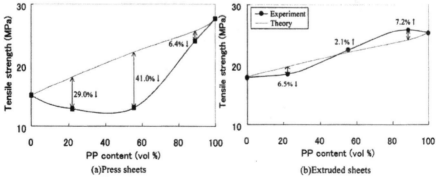

(a)Press sheets   (b)Extruded sheets
Figure 3 Relationship between PP content and tensile strength
of the press and extruded PP/PCL blended sheets.

3-2. Morphology

Figure4 shows the SEM photographs of the PP/PCL=8/2 blend sheets manufactured by both methods. The PCL components of the heat pressed sheets were dissolved by the chloroform, and observed as black spherical ditches that the maximum size was 8 μ m and minimal size was 0.3 μ m. In case of the extrusive sheets, the PCL components at the skin layer exposed the shear flow formed long fiber along the flow direction. The number of the droplets gradually increased in the core layer, attributed to the low shear flow and the relaxation of orientation resulting from quenching rate difference. Figure5 shows the SEM photographs of the both PP/PCL=5/5 blend sheets. In the press sheets, the PCL domains of spherical shape with various sizes were dispersed in the PP matrix. The particle size was so larger than 8/2 blends as shown in figure4-(a) because of domain coalescence. The morphology of the extrusive sheets was the co-continuous structure in both skin and core layers. Figure6 shows the morphology of the PP/PCL=2/8 blend sheets, where the PCL was matrix and dissolved by chloroform, so that the PP dispersive phase was seen in this photographs. The pressed sheets were

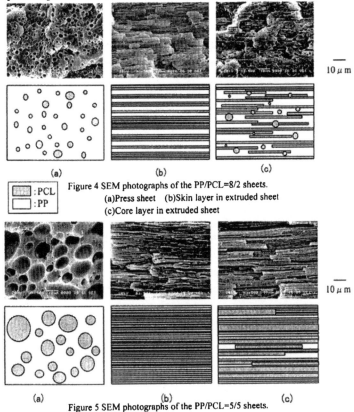

10 μ m

(a)   (b)   (c)

:PCL
:PP

Figure 4 SEM photographs of the PP/PCL=8/2 sheets.
(a)Press sheet   (b)Skin layer in extruded sheet
(c)Core layer in extruded sheet

10 μ m

(a)   (b)   (c)

Figure 5 SEM photographs of the PP/PCL=5/5 sheets.
(a)Press sheet   (b)Skin layer in extruded sheet
(c)Core layer in extruded sheet

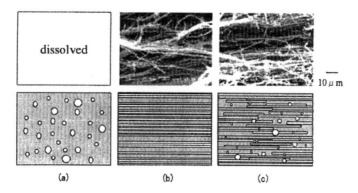

Figure 6 SEM photographs and expected schematic diagram of the PP/PCL=2/8 sheets.
(a)Press sheet   (b)Skin layer in extruded sheet
(c)Core layer in extruded sheet

completely dissolved by the chloroform.   Therefore it is considered that the PP component in these sheets must be distributed with very fine sphere shape.   The expected schematic diagram is shown in Figure6(a).   The fine continuous fiber reinforcements were clearly dispersed from skin to core layer in the extruded sheet as shown in Figure6(b)(c).   It can be seen that the mostly diameter of the continuousfibers was less than 1μ m.   From these observation results, the minor component in the PP/PCL blend sheets was deformed to continuous long fiber as the result of shear and elongational flow, and the continuous long fiber effectively reinforced the matrix as mentioned in previous chapter.   The developing mechanism of the fiber formation in this processing is that the dispersive components are compounded and fixed as droplets by the screw compounding, and the shear flow in slit type die transform these droplets to the long fibers.   Furthermore, the elongational flow between the die lip and the quenching zone accelerate the fiber formation.

3-3. Evaluation of crystallinity

The DSC thermograms of the PP/PCL=8/2 sheets manufactured by both methods as shown in Figure7.   The shape of the PCL endothermic peak at near 60℃ was different in the processing methods.   The press sheets has shoulder peak at 55.6℃ except the main peak resulting from the existence of micro crystal structure.   The crystal of extrusive sheets is homogeneous structure so that the peak shape was smooth.   Table1 summarized the melting enthalpy data(Δ H) from DSC analysis.   In case of the PP/PCL=8/2, the ΔH_{PCL} of the press and the extrusive sheets were 75.21J/g and 62.14J/g respectively.   The deformation from the particle to the fiber shape of the PCL dispersive phase in the extrusion make   great influence on the growth of ΔH_{PCL}.   On the other hand, the shape of the PP endothermic part and the ΔH_{PP} were almost same, namely, the crystal structure and the crystallinity of the matrix are not affected by shear and elongational flow in extruder.   In case of the PP/PCL=5/5, the shoulder peak of the PCL endothermic part was observed in both processing sheets.   This phenomena of little increase of the ΔH_{PCL} and the ΔH_{PP} suit that the dispersive phase formed the co-continuous structure as shown in Figure6.   The PP fiber reinforcements were

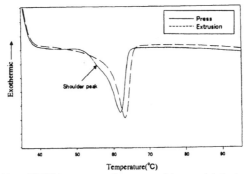

Figure 7 DSC thermograms of the press and the extruded sheets.

unit:Jg⁻¹

	PCL		PP/PCL=2/8		PP/PCL=5/5		PP/PCL=8/2		PP	
	PCL	PP	PCL	PP	PCL	PP	PCL	PP	PCL	PP
(a)press	71.30	×	70.81	74.33	71.58	78.69	62.14	81.04	×	78.15
(b)extrusion	76.40	×	71.35	81.30	74.95	79.39	75.21	78.44	×	78.75
(b)-(a)	5.10	×	0.54	6.97	3.37	0.70	13.07	-2.60	×	0.60

Table 1 Melting enthalpy of each blend sheets

constituted in the extrusive PP/PCL=2/8 sheets that the matrix is the PCL.   The ΔH_{PP} of the fiber reinforcements in the extrusive sheets was also larger than that of press sheet. Some former study, for instance, proved that crystallinity of isotactic polypropylene under shear flow was grown as the results of nucleus increase induced by shear flow [11], [12].   In the PP/PCL blend system, the high crystallinity of the fiber dispersive phase was performed, because shear and elongational flow efficiently convey the dispersive phase through the interface between matrix/dispersion by special matching of PP and PCL. The DSC analysis give the information that the crystallinity of the fiber reinforcement is developed by shear and elongational flow.

4. CONCLUSION

In this study, the effects of the shear and elongational flow in the different processing on the mechanical properties and morphology of PP/PCL blend which is one of in-situ fiber reinforced composites was investigated.   The main results might be summarized as follows;
1. The shear and elongational flow promoted the dispersivre phase to transform the long fiber reinforcements.
2. In the PP/PCL blends, the component with low ratio formed the long fiber reinforcement in the matrix.
3. The tensile properties of the extrusive sheets, such as moduli, strength and ultimate strain, show multiple effects remarkably.
4. The enthalpy of fiber components increased in contrast with sphere shape dispersion. The obtained results in these analyses were that the high crystallinity was achieved in

fiber components.

REFERENCES
(1)L. A. Utraci, "Polymer Alloys and Blends" Hanser Publishers. (1989)
(2) K. Kitagawa, T. Semba and H. Hamada: J. Material Sci. Japan 47, 12(1998)
(3) P.Cassagnau, R.Fulchiron, and A.Michel: J. Polym. Sci., PartB, Vol.36,No.14(1998)
(4)A.Monticciolo, P.Cassagnau and A.Michel: Polym. Eng. Sci., Vol.38, No.11(1998)
(5)M.F.Boyaud, P.Cassagnau, A.Michel and M.Bousmina: Polym. Eng. Sci., Vol.41, No.4(2001)
(6)I.Pesneau, P.Cassagnau and A.Michel: J. Appl. Polym. Sci., Vol82, No.14(2001)
(7)T.Semba, K.Kitagawa and H.Hamada: Proc. 6th Japan International SAMPE symposium, Oct. 26-29,(1999)
(8)T.Semba, K.Kitagawa, H.Hamada, M.Kwazuma and K.Maeda: 11th Annual Meeting of The Jap. Society of Polym. Processing,(2000)
(9)J.M.H.Jansen and H.E.H.Meijer: J. Rheol. Vol.37, No.4 (1993)
(10)M.A.Huneault, Z.H.Shi and L.A.Utracki: Polym. Eng. Sci. Vol.35, No.1 (1995)
(11)S.Liedauer, G.Eder, H.Janeschitz-Kriegl, P.Jerschow, W.Geymayer and E.Ingolic: Int. Polym. Process. Vol.8, No.3(1993)
(12)G.Kumaraswamy, AM.Issaian and JA.Kornfield: Macromolcules, Vol.32(1999)

Thermal Conductivity of plain weaves fabric
which consists of high strength plain fiber

Yoshihiro Yamashita, Hiroaki Yamada, Akira Tanaka and Sueo Kawabata
Department of Materials Science, The University of Shiga Prefecture
2500 Hassaka, Hikone 5228533, Japan

1.Introduction

The thermal conductivity and the moisture transfer characteristic of the cloth are deeply related to the comfortability of clothes. Recently, the high-strength fiber is used also for the fireman uniform, protective clothing, and the rider suit. The thermal behavior analysis of the structure, which consists of the cloth under a severe environment, is important. In addition, the cloth, which consists of the high-strength fiber, is often used also for the composite material. For example, the diffusion of heat by the cloth composite substrate such as cellular phones is an important problem. However, the thermal conductivity of the composite material which uses the cloth has not been elucidated enough. We examined the thermal conductivity in the transverse direction of the yarn and the direction of the thickness of plain weave fabric and plain weave fabric/resin composite from the structure model and the measurement. The plain weave fabric was used in general widely for the simplest geometrical pattern. Kawabata found fibers of variety having strong anisotropy in the fiber-axis direction and the transverse direction like having shown the coefficient of thermal conductivity of the fiber with Table 1. We elucidated the effect that anisotropy of the coefficient of thermal conductivity of the fiber caused for a yarn structure material. We obtained the coefficient of thermal conductivity of the transverse direction of the yarn by first using the coefficient of thermal conductivity of the fiber.

Table 1 Heat conductivity of fiber $(Wm^{-1}K^{-1})$

Fiber	longitudinal λ_{FL}	transverse λ_{FT}	$\lambda_{FL} / \lambda_{FT}$
Carbon T-300	7.81	0.675	11.9
Kevlar 29	3.33	0.181	18.4
E-Glass	2.25	0.509	4.42

2.Experimental

2.1 Device

KES-F7 Thermo Labo type II made of the Kato tech Ltd. was used to measure the coefficient of thermal conductivity to the transverse direction of the yarn and the direction of the thickness of the plain weave fabric and this composite structure. The area of the hot plate of "BT-Box" used 2.5mm x 2.5mm in the measurement of the coefficient of thermal conductivity to the transverse direction of the yarn and 8mm x 8mm was used for the measurement of the epoxy and silicone resin composites. The sample is put on the stage of KES handy compression tester, and it is compressed. The thickness of the cloth was obtained by the extrapolation to the value of load 0. Capacity of the load cell is 20Kg.

2.2 Yarn sample

Kevlar 29 used four kinds of cloths with the different number of the fiber that composed the yarn; therefore, the density and the thickness of the cloth are different. The coefficient of thermal conductivity in the part where warp and weft intersected was measured. These cloths made the composite material with epoxy resin and the silicone resin for heat radiation.

2.3 Composite materials

2.3.1 Epoxy resin matrix

Epikoto819 and the triethylenetetramine are mixed by a percentage by weight of 10:1. The cloth was washed with the acetone, and the resin was spread, and the deairing was done. The resin post-cured for one hour at 100°C after the hardening a day in the room temperature.

2.3.2 Silicone resin matrix

To mix made of KE-1223 and CLA-3 (Shinetsu Chemical Industry Ltd.) by percentage by weight 50:1, and to lower the viscosity, the toluene of 10wt% is added and diluted for the resin. After the resin had been spread on the cloth, the deaeration was done. The resin post-cured for one hour at 80°C after the hardening a day in the room temperature.

3.Theory

The analogy to the electric circuit was tried for the heat transfer analysis of the cloth. The structure model of the yarn and the plain weave cloth was made, and it was made a circuit. The formula of the coefficient of thermal conductivity to the transverse direction of the yarn, the direction of the thickness of the plain weave crossing, and the direction of the thickness of plain weave crossing/resin composite was led by using this circuit. Next, those coefficients of thermal conductivity were measured. The coefficient of thermal conductivity led from the model theory was compared with it from the measurement and considered. Moreover, heat flow in the plain weave fabric was analyzed by the finite element method.

3.1 Experimental theory of coefficient of thermal conductivity

Thickness d (m) of the sample is measured by using KES handy compression tester. The thickness was measured while compressing the plain weave fabric, and the thickness in no-load was presumed. Next, Water Box of KES-F7 Thermo Labo type II was kept T_w (25 °C) with the constant temperature water circulation device. Temperature gradient DT ($=T_B-T_W$) of BT Box (T_B°C) and Water Box (T_W°C) become about 10°C. Heat flow q (Watt) in area A (m^2) of the sample and BT Box temperature T_B°C are measured. Here, the value of q is an average of the measurement between 1 min in the stationary state. The weight of BT Box was changed, and the heat flow in no-load of the plain weave fabric was presumed. Coefficient of thermal conductivity λ (Wm^{-1}K^{-1}) is calculated from the following expressions. Heat flow assumes the coordinate system in the direction of X.

$$\frac{q}{A} = -\lambda \frac{\partial T}{\partial x} = -\lambda \frac{T_W - T_B}{d} \qquad (1)$$

The coefficient of thermal conductivity l becomes equation (2) from equation (1).

$$\lambda = \frac{qd}{A\Delta T} \qquad (2)$$

3.2 Structure model of yarn

The structure model to obtain the coefficient of thermal conductivity of the transverse direction of the yarn is described. It was assumed that the coefficient of thermal conductivity in the fiber-axis direction of the yarn and that of the single fiber were the same. And, it was assumed that the coefficient of thermal conductivity to the transverse direction of the yarn was the total of thermal conductivity's of the single fiber and air (that of the matrix resin in case of the composite).

3.3 Modeling of yarn section

The section of the fiber is modeled by the square. The fiber section of D in the diameter was modeled by the square of the same area (equation (3) and figure 1).

$$C = \sqrt{\frac{\pi}{4}D^2} = \frac{\sqrt{\pi}}{2}D \qquad (3)$$

Fig.1   The section of the fiber is modeled by the square.

In addition, the part of the fiber and the air of the yarn were modeled as follows. The section of the yarn was observed with the stereoscopic microscope because the section of an actual yarn was flat and the ratio in the air part was examined. V_f is a packing fraction of the fiber in the yarn. To make the model easy, it was assumed $b=a$. When the 3-D shape shows the model, it is shown in figure 2. The length of the depth h_3 is assumed to be 1.

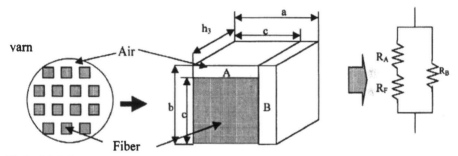

Fig.2   Modeling of yarn section

3.4 Calculation of thermal resistance of yarn

The thermal resistance of the yarn is calculated by using the same model. Thermal resistance R_Y of model structure body, R_A, R_B, and R_{TF} are the thermal resistance of air and the fiber. λ_A, λ_B, and λ_{FT} are assumed to be air space A and B, a coefficient of thermal conductivity of the transverse direction of the fiber respectively here. Finally, coefficient of thermal conductivity to the transverse direction of the yarn λ_{YT} was obtained.

$$\frac{1}{R_Y}=\frac{1}{R_B}+\frac{1}{R_A+R_F} \qquad (4)$$

$$\lambda_{YT}=\frac{b}{R_Y a h_3}=\frac{k}{R_Y h_3} \qquad (5)$$

3.5 Structure model of plain weave fabric

To calculate the coefficient of thermal conductivity to the direction of the thickness of the plain weave fabric, an easy structural model was assumed. This structure model consists of two elements concerning heat conduction of the cloth. This plain weave fabric model combined the thermal conductivities in the direction of the yarn of the transverse direction and the parallel. This model considers heat flow to both directions of the yarn (model 1, figure 3). Another model considered only heat flow to the transverse direction of the yarn (model 2). Heat conduction can do the analogy with the electric circuit. Heat conduction of the repeat unit organization of the plain weave fabric can be replaced with the network circuit like figure 3 by consisting of the same fiber as the warp and the weft and assuming that the fiber volume fraction of the yarn is also equal. Here, r and rL are

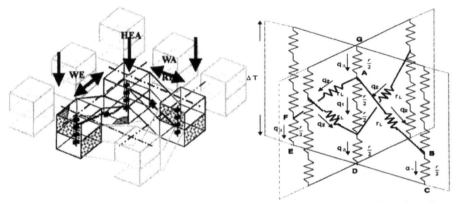

Fig.3   Plain weave fabric model combined the thermal conductivities in the direction of the yarn of the transverse direction and the parallel.

thermal resistance in a transverse direction and a parallel direction of the yarn respectively. The heat flow in the direction of the thickness was calculated from this structure model. The analogy to the heat flow of the Kirchhoff's law was used to calculate. The heat flow which flows in the direction of the thickness of the cloth can be shown by the equation (6). Therefore, the coefficient of thermal conductivity to the direction of the thickness of the cloth becomes like the equation (7).

$$q_3 = \frac{\Delta T}{2r} \cdot \frac{(1 + \frac{2r}{r_L})}{(1 + \frac{r}{r_L})} \tag{6}$$

$$\lambda_{WT} = \frac{q_3 \, d}{A \Delta T} \tag{7}$$

3.6 Calculation of coefficient of thermal conductivity in direction of thickness of

fabric and composite material

It is necessary to consider the gap between the textures of the plain weave fabric. Moreover, the thermal performance in the air part of the structure model of the yarn is replaced with that of the resin for the composite material. In addition, because the matrix part exists the top and bottom of the cloth slightly, it is considered for the composite material (equation (8)).

$$R_c = R_4 + \frac{R_1 R_2 R_3}{R_1 R_2 + R_2 R_3 + R_3 R_1} \quad , R_2 = R_5 + R_6 \quad (8)$$

3.7 Heat transfer analysis by finite element method

The heat transfer analysis was calculated by using MSC-NASTRAN. The coefficient of thermal conductivity of the measured yarn was used for anisotropy of the material.

4.Result and discussion

4.1 Coefficient of thermal conductivity of transverse direction of yarn

Coefficients of thermal conductivity of the transverse direction of an actual yarn were compared with it calculated from the structure model in the section of the yarn. The coefficient of thermal conductivity of air λ_a is 0.026 W m^{-1} K^{-1} at 25°C. λ_{FT} used the value, which had been obtained with Table.1. The calculation value of the yarn of three kinds of different fibers is almost equal to measurements. Moreover, the result of the calculation and the measurement was almost the same values though the number of the fiber was different. Therefore, it is thought that the theoretical formula led from the structure model of the yarn made by this research is correct.

4.2 Coefficients of thermal conductivity in direction of thickness of plain weave fabric

Coefficients of thermal conductivity in the direction of the thickness of the plain weave fabric were measured and compared with it calculated from the structure model. Measurements were the values between model 1 and model 2. Moreover, these were values that were closer to the value of model 1 than model 2.This result shows that there is obviously heat flow to parallel to the yarn direction in the plain weave fabric. The calculation value reached almost the same value as measurements though the density of the yarn and the kind of the yarn was different. Therefore, it is thought that the structure model of the plain weave fabric made by this research is right.

4.3 Coefficients of thermal conductivity to direction of thickness of plain weave fabric/resin composite

The coefficient of thermal conductivity of epoxy resin and the silicone resin was respectively 0.200 W m^{-1} K^{-1} and 0.900 W m^{-1} K^{-1}. The value of measurements of the plain weave fabric, which uses epoxy resin and the silicone resin for the matrix, is almost equal to model 1. That is, even the composite material could confirm heat flow to the direction of the yarn. The calculation and the measurement were almost the same values though the kind and the density of the yarn was different. It is thought that the structure model of the plain weave fabric made by this research can adjust also to the composite material, and the theoretical formula to obtain the coefficient of thermal conductivity is right. Moreover, it seems that this formula can use the resin with a different coefficient of thermal conductivity.

4.4 Relation between volume fraction of fiber and coefficient of thermal conductivity

The relation between the fiber volume fraction and the coefficient of thermal conductivity was shown in figure 5. The number of coefficients of thermal conductivity of the plain weave fabric increases as the fiber volume fraction increases because the coefficient of thermal conductivity of air is smaller than that of the fiber. The coefficient of thermal conductivity of the composite material does not depend on the volume fraction of the fiber when epoxy resin with almost the same coefficient of thermal conductivity as the plain weave fabric is used for the matrix. However, it has decreased when the silicone resin with higher coefficient of thermal conductivity than the plain weave fabric is used. That is, the coefficient of thermal conductivity to the direction of the thickness of the plain weave fabric and the composite material is proportional to the fiber volume fraction. The coefficient of thermal conductivity of Kevlar 29 unwoven cloths/epoxy resin composite was measured for the comparison. There is no directionality in the plane because this unwoven cloth has distributed a short fiber at random. This was corresponding to the calculation result of model 2 who did not contain heat conduction to the direction of the yarn. It could be confirmed that the thermal conductivity to the direction of the yarn of the plain weave fabric contributed to heat conduction to the direction of the thickness.

4.5 Effect that coefficient of thermal conductivity anisotropy of fiber

The influence that ratio (rT/rL) of the thermal resistance of the direction of the yarn and the transverse direction of the yarn exerted on the heat flow in the direction of the thickness of the cloth was examined. The heat flow to the direction of the thickness of the plain weave fabric was at most twice though was large the coefficient of thermal conductivity in the direction of the yarn. This result shows that heat conduction anisotropy of the fiber does not strongly contribute to the coefficient of thermal conductivity to the direction of the thickness.

4.6 Heat transfer analyses of plain weave fabric by finite element method

The heat flow in the direction of the thickness when the coefficient of thermal conductivity of the transverse direction of the yarn was fixed, and the coefficient of thermal conductivity in the direction of the yarn was changed in the Kevlar plain weave fabric. It has been understood that the heat flow in the direction of the yarn contributes to the coefficient of thermal conductivity to the direction of the thickness of the plain weave fabric. However, the contribution was a little as well as the result in the structure model calculation.

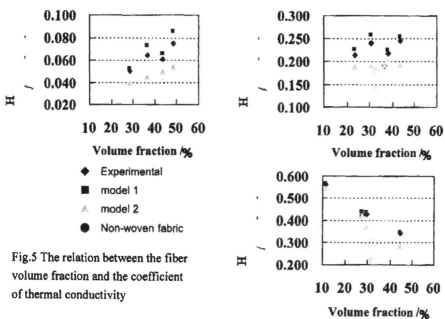

Fig.5 The relation between the fiber volume fraction and the coefficient of thermal conductivity

5.Conclusion

It has been understood that heat not only flows to the transverse direction of the yarn but also flows in the direction of the yarn in the plain weave fabric. However, the influence that heat flow to a parallel direction of the yarn exerted on heat flow to the direction of the thickness of the plain weave fabric became only twice when assuming that heat did not flow in the direction of the yarn at all.

Void evaluation of polyurethane foam materials
by image processing technology

Akihiko GOTO*, Kazumi YAMAGUCHI** and Hiroyuki HAMADA***

*Department of Information Systems Engineering, Osaka Sangyo University
3-1-1 Nakagaito, Daito, Osaka 574-8530, JAPAN
** Industrial Technology Center of Wakayama Prefecture
60 Ogura, Wakayama 649-6261, JAPAN
*** Division of Advanced Fibro-Science Graduate School, Kyoto Institute of Technology
Matsugasaki, Sakyo-ku, Kyoto 606-8585, JAPAN

ABSTRACT

We have been investigated the heterogeneity of polymeric composite materials. Distributions of fiber length and fiber direction in fiber reinforced plastics are very importance factor to evaluate materials. By image processing technology, evaluation methodology of these factors has examined.

In this study, we attended to the void evaluation of polyurethane foam materials. Void state is significant factor of the heterogeneity. Several kinds of foam materials with different foam states were employed. Each specimen was cut three kinds of part. There were bottom part, middle part and top part, respectively. Image of these cross section were scanned in the computer. Void feature was extracted as white pixels by binary image. Here, influence of threshold on the binary image was investigated, so that the optimum threshold for binary image was attended to be clears the void shape. The algorithm of extraction for geometric information of void was examined. The algorithm measured the void shape and the distribution of void size and direction, so that this information was evaluated. Moreover, we attended to the correlation between geometric information of void and location in detail.

INTRODUCTION

Various properties of foam materials are depended upon the size and state of void. Internal void state of foam material is generated probability when the molding. However, it is difficult to grasp the internal state form molding condition. Then, internal observation is often done by image processing. When roughly dividing, there are two kinds of image processing methods. One is two-dimensional image processing and the other is three-dimensional image processing. In the case of the former, there are methods of using the microscope with CCD camera, image scanner or film scanner and so on. On the other hand, in the case of the latter, there are methods of using tomogram by CT scanning. Moreover, statistical processing is done, the void cell is modeled, and the inside is observed.

Here, we aimed at the construction of the system to understand the internal state of the form material with easily and cheap in the molding factory. The rigid polyurethane was employed. The image of surface of the foam material was taken into the personal computer by scanner. Geometric information of void was extracted from the 2D image. Evaluation of size and shape of void was carried out.

ANALYTICAL PROCEDURE

Objective material

Objective foam materials were molded by mixed polyether-based polyol, silicon-based bubble adjustment agent, foaming agent, and so on. The size of material was 200 mm x 200 mm in bottom, about 200mm in height. Figure 1 shows an example of the appearance of the foam materials. The foam materials were divided into three kinds of area along the foaming direction. There were top part, middle part and bottom part. In each part, samples were cut out as a cube of one side 50 mm, so that we called Type-1, 2 and 3, respectively. Figure 2 shows the analytical part of a block of the foam.

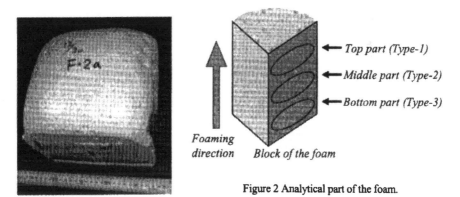

Top part (Type-1)

Middle part (Type-2)

Bottom part (Type-3)

Foaming direction   *Block of the foam*

Figure 2 Analytical part of the foam.

Figure 1 Appearance of the foam material.

Image processing

Six surfaces of cube were taken into the computer as digital images by scanner, so that geometric information of void in each surface was extracted. Resolution of scanner was 200 dpi. The size of analytical image was 320 pixels x 240 pixels. This corresponds to 40mm x 30mm in an actual size. One pixel was about 0.0156 mm^2. Figure 3 shows scanned original images, which were top part (Type-1) and bottom part (Type-3).

Firstly, the taken image gave the binary image processing. Threshold to make the binary image was varied from 0 to 255. Here, void was identified as a white pixel. In the case of each threshold, total white pixel and the number of the lump of white pixel were measured. When the number was maximum value, the threshold was noticed. Secondary, in the threshold, coordinates of lump of white pixel were measured to get the geometric information of void. In our previous investigation, we are obtaining an interesting result. In the case of fiber reinforced plastics, measured total white pixels means fiber volume fraction in the threshold. We paid attention to this threshold in the form material based on this result. Figure 3 shows flowchart of algorithm to find white pixel. Raster scan is executed for the analytical image. It searches for a white pixel. If it is possible to search for a white pixel, continuity of white pixel in surrounding adjacent points is investigated.

Thus, the distribution of the void size and the deformation of the void shape were considered based on the coordinates value of the measured white pixel.

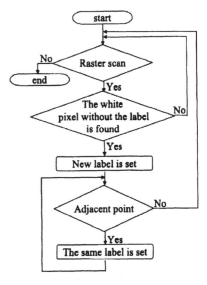

Figure 3 Flowchart of algorithm to find white pixel.

RESULTS & DISCUSSIONS

Figure 4 shows an example of analytical images. These are Type-1 and 3. After the tone of these images reverses, the threshold was changed one by one and the number of lump of white pixel was measured. Variation of the number with varying threshold is shown in figure 5. Threshold of peak of the number was noticed. Figure 6 shows binary images in the threshold. In such a binary image, Type-1 and 3 both showed about 19% in shares of the number of white pixels to the number of total pixels of an analytical image. The share of white pixels of both is the same. However, it is clear in each image that the size of void is different.

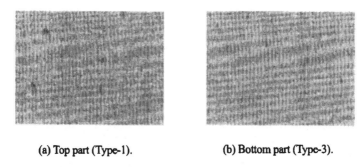

(a) Top part (Type-1).   (b) Bottom part (Type-3).

Figure 4 Analytical images.

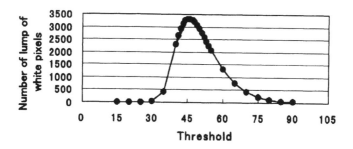

Figure 5 Variation of the number of lump of white pixels with varying threshold (Type-3).

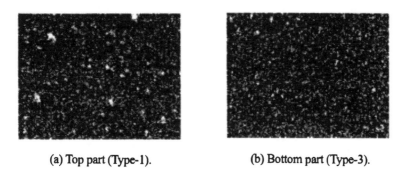

(a) Top part (Type-1).   (b) Bottom part (Type-3).

Figure 6 Binary images.

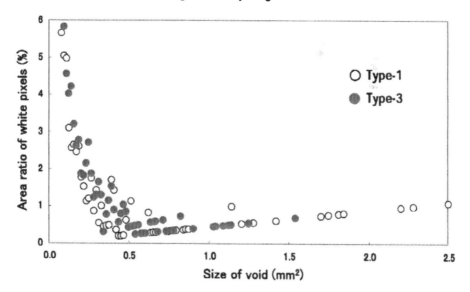

Figure 7 Relationship between area ratio of white pixels and size of void.

Then, the size of each white pixel was totaled in the binary image of both. The proportion of the lump of each white pixel to the number of total white pixels in this threshold was calculated and evaluated respectively. The relation between both is shown in figure 7. The share of void of a very small size showed similar tendency of both type-1 and 3. On the other hand, void of the largest size in Type-1 was about 1.5 times Type-3 and shares of about 1%. When the size of void is about $0.4mm^2$, it is clear that area ratio region of white pixels of Type-1 is wider than that of Type-3. Thus, it tends to include both size of large and small diameter of void in top part of foaming direction. Therefore, the difference of the share is large to the change in the size of void. However, in bottom part of foaming direction, many very small sizes of void are included. The difference of the share is small to the change in the size of void

CONCLUSIONS

It is considered that the state of void included in the foam material can be evaluated from the above-mentioned by doing such an image processing. In addition, it pays attention to the void shape in the foaming direction, the directionality is measured, and distribution is clarified. It pays attention to the deformation of void, and the examination of the correlation with mechanical properties is advanced.

Fiber Orientation and Mechanical Properties of Injection Molded CF/LCP Thin Plates

Akihiro FUJITA and Fumiaki BABA,
Advanced Technology R&D Center, Mitsubishi Electric Corporation
8-1-1, Tsukaguchi-honmachi, Amagasaki, Hyogo 661-8661, JAPAN

Chimyon GON, Daisuke WATANABE, Asami NAKAI and Hiroyuki HAMADA
Advanced Fibro-science, Kyoto Institute of Technology
Matsugasaki, Sakyo-ku, Kyoto 606-8585, JAPAN

ABSTRACT

Fiber reinforced plastics formed a multi layer structure by injection molding. To estimate the mechanical properties of injection molded articles, sliced technique are applied. Injection molded plates with 1mm thickness under different injection speeds of Liquid Crystalline Polymer (LCP) and carbon fiber reinforced LCP composite (CF/LCP) were sliced at every 100μm layer and measured the dynamic modulus. Cox formula and laminate theory on the based on the fiber orientation and modulus of each LCP layer were applied to calculate the modulus of the CF/LCP multi layer plate.

It could be confirmed that the injection molded CF/LCP plates can be treated as general laminated composite, and distribution of the modulus in the injection molded CF/LCP can be evaluated by the sliced technique. Comparing the calculated modulus to the experimental modulus in MD (machine direction) and TD (transverse direction) specimens, it is concluded that the degree of local orientation of LCP near fiber in CF/LCP is different from that of LCP without fiber. Moreover, it was cleared that the fiber orientation contributed to difference of the modulus near the surface layer due to the injection speed by evaluation of fiber orientation and modulus of sliced specimen.

INTRODUCTION

Injection molded thin plate of Liquid Crystalline Polymer (LCP) has unique structure and mechanical properties. On account of shear force and elongation force in injection molding, especially near region of mold wall, LCP aligns, and consequently it exhibits anisotoropic property [1-7]. In order to control the anisotropy in industrial process, the fillers, fibers were filled. LCP molded parts have been used as various parts for electronic applications, because of the high strength and stiffness, superior damping properties, high heat resistance, low coefficiency of thermal expansion and superior process-ability namely low viscosity.

The effects of gate figure and injection speed on the multi-layer structure, fiber orientation, mechanical properties, coefficient of linear expansion in injection molded carbon fiber reinforced LCP (CF/LCP) thin plate have been clarified [8-11]. Moreover, the estimation method of modulus has been examined, taking account of layer structure and fiber orientation [12]

The most important characteristic of injection molded CF/LCP thin plate is that it has layer structure in which fiber orientation and length were different. Also the bending fracture behavior was similar to the laminated composites such as carbon reinforced epoxy resin and so on. Therefore, in order to estimate modulus of thin plates fabricated by CF/LCP, the concept of lamination should be used. The distribution of modulus in the thickness direction has not been quantitatively measured in the injection molded thin plate.

In this paper, injection molded LCP and CF/LCP thin plates with two different injection speeds were sliced at every 100μm layer, and the modulus was measured with Dynamic Mechanical Spectrometer (DMS). Moreover the modulus of the CF/LCP thin plate was calculated by Cox formula and laminate theory on the based on the fiber orientation and distribution of modulus of LCP in thickness direction. Calculated value was compared with experimental value, and the effect of carbon fiber on orientation of LCP was examined.

EXPERIMENTAL PROCEDURES

Materials

Resins used in this study were a neat LCP (VECTRA A950, Polyplastics Co.) and CF/LCP (VECTRA A230, Polyplastics Co., Vf=30wt%, average fiber length is about 200μm). These two kinds of pellets were mixed under dry condition, to obtain 20wt% of carbon fiber content. Neat LCP (A950) and reinforced LCP (A220) specimen were injection molded for experiments.

Injection molding

Mold used in this study was plate shape with film gate as indicated in Fig.1 Dimension of plates was 70mm (length) x 55mm (width) x 1.0mm (thickness). A reciprocating screw injection molding machine with a hydraulic accumulator system (V110/75V, Sumitomo Heavy Machinery .Co) was used. Cylinder temperature was 320°C, mold temperature was 100°C, and injection speeds were 150mm/sec. (68cc/sec., low speed) and 300mm/sec. (136cc/sec., high speed).

Measurement of modulus by sliced specimen

A220 and A950 specimen were cut out in parallel (machine direction) and perpendicular (transverse direction) to machine direction of plate as indicated in Fig.1. The specimen which is parallel to machine direction is named 'MD specimen', and the specimen which is perpendicular to machine direction is named 'TD specimen'. The specimens were cut into with 9mm width.

MD and TD specimen were sliced at every 100μm of thickness by microtome (SM2500S, Leica micro systems co.). The standard slicing conditions were as follows; the speed was 5mm/s in MD and TD specimen of A220, 3mm/s in MD, 10mm/s in TD of A950. The tool angle was 40°. Storage elastic modulus of each sliced specimen was measured by viscoelastometer (DMS210 Seiko electronics industry) in frequency of 1Hz.

Evaluation of fiber orientation

Evaluation of fiber orientation was performed with each sliced specimen of A220. The surface of each sliced specimen was polished, and optical microphotographs of polished surface were taken. Image processing was applied in the center region (3x4mm) on molded plate as indicated with black square in Fig.1. The conversion of picture to monochrome was performed to distinguish the fiber clearly and the threshold was important parameter. Threshold was calculated on the assumption that the rate of white region was fiber content. The measuring routine was used, which is called as measurement of maximum diameter. The fiber length and orientation angle of each fiber could be calculated and fiber orientation and length distribution were obtained from these data.

RESULTS AND DISCUSSION

Verification of modulus in sliced specimen by laminate beam theory

In order to verify that the injection molded CF/LCP can be treated as general laminated composite, modulus of the whole specimen was calculated by laminate beam theory using each modulus with sliced specimen, and compared with experimental value measured by three point bending test.

Fig.2 shows the relationship between modulus and depth from surface in injection molded A220 plate. Depth from surface of the sliced specimen in the molded plates is the ratio of distance from surface to depth of molded plate. 0% means the surface of molded plate, and 50% means the center of thickness. Here, validity of this method by sliced specimen was verified by modulus of molded plate in the low injection speed.

It was assumed that MD and TD specimen has symmetry in thickness direction. The formula for laminate beam theory is shown as follows;

$$EI = \sum E_i \left(A_i z_{0i}^2 + I_{0i} \right) - z_0^2 \sum E_i A_i \qquad (1)$$

where, EI: bending rigidity of laminated plate, E_i: modulus of ith layer, A_i : cross section area of ith layer, z_{0i}: distance between any axis and neutral axis of ith layer, I_{0i}: second moment of area against neutral axis of ith layer, z_0: distance between any axis and neutral axis of laminated plate.

For both MD and TD specimen, calculated values (MD: 28.8GPa, TD: 3.6GPa) were well agreement with experimental values (MD: 28.1GPa, TD: 3.4GPa) From this result, it is considered that there was no effect of relaxation for residual stress etc on elastic property, when specimen was sliced. It could be confirmed that the injection molded CF/LCP can be treated as general laminated composite.

Distribution of modulus

Bending modulus (28.1GPa) of injection molded A220 plate in the low injection speed was higher than that (23.4GPa) of the molded plate in the high injection speed in MD specimen, whereas effect of the injection speed on the bending modulus did not appear in TD specimen (low: 3.4GPa, high: 3.7GPa).

Fig.3 shows the relationship between modulus and depth from surface in molded A950 plates. In MD specimen of the molded A220 plate, as shown in Fig.2, the modulus decreased from 0% (surface) to 50% (center) in the low speed specimen, whereas it exhibited maximum value around 15% in the high speed specimen. Difference of distribution of modulus due to the injection speed was exhibited near the surface. In TD specimen, modulus was constant up to 20%, and increased from 20% to 50%. Distribution of modulus in the low speed specimen was similar to that of modulus in the high speed specimen. On the other hand, modulus of the molded A950 plate decreased in MD specimen and increased in TD specimen with increasing of depth, regardless to injection speed, as shown in Fig.3. Therefore, it is considered that difference of the modulus near the surface leads to difference of the bending modulus due to the injection speed in MD specimen of the molded A220 plate.

Prediction of modulus

The modulus of sliced specimen for injection molded CF/LCP thin plate was predicted by formula Cox and laminate theory. Fig.4 shows flowchart for prediction of modulus. Cox equation is expressed as follows;

$$E_{iL,T} = \eta_0 \eta_{li} E_f V_f + E_{mi}(1 - V_f)$$

$$\eta_{li} = 1 - (\tanh \frac{1}{2}\beta \ell i) / \frac{1}{2}\beta \ell i, \quad \beta = \{\frac{2G_m}{E_f r^2 \ln(R/r)}\}^{\frac{1}{2}} \tag{2}$$

E, Ef, Emi: modulus of each layer, fiber, matrix, *Vf*: fiber volume fraction, η_0: coefficient of fiber orientation, ℓ_i: fiber length in each layer, G_m: shear modulus of matrix, R: half of distance between fibers, r: radius of fiber. Here fiber length distribution was obtained by image processing for sliced specimen, and modulus of A950 was used for *Emi* as shown in Fig.3. Consequently, E_{iL} and E_{iT} were calculated.

Fig.5 shows one of the optical microphotograph of the polished surface. Fig.6 shows the result of evaluation for fiber orientation by image processing. Axis of abscissa shows fiber orientation angle to flow direction of resin every 10 degree. Axis of ordinate shows the coefficient of fiber orientation, which was ratio of sum of fiber length in each fiber orientation angle to total fiber length. It was assumed that each sliced specimen with distribution of fiber orientation angle was laminate plate consists of unidirectional plate with each fiber orientation angle. The coefficient of fiber orientation in each orientation angle was converted into the rate of thickness of unidirectional plate in laminated plate. The modulus of each layer was calculated using the modulus of unidirectional plate and the thickness by laminate theory.

Fig.7 shows relationship between the calculated modulus, the experimental modulus in MD, TD specimen and the depth from surface in the low injection speed. In the result of MD specimen, calculated value showed same tendency with experimental value. All of calculated value was higher than experimental value in same depth. Whereas, in TD specimen, calculated value also showed same tendency with experimental value, but all of calculated value was lower than experimental value in same depth. This difference is caused by the degree of local orientation of LCP near fiber.

Effect of injection speed on distribution of modulus

In MD specimen of the injection molded CF/LCP plate, difference of distribution of modulus due to the injection speed was exhibited near the surface, as shown in Fig.2. Therefore, comparing to the fiber orientations of low and high speed specimens, as indicated in Fig.8, fiber orientations of 0° to 10° in the low speed and 20° to 30° in the high speed were dominant.

Moreover, modulus of each layer was calculated by using estimation method suggested in the previous section on the based on these fiber orientations. Fig.9 shows relationship between the calculated modulus and the depth from surface in MD specimen under both low and high injection speeds. Difference of the calculated modulus near the surface was same as the experimental result. From this result, it was cleared that the fiber orientation contributed to difference of the modulus near the surface due to the injection speed.

CONCLUSION

In this study, distribution of the modulus in injection molded LCP and CF/LCP thin plates with two different injection speeds was investigated by using sliced technique at every 100μm layer Moreover the modulus of the CF/LCP thin plate was estimated by Cox formula and laminate theory on the based on the fiber orientation and modulus of each LCP layer.

Consequently, it could be confirmed that the injection molded CF/LCP can be treated as general laminated composite, and distribution of the modulus in the injection molded CF/LCP can be evaluated by the sliced technique. Comparing the calculated modulus to the experimental modulus in MD and TD specimens the degree of local orientation of LCP domain near fiber is different from that of LCP without fiber. Moreover, it was cleared that the fiber orientation contributed to difference of the modulus near the surface due to the injection speed by evaluation of fiber orientation and modulus of sliced specimen.

REFERENCES

1. A.E.Zachariades and R.S.Porter, "High Modulus Polymers", Marcel Dekker Inc., 1987.
2. G.Menges,T.Schacht, H.Ecker and S. Ott, Inter. Polymer Processing, Vol.2, 1987, p.77.
3. K.Engberg, O.Stromberg and J.Martinsson, Polymer Eng. Sci., Vol.34, 1986, p.1336.
4. S.Kenig, Polymer Composites, Vol.7, 1986, p.50.
5. T.Weng, A.Hiltner and E.Bear, J. Mat. Sci., Vol.21, 1986, p.744.
6. R.S.Bay and C.L.TuckerIII, Polymer Composites, Vol.13, 1992, p.332.
7. S.Kenig, B.Trattner and H.Anderson, Polymer Composites, Vol.9, 1988, p.20.
8. A.Fujita and F.Baba, Proceedings of the Second Joint Canada-Japan Workshop on Composites, 1998, p.3.
9. A.Fujita, H.Ishida, E.Tanigaki, H.Hamada and F.Baba, Proceedings of the Third Joint Canada-Japan Workshop on Composites, 2000.
10. A.Fujita, A.Yamada, F.Baba, H.Ishida, E.Tanigaki and H.Hamada, JSPP'00, 2000, p.329.
11. A.Fujita, F.Baba and H.Hamada, JSPP Sympo'00, 2000, p.293.
12. A.Fujita, F.Baba, A.Goto, H.Ishida, A.Nakai and H.Hamada, JSPP'01, 2001, P.203.

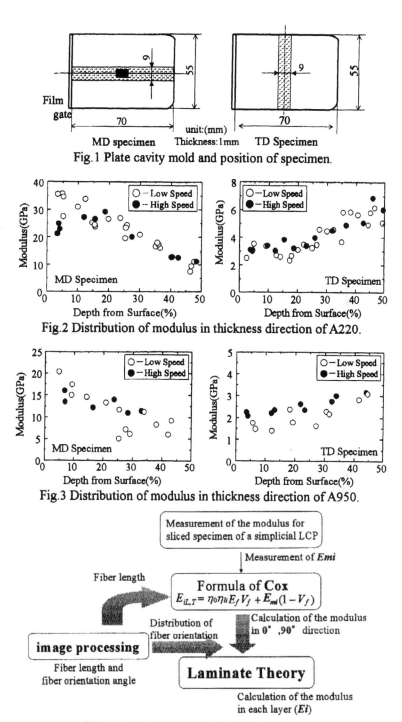

Fig.1 Plate cavity mold and position of specimen.

Fig.2 Distribution of modulus in thickness direction of A220.

Fig.3 Distribution of modulus in thickness direction of A950.

Fig.4 Flowchart for prediction of modulus.

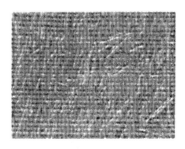

Fig.5 Microphotograph of polished surface.

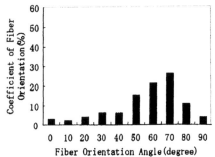

Fig.6 Distribution of fiber orientation.

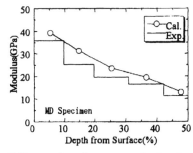

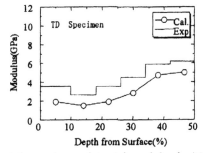

Fig.7 Comparison between calculated modulus and experimental modulus in A220.

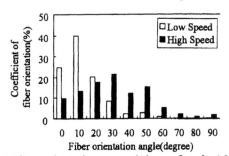

Fig.8 Fiber orientation around the surface in A220.

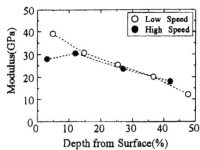

Fig.9 **Comparison of calculated modulus between low and high injection speed in A220.**

Microscopic observation of initial fracture in textile composites

Toshiko Osada
Graduate school,
Kyoto Institute of Technology
Matsugasaki, Sakyo-ku, Kyoto 606-8585, Japan
b5330033@ipc.kit.ac.jp

Asami Nakai
Advanced fibro-science
Kyoto Institute of Technology
Matsugasaki, Sakyo-ku, Kyoto 606-8585, Japan
nakai@ipc.kit.ac.jp

Hiroyuki Hamada
Advanced fibro-science
Kyoto Institute of Technology
Matsugasaki, Sakyo-ku, Kyoto 606-8585, Japan
hhamada@ipc.kit.ac.jp

ABSTRACT

In this study, initial fracture in textile composites was considered. Tensile testing of textile composites was performed and knee point on stress-strain curve was identified by least squares method. In situ observation of initial fracture in textile composites was also conducted. The relationship between initial fracture and knee point was considered Initial fracture in woven fabric composite was transverse crack in weft fiber bundles, that is, fracture at the interface inside of fiber bundle. Initial fractures in Non-cut and Cut specimens of braided composites were filament fracture and delamination at the fiber crossing part, respectively. Initial fracture in wale specimen of knitted composite was transverse crack at the loop head, course specimen was fracture at the fiber crossing part. Each fracture reflect interfacial property of around a filament, inside of fiber bundle, around a fiber bundle, that is, fracture at the crossing part.

INTRODUCTION

Utilizing well-developed textile technologies such as weaving, knitting, and braiding, structural preforms may be produced with fiber architectures such that the continuous orientation of fibers is not restricted at any point, resulting in superior mechanical properties. Textile composites have mainly been investigated from two points of view; development of 3-D textile fabrication techniques and evaluation of mechanical properties.

A characteristic property of textile composites is a knee point on the stress-strain curve of tensile test as shown in Fig.1. Knee point appeared at elastic limit, so that knee point can be defined as a dropping off point from initial elastic line. It is possible that the initial fracture occur at the knee point. Knee point found on the stress-strain curve of tensile test is considered to be used for quantitative evaluation of the initial fracture. When the knee point can be determined quantitatively, stress or strain at the knee point can be used for structural design. Namely, knee point can be used designing stress of safety factor of 1, because any fracture does not occur until the knee point. Moreover, since strength is considered to keep same value after the fatigue test under the knee point stress as virgin specimen, knee point stress seems to allowable stress for long life span.

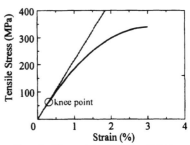

Figure 1   Knee point on the stress-strain curve.

In this study, initial fracture in textile composites was considered. Tensile testing of textile composites was performed. Knee point on the stress-strain curve was identified by least squares method. In-situ observation of initial fracture in textile composites was also conducted. From these results, the relationship between initial fracture behavior in textile composites and knee point was considered.

INITIAL FRACTURE IN WOVEN FABRIC COMPOSITE

Materials

Materials used in this study were plain glass woven fabrics (WE18W: Nitto Boseki Co., Ltd., Japan) as reinforcement with 0.4 wt% acryl silane coupling agent and vinyl ester resin (R-806; Showa Highpolymer Co., Ltd., Japan) as the matrix resin. The room temperature catalyst used was 0.7 phr methylethylketoneperoxide (MEKPO). One ply woven fabric composite was fabricated for tensile test to detect the onset and growth of initial fracture clearly.

Tensile testing was performed at a tensile speed of 1 mm/min using an Instron universal testing machine (model 4206). In order to know the onset and growth of initial fracture, during the tensile tests, the testing machine was periodically stopped and observation of initial fracture in weft fiber bundles using optical microscopy was performed.

Experimental results

Figure 2 shows stress-strain curve.   Knee point can be identified by least squares method.   Knee point stress and strain were 97.5MPa, 0.4% respectively.   Figure 3 shows the fracture aspect by optical microscopic observation.   At the knee point strain (0.4%) fracture cannot be seen, then, transverse cracks appeared in weft fiber bundles at 1.0%.

Fracture process was considered as follows; the initiation of transverse crack appeared at the knee point and then the crack propagated throughout the weft fiber bundle as shown in Fig.4.   Crack at the knee point is considered to be too small and closed when the load was unloaded to 0.   It seems that the observed crack in Fig.3 was the crack which propagated throughout the weft fiber bundle.

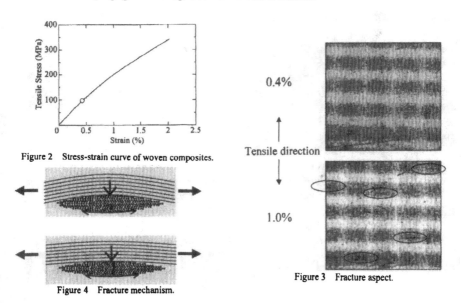

Figure 2   Stress-strain curve of woven composites.

0.4%

Tensile direction

1.0%

Figure 4   Fracture mechanism.

Figure 3   Fracture aspect.

INITIAL FRACTURE IN BRAIDED COMPOSITE

Materials

Materials used were glass fiber (ER575 F-165N, $17\,\mu$ m, 1000 filaments, Nippon Electrical Glass Co., Ltd.) as reinforcement, and epoxy resin (EPOMIK R140, Mitsui

Petrochemical Industry Co., Ltd.) as matrix resin together with 50% of a hardening catalyst, polyamidamine.

Two types of braided composite were used in order to investigate the effect of fiber continuity at the side edge on the fracture behavior of composites: the first type was a normal flat braided composite ("Non-cut" specimen), whilst the second type was a flat braided composite whose fiber bundles at the side edge was cut ("Cut" specimen). Schematic drawing of both specimens are shown in Fig.5. Tensile testing was carried out using an Instron Universal Testing Machine (Type 4206) at a cross-head speed of 0.5 mm/min.

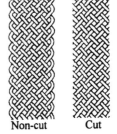

Non-cut   Cut

Figure 5   Braided fabric.

In order to clarify the microscopic fracture process for these braided composites, two types of observation were performed; one was the observation of filament fracture in fiber bundles using optical microscopy, the other was replica observation at the surface edge of the "Cut" specimens. For the replica observation, the surface at the one side edge of the specimen should be polished, whilst that of the other side was not polished, but instead cut by a diamond saw in order to make the specimen symmetrical. Tensile tests were performed with a cross-head speed of 0.5 mm/min. During tensile testing, the experiment was periodically stopped and the number of filament fractures in fiber bundles were counted using optical microscopy. In the same way, the testing machine was periodically stopped and the polished surface edge was replicated on a replica film (acetyl cellulose film) with methyl acetate as a solvent. The replicated film was then observed using optical microscopy.

Experimental results

Figure 6 shows stress-strain curves of specimens. Non-cut specimen has twice tensile strength as cut specimen, because of continuity of fiber bundle. Knee point could be identified. Knee point stress of Non-cut and cut specimen was 315MPa and 123MPa, strain was 1.40% and 0.69% respectively.

Figure 7 shows an example of fractured filaments observed in the Non-cut specimen. Relationship between number of cracks and strain was shown in Fig.8. The number of filament fractures increased with increasing composite strain for "Non-cut" specimens, while scarcely increased for "Cut" specimens.

Figure 9 illustrates the onset and growth of delamination between fiber bundles at the surface edge of "Cut" specimens obtained by replica observation ((a) strain=0.5%, and (b) 0.9%). In the "Cut" specimens, initial microscopic damage was delamination of the fiber bundles at the center of the fiber crossing region (Fig.9 (a)). The delamination progressed along the fiber crossing region and then around the fiber bundles. As the tensile stress was increased further, the delamination reached the specimen surface (Fig.9 (b)) and the fiber bundles were eventually pulled out of the matrix.

The strain at which filament fracture was observed in Non-cut specimen(1.4%) and knee point strain(1.40%) were well agreed.   On the other hand, in Cut specimen, the strain at the initiation of delamination(0.6%) is agreed with knee point strain(0.69%).

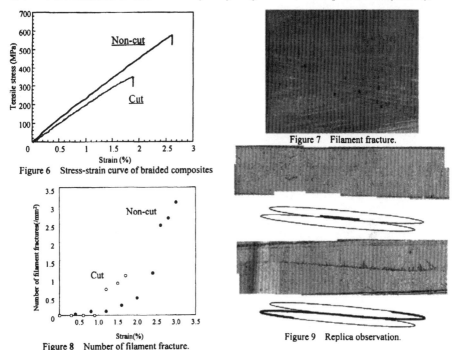

Figure 6   Stress-strain curve of braided composites

Figure 7   Filament fracture.

Figure 8   Number of filament fracture.

Figure 9   Replica observation.

INITIAL FRACTURE IN KNITTED COMPOSITE

Materials

Materials used were glass fiber (D B450 1/2 44S Y23, Nippon Electrical Glass Co., Ltd.) as reinforcement, bisphenol-A type epoxy resin (Epikote 828, Japan Epoxy Resin Co., Ltd.) as matrix resin and triethylenetetramine as the hardener.   Plain weft knitted fabric as shown in Fig. 10 was fabricated by the weft knitting machine.   The plate with 1 ply of knitted fabric was prepared by hand lay-up method, and cured for 48 hours at room temperature, followed by post cure for 2 hours at 100°C.

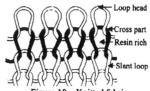

Figure 10   Knitted fabric

Tensile testing

Tensile testing of knitted composites was performed by using an Instron universal testing machine (type 4206).   Dimensions of the specimen were 1 mm in thickness, 200 mm in length, and 20 mm in width.   Tensile specimens were cut along with both wale and course direction.

Woven and braided fabric consists of aggregation of fiber bundle cross sections. On the other hand, knitted fabric composites has loop structure, there are not only fiber crossing part, but also fiber bundle region which composed the loop and resin rich region from microscopic viewpoint and are considered to be not homogeneous but heterogeneous.   It is important to decide which part of strain should be measured on the specimen.   Therefore, tensile testing of knitted composite was performed with 1mm of strain gage length (Kyowa Electronic Instruments Co., Ltd.).   Figure 11 shows position of strain gages, in wale and course.   As shown in Fig. 11 strain gages with 1mm were attached at the four different locations.   Position of strain gages were at the loop head(A), the cross part(B), the resin rich region(C), and the slant loop(D).

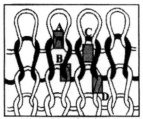 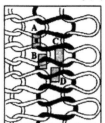

Figure 11   Position of strain gages, wale.   Figure 11   Position of strain gages, course.

Knee points on the global stress-local strain curves were determined using least squares method.   The fracture behavior was observed by using two different methods. The first method was that during the tensile tests, the testing machine was periodically stopped and unloaded to 0, and observations using optical microscopy were performed. The other was that the aspects of specimen during tensile test were recorded as digital image by using CCD camera continuously.   From these images the initial fracture could be identified.

Experimental results

Figure 12 shows stress-strain curves of the knitted composites in the both wale and course direction.   Knee points were found on the stress-strain curves shown in Fig.12. Knee point stress and strain were shown in Table 1, they showed different value in all position.   In wale specimen, knee point stress and strain at loop head were lower value (stress: 14.59MPa, strain: 0.18%) than the others.   It is estimated that initial fracture occurred at the loop head.   On the other hand, in course specimen, knee point stress and strain at the cross part were lower value (stress: 9.26MPa, strain: 0.19%) than the others. It is estimated that initial fracture occurred at the cross part.

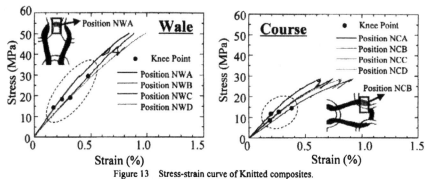

Figure 13   Stress-strain curve of Knitted composites.

Table 1   Knee point stress and strain.

	Knee stress (MPa)	Knee strain (%)		Knee stress (MPa)	Knee strain (%)
Loop head	14.59	0.18	Loop head	12.19	0.21
Cross part	18.97	0.26	Cross part	9.26	0.19
Resin rich	31.05	0.51	Resin rich	14.89	0.38
Slant loop	19.12	0.32	Slant loop	12.19	0.26

The photographs of fracture aspect of wale specimen at the 0.3% strain in wale direction obtained by using optical microscope were shown in Figure 13.   The aspect of the specimen before test was also shown as a reference.   The strain value in Fig.13 was obtained from the strain gages at the loop head, because it was expected that the initial fracture might occur at the loop head in wale direction.   Although the knee point was found at 0.18 % by using the least squares method as described above, any fracture could not be observed until 0.3 % strain.   According to Fig.13 two different fractures were obtained; those are loop head and cross part.   Loop head was transverse crack inside of fiber bundle, and cross part was delamination at the fiber crossing part.   Those micro fractures could not seen at lower strain than 0.3 %.   It was supposed that micro fractures were closed during unloaded process.   From these observations both or either of loop head and cross part should be initial fracture in wale direction.   From the observation with CCD camera, initial fracture occurred at loop head at about 0.2% strain, although clear photograph cannot be obtained from the CCD camera.

Similarly, Figure 14 shows photographs of fracture aspect at strain 0.3 % of the strain gages with 1 mm gage length using optical microscope in course specimen.   At the 0.3 % strain, two types of fracture were observed; one was the delamination at the fiber cross part, the other was the crack inside of fiber bundle at the slant loop.   Here, knee point strain at the cross part was 0.2 % on stress-strain curves.   At the 0.2% strain fracture was not observed, however, initial fracture observed by CCD camera during the tensile test.   Initial fracture in course was not transverse crack, but delamination at the fiber cross part.

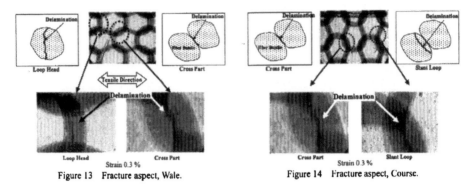

Figure 13   Fracture aspect, Wale.   Figure 14   Fracture aspect, Course.

INTERFACIAL PROPERTIES IN TEXTILE COMPOSITES

Figure 15 shows a classification of fiber/matrix interfaces; Interface around a single fiber filament, inside of a fiber bundle, around a fiber bundle, and between lamina.

Initial fracture of woven fabric and knitted wale specimen, transverse crack occurred. The transverse crack is a fracture at the interphase inside of fiber bundle. On the other hand, initial fracture of the braided-cut specimen and knitted course specimen, initial fracture occurred at the fiber crossing part which correspond to interface around a fiber bundle. In case of braided, Non-cut specimen, multiple filament fracture can be observed. Multiple filament fracture depends on the interfacial property around single fiber filament.

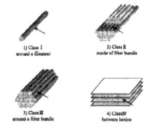

Figure 15   Classification of interface

Namely, initial fracture at the knee point occurred at the interface, so that, it is considered that evaluation of knee point leads to the evaluation of interface. From the results, guideline for design of interface was clarified.

As mentioned before, the knee point stress might be used as designing stress with safety factor of 1 because any fracture occurred until the knee point. Therefore, designing of knee point stress or strain is considered as one of the guidelines for design of composites. If interfacial property can be controlled, consequently, design of composite can be realized.

CONCLUSION

In this study, initial fracture in textile composites, that is, woven, braided and knitted fabric composites was considered with knee point and in-situ observation. Initial fracture in woven fabric composite was transverse crack in weft fiber bundles, that is, fracture at the interface inside of fiber bundle. Initial fractures in Non-cut and Cut specimens of braided composites were filament fracture and delamination at the fiber crossing part, respectively. Initial fracture in wale specimen of knitted composite was transverse crack at the loop head, course specimen was fracture at the fiber crossing part.

Lifetime prediction in anti-corrosive glass fiber reinforced composite pipe under acidic condition for sewage application

H.HAMADA, M.MIZOGUCHI and T.SHIMIZU
Advanced Fibro-Science
Kyoto Institute of Technology
Matsugasaki, Sakyo-ku, Kyoto
606-8585,Japan
hhamada@ipc.kit.ac.jp

Y.FUJI
Seikow Chemical Engineering & Machinery Ltd.
Sioe 3-1-16 Amagasaki
661-0976,JAPAN
yfuji@gold.ocn.ne.jp

ABSTRCT

In this study, ECR-glass fiber which has been recognized as anticorrosive glass, reinforced GFRP pipe was investigated by the stress relaxation test under water and acid conditions. The application of the time-temperature superposition principle (TTSP) was discussed in order to establish the lifetime prediction method. TTSP could be applied so that a smooth master curve was obtained under water conditions. Accordingly the stiffness of the pipe after 100years can be predicted. The test result under acid conditions was similar to that under water. The acid stress corrosion which is very common phenomenon in the case of E glass fiber reinforced plastics was not observed. The design method of the pipe using the result of stress relaxation test was discussed.

INTRODUCTION

It has been 1 century since sewerage, which was used as storm drain, was built in Tokyo in Japan. The role of sewerage pipes has been expanded, and the principal purpose of the application is the water quality preservation at present. The saturation level of sewerage reached 56% all over Japan, and particularly it reached more than 96% in the great cities where the populations are more than one million according to the statistics in 1997. Thus, the pipes have been used more than 50 years and the decay of the pipes became severe problems. Due to the increase of traffic, the decay leads failures and fractures, and consequently immersion of water, the obstruction of pipes and craving of roads sometimes occur.

The condition of sewerage has been changed and it is under high concentration acid condition in the sewerage pipes, because the bacteria which changes hydrogen sulfide to sulfuric acid exists. Therefore, the sewerage pipes are exposed to acid environment, and the pipes corrode severely. GFRP which has high mechanical properties and anticorrosive performance has been paid attention as sewerage pipes and rehabilitation of the pipes. However, in the case that GFRP is applied to sewerage pipes, there are a lot of problems. When GFRP is used under acid condition, the acid stress corrosion behavior, which the degradation of E-glass fiber causes, appeared. E-glass fiber corrodes rapidly under acid conditions. Therefore, the use of anticorrosive glass such as C-glass and ECR-glass is recommended in acidic environments, and it is indispensable in the sewerage field. Table 1 shows the contents of E-glass and ECR-glass, Figure 1(a) shows the SEM photograph of E-glass fiber after immersion for 1 day, and Figure 1(b) shows the SEM photograph of ECR-glass fiber after immersion for 30 day. As shown in this figure, on E-glass fiber, the fracture called spiral crack appeared for short time, on the other hand on ECR-glass fiber, the fracture is not observed even after longer-term immersion.

One more important problem is the guarantee of the lifetime more than 50years for sewerage pipes. There are some problems in the method for lifetime prediction established

Table 1   Contents of E-glass and ECR-glass.

	E	NCR	
SiO_2	52-56	56-62	
AL_2O_3	12-16	9-15	
CaO	16-25	17-25	
MgO	0-6	0-5	
B_2O_3	8-13	-	
Na_2O+K_2O	-	0-1	
TiO_2	-	0-4	
ZnO	-	0-5	%

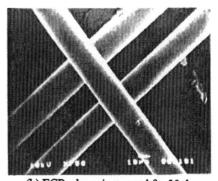

(a) E-glass immersed for 1 day
in sulfuric acid

(b) ECR-glass immersed for 30 day
in sulfuric acid

Figure 1 SEM photograph of E-glass and ECR-glass immersed in acid.

in JIS K 7034. Although the widely accepted method of lifetime prediction is reaction kinetics of Arrhenius and the time-temperature superposition principle (TTSP) in the polymer materials filed, it is necessary to discuss whether the methods can be applied for GFRP and GFRP pipes with chemical degradation.

A purpose of this study is to establish lifetime prediction method for ECR-glass fiber, reinforced GFRP pipes. The stress relaxation behavior of the pipes was investigated by the stress relaxation test under water and acid conditions and TTSP was applied.

Problems of lifetime prediction method at present

The evaluation method of the corrosion resistance of GFRP pipes is defined in JIS K 7034 (Figure 2). In this method, the corrosion resistance of deformed pipe in chemical

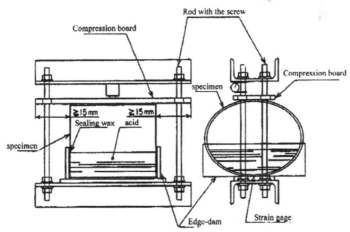

Figure 2   The evaluation method of corrosion resistance of GFRP-pipe in JIS K 7034.

environment is evaluated. The method includes the acceleration of degradation by the deformation. The constant deflection is given to the pipes in which chemicals are injected, and the time when the leak failure occurs is measured. After the test, the relationship between the time and the deflection is obtained. The regression line is determined, and then the deflection after 50 years can be obtained from the extended line. Compared with the permission deflection used in the design, it is confirmed that the value is larger than the permission deflection. Consequently the pipes are guaranteed to be used for 50 years. In this method, it is necessary to carry out the long-term test more than one year. Moreover because the lifetime is predicted by the simplex extension line of the regression line, it is doubtful whether it represents the complex fracture behavior in composite materials. Furthermore, the data is necessary to be obtained by public organization, and the cost of the long-term test rises. Therefore, the method to predict the lifetime of GFRP accurately in short-term has been demanded.

Materials and Experimental Method

In FW-GFRP pipe, Isoide unsaturated polyester was used as matrix and ECR-glass fiber as reinforcement fiber. The winding angle of this pipe was ±10degree, the number of layer was 7, and the fiber volume fraction was 40%. The dimensions of these pipes were as follows; internal diameter was 200mm, thickness was approximately 4.0mm, and width of the pipe was 50mm. The sewerage pipe was generally designed under the permission deflection ratio at 5%. Here, the deflection ratio is determined by dividing applied the deflection by the internal diameter of the pipe. In this study, the stress relaxation test was carried out under the deflection ratio at 2.5%, 5.0% and 10%. The stress relaxation test was carried out in 30°C, 50°C and 70°C under water and 1N sulfuric acid conditions. First, GFRP pipe was set on the

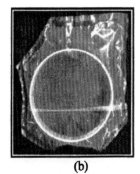

(a)   (b)

Figure 3   Pictures of stress relaxation testing system.
(a) Stress relaxation testing machine   (b) Pipe in PE bag

stress relaxation testing machines as shown in Figure 3(a).   The testing machines and the pipes were put in a constant temperature water bath.   After one hour, the deflection ratio of 2.5%, 5.0% and 10% was applied to the pipe, and the change of load was measured for a week.   In testing under the acid conditions, the pipe and the sulfuric acid were put in the polyethylene bag as shown in figure 3(b).   The edge of the pipe was exposed to the acid solution.

RESULTS AND DISCUSSION

Figure 4 shows the results of the stress relaxation test under water conditions at the deformation ratio 10% at 30°C, 50°C and 70°C.   Here, the load was transformed to the stiffness by using the following equation,

$$S = \frac{f \times F}{L \times y}.$$

Where, S ; Stiffness (kN/m^2)
  f ; Coefficient of deflection

$$f = \left[1860 + \left(2500 \times \frac{y}{d_m} \right) \right] \times 10^{-3}$$

  d_m ; Diameter (m)
  L ; Length of specimen (m)
  F ; Load (N)
  y ; Deflection (m)

This equation is defined in JIS K7035.   Next, the time-temperature superposition principle (TTSP) was applied.   The data for the reference temperature T_0=30°C are held fixed, and the

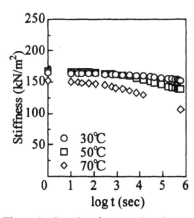

Figure 4   Results of stress relaxation test under water condition. (Deformation ratio = 10%)

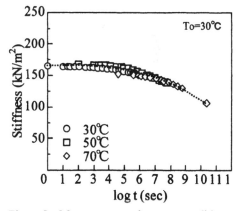

Figure 5   Master curve under water condition. (Deformation ratio = 10%)

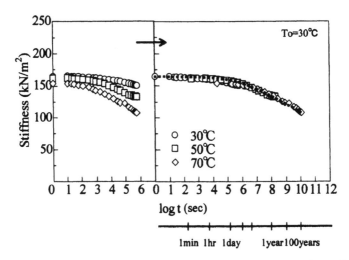

Figure 6   Master curve under acid condition.   (Deformation ratio = 10%)

other curves (50˚C, 70˚C) are shifted horizontally along the time axis until the points coalesce to the single curve.   Thereby a smooth curve, so called master curve, is obtained as shown figure 5.   The defect could not be observed by eye in the pipe after the tests.

Figure 6 shows the result of the stress relaxation test under acid conditions at the deflection ratio 10% at 30˚C, 50˚C and 70˚C.   The test results were similar to that under water conditions.   By using TTSP, one smooth master curve could be obtained.   Figure 7 shows the relationship between the time-temperature shift factors $a_{T_0}(T)$ when shifting the plots horizontally and reciprocal of the absolute temperature 1/T under water and acid conditions.   As shown in this figure, the two shift factors of water and acid conditions represent linear, and they agree well with following Arrhenius equation,

$$\log a_{T_0}(T) = \frac{\Delta H}{2.303G}\left(\frac{1}{T} - \frac{1}{T_0}\right).$$

Where G is gas constant ($8.314 \times 10^3 [kJ/mol \cdot K]$).   The calculated activation energy ΔH was $198 kJ/mol \cdot K$.

Next comparison with the stress relaxation behaviors at each the deflection ratio under water and acid conditions is shown in figure 8.   The stiffness ratio - logarithmic time curves at each the deflection ratio were corresponded on the curves of each temperature.   Here, the stiffness was normalized by dividing by the initial stiffness.   In addition the stiffness ratio - logarithmic time curves under acid conditions were similar to that under water conditions.   Figure 9 shows the picture of edge of the pipe after the test under water and acid conditions.   No defect was observed under both conditions.

As mentioned above, on the pipe reinforced by ECR-glass recognized as anticorrosive glass, there are no differences between the stress relaxation behaviors under water and acid conditions. Furthermore, TTSP can be applied, and the stiffness of the pipe after 100years can be predicted under both conditions. Therefore these results suggest a possibility to estimate the stress relaxation behavior under acid by using of the results under water conditions.   This method is very simplified to predict compared with JIS K7034, for which it takes a year to predict the deflection to failure after 50years.

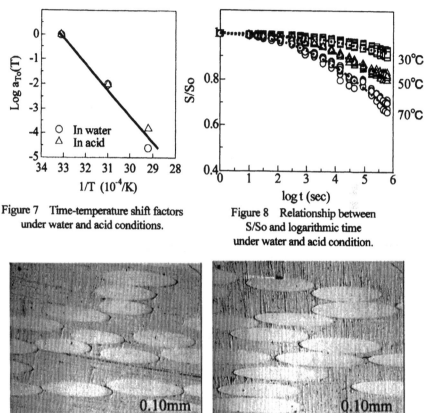

Figure 7   Time-temperature shift factors under water and acid conditions.

Figure 8   Relationship between S/So and logarithmic time under water and acid condition.

Under water condition   Under acid condition
Figure 9   The picture of edge of pipe after test under water and acid conditions

Since it was clarified that the deflection in 100 years could be obtained from the stress relaxation test and TTSP, the application of the results to the design of the pipe was considered. In the design of sewerage pipe, the permission deflection is normally calculated.   S (t), at a certain time can be calculated by the master curve obtained from the stress relaxation test and the TTSP. The actual load such as soil pressure which applied to the pipe is constant, so that the deflection, Y (t), at a certain time can be calculated by using S (t).   It should be confirmed that the deflection after 100years, Y (100), is less than the permission deflection, which is 5%

Y (100) < permission deflection (5%)

Thus, according to calculation of the load of the pipe after 100years by stress relaxation test, the safety of the pipe could be predicted.

CONCLUSION

In order to apply the time-temperature superposition principle for the lifetime prediction of the GFRP pipe, the stress relaxation test curried out under water and acid conditions.   As a consequence, following things were clarified.

1.   On the pipe reinforced by ECR-glass recognized as anticorrosive glass, TTSP can be applied, and the stiffness of the pipe after 100years can be predicted under both water and acid conditions.
2.   It is possible to estimate the stress relaxation behavior under acid conditions by using of the results under water conditions.
3.   According to calculation of the load of pipe after 100years by the stress relaxation test, the safety of the pipe could be predicted.

REFERENCE

1) Isao Kinoshita, Journal of the society of materials science, Japan, vol.47, No.10, 1031-1040 (1998) In Japanese
2) Jones, F.R., Rock, J.W. and Wheatley, A.R., Composites, 14 3, 262-269 (1983)
3) Carswell, W.S., ASME Petroleum Division, 24, 105-110 (1988)
4) Hogg, P.J., Hull, D. and Legg, M.J., Composite Structures, 106-122 (1981)
5) Hogg, P.J., Composites, 14, 3, 254 (1983)
6) Hogg, P.J., Composites Science and Technology, 38,23-42 (1990)
7) Fujii,Y., Kubota,T., Maekawa,Z., Hamada,H., Murakami,A. and Yoshiki,T., The proceedings of the 36th Japan Congress on Materials Reserch; 234-239 (1993)
8) Yasushi Miyano, Masayuki Nakada, Michael K. M^{C}murray, Rokuro Muki, Material system, 17,63 (1998) In Japanese
9) JIS K 7034, 758 (1998)

Fabrication and mechanical property of long fiber reinforced thermoplastic composites using braiding technique

A.Nakai*, H. Hamada and D.Nakaami
Advanced Fibro Science
Kyoto Institute of Technology
Matsugasaki, Sakyo-ku
606-8585, Japan
*nakai@ipc.kit.ac.jp

ABSTRACT

Two types of unidirectional carbon fiber reinforced thermoplastic composites, that is, CF/PPS (polyphenylene sulfide) and CF/PA66 composites, were selected to investigate the correlations between the impregnation property and mechanical properties. Unidirectional carbon fiber reinforced thermoplastic composites were fabricated by compression molding using " Micro-braided yarn " that is composed of reinforcement fiber and matrix fiber. 3 point bending test and the observation of impregnation state were also performed to examine the influence of impregnated states on the bending properties. As a result, the bending modulus and strength increased, in accordance with the impregnation states. The bending modulus of CF/PA66 composites was almost as same as CF/PPS composites on each impregnation states. On the other hand, the bending strength of CF/PA66 was constantly higher than CF/PPS on each impregnation states because of its high fracture toughness and good adhesion between carbon monofilaments and PA66 resin compared to PPS resin.

INTRODUCTION

There has been an increasing interest in the use of thermoplastic matrices for the continuous fiber reinforced composites. High ultimate strain, high fracture toughness, better impact tolerance, short molding cycle time, recycle ability and the ability to be re-melted and reprocessed are examples of the reason. However, thermoplastic matrices have some problems. One of that is the difficulty in impregnating resin into reinforcing fiber bundles because of its high molten viscosity, and the prepregs for thermoplastic composites have less drapability. In order to overcome these difficulties, various intermediate materials have been developed; for example, Commingled yarn , Powder coated yarn and Co-woven fabric. These intermediate materials have good impregnation property and drapability, however, the cost is high in most cases, and these materials are special so that combination of reinforcement fiber and matrix is limited; the most case is GF/PP. In the case of Commingled yarn, it is possible that reinforcement fiber bundle is damaged during the process of making reinforcement fiber bundle open, and this process is indispensable to making Commingled yarn.

Therefore, we have been developing the intermediate material called " Micro-braided yarn " for fabrication of continuous fiber reinforced thermoplastic composites. Micro-braided yarn is composed of the reinforcement fiber bundle that is straightly inserted in the center ('axial fiber') and the matrix resin fiber bundles that are braided around reinforcing fiber bundle, as shown in Figure 1. Due to this construction, the matrix resin can be located near the reinforcement fiber bundles. Micro-braided yarn

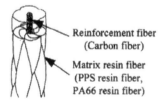

Reinforcement fiber
(Carbon fiber)

Matrix resin fiber
(PPS resin fiber,
PA66 resin fiber)

Characteristics of Micro braided yarn

• Micro braided yarn can be treated as a single fiber bundle.

• Damage to reinforcement fiber bundle is small in making Micro braided yarn .

• Various resin as matrix can be adopted.

Figure 1. Schematic diagram and characteristic of Micro braided yarn

Figure 2. Photograph of tubular braiding machine

is braided by tubular braiding machine, as shown in Figure 2, and another processing is not necessary. Thus, the damage to reinforcement fiber bundle is decreased in making Micro-braided yarn, compared to making Commingled yarn, and the reduction of cost for fabrication of long-fiber reinforced thermoplastic composites would be possible. Additionally, various matrix resin can be adopted to Micro-braided yarn in case that resin can be made into fibrous shape. Micro-braided yarn have great drapability because of using dry thermoplastic fiber, and can be treated as a single fiber bundle, so that textile processed goods can be easily fabricated. The dimension of Micro-braided yarn depends on the volume and number of both reinforcement and matrix fiber bundles and the orientation angle of matrix fiber bundle. Selecting these factors can make various dimension of Micro-braided yarn, that is, the impregnation distance and fiber volume fraction of Micro-braided yarn can be selected.

In this study, we fabricated two types of unidirectional carbon fiber reinforced thermoplastic composites; CF/PPS (polyphenylene sulfide) and CF/PA66 composites using Micro-braided yarn. 3 point bending test and microscopic observation in the fiber-axis direction was performed to investigate the correlations between their impregnation properties and mechanical properties.

EXPERIMENTAL

Materials and specimen fabrication

The materials used in this study were carbon fiber bundles (T300B, 12000 filaments, TORAY Co., Ltd.) as reinforcement, PPS resin fiber bundles (100 filaments, 400 denier, TORAY Co., Ltd.) and PA66 resin fiber bundles (TORAY Co., Ltd.) as matrix resin. Matrix resin fiber bundles were braided on carbon fiber bundle by tubular braiding machine with 16 spindles, as to be about 42% in fiber volume fraction of each Micro

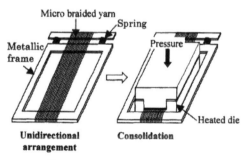

Figure 3 Schematic diagram of fabrication method

Table1.Processing Conditions

	CF/PPS	CF/PA66
Temperature (°C)	310	290
Pressure (MPa)	4.0	
Time (min.)	3,5,10,20,40	
Cooling	quenched	

braided yarn. The diameter of each Micro braided yarn is about 1mm. Unidirectional CF/PPS and CF/PA66 composites were fabricated using compression molding with the change of processing condition, as shown in Figure 3; first, to array and tense the micro-braided yarn properly, the micro-braided yarn was wound at 16 times parallel onto metallic frame that has adjust mechanism with the spring, which can adjust the tension caused by the thermal shrinkage of the carbon fibers. Next, the metallic frame wounded with yarn was placed in a heated die and pressed by compression molding machine. The heated die was quenched in ice water at 0°C under appropriate pressure. Processing conditions are shown in Table 1. The compression time (T_c) means the period during which the compressive pressure was maintained. Each molding temperature was 35 °C higher than each melting temperature of matrix resin.

Experimental procedures

Microscopic observation of unload specimens in the fiber-axis direction was performed with optical microscopy to investigate impregnation states. 3-point bending test was also performed using Instron Universal Testing Machine (Type 4206) at crosshead speed of 1mm/min. The specimens for 3 point bending test were cut along the fiber-axis direction. Geometry of the specimens was 1.1~1.3mm in thickness, 20mm in width, and the span length was 35.2mm. After bending test, microscopic observation of loaded specimens was   performed with optical microscopy.

RESULT AND DISCUSSION

Microscopic observation

Figure 4 shows the cross-sectional photograph of unloaded specimens of T_c =5 and 40min in the fiber-axis direction. The circular region shown in the photograph is a carbon fiber bundle. The region including a lot of void inside of the carbon bundle was defined as unimpregnated region, and volume ratios of unimpregnated region to whole specimen was defined as Unimpregnated volume ratio (V_{unimp}). In both case of CF/PPS and CF/PA66, as T_c became longer, unimpregnated region decreases according to the impregnation of matrix resin into carbon fiber bundles.

Relationship between V_{unimp} and T_c is shown in Figure 5. V_{unimp} of CF/PPS and CF/PA66 rapidly decrease until T_c =10min and 20min, and become almost constant in T_c =10min and 20min respectively. The constant value of V_{unimp} is not zero, and that of CF/PA66 is bigger than CF/PPS. Thus, it is considered that the impregnation of CF/PPS and CF/PA66 has almost finished until T_c =10min and 20min. Compared to CF/PPS and CF/PA66, the impregnation time of CF/PA66 is as two times longer than CF/PPS. This

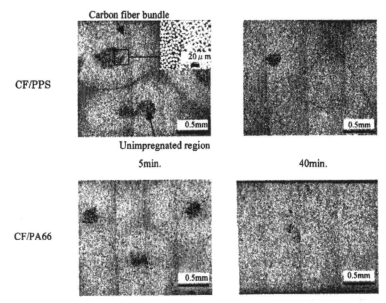

Figure 4 Cross-sectional photograph of the
unloaded specimens (Tc=5,40min.)

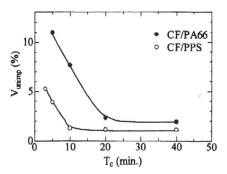

Figure 5 Relationship between V_{unimp} and T_c

result indicated that CF/PPS is superior to CF/PA66 in impregnation property.

3 point bending test

    The bending specimens after fracture are shown in Figure 6. In the case of CF/PPS of T_c =5min, not fiber breakage on tension side but fiber breakage with kink on compression side was observed. In the case of T_c =40min, both fiber breakage with kink on compression side and fiber breakage on tension side were observed, as shown in Figure 6 (a). In the case

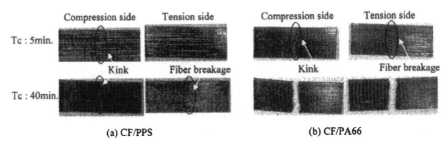

Figure 6 Bending specimens after fracture

of CF/PA66 of T_c =5min, both fiber breakage with kink on compression side and fiber breakage on tension side were observed. In the case of T_c =40min, the specimen was divided into two pieces, as shown in Figure 5 (b). Fiber breakage on tension side was observed in each T_c.

Figure 7 shows typical bending stress-displacement curves of CF/PPS and CF/PA66 on each T_c. In the case of CF/PPS, the bending stress gradually decreased after it reached a maximum value, especially in case of short compression time (Figure 7 (a)). According to the optical microscopic observation during bending test, the cause of decrease in bending stress before fracture would be fiber breakage with kink on compression side. On the other hand, in the case of CF/PA66, the fiber breakage on tension side occurred at maximum bending stress, and bending stress decreased brittlely. The fracture behavior of CF/PA66 tended to be brittle compared to CF/PPS as a whole. In addition, when T_c was short, the fracture behavior tended not to be brittle in comparison to that in long T_c in both cases of CF/PPS and CF/PA66.

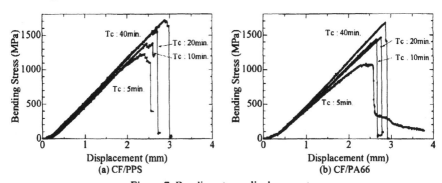

Figure 7. Bending stress-displacement curves

Following bending specimens after fracture and stress-displacement curves, it is considered that the fracture behavior of CF/PA66 tended to be brittle with fiber breakage on tension side in comparison to CF/PPS.

The effect of V_{unimp} on bending properties was also investigated. Relationship between the normalized bending modulus (E_{exp}/E_{theo}) and V_{unimp} is shown in Figure 8. Here, the experimental bending modulus was normalized by the theoretical modulus using rule of mixture.

$$E_{theo} = E_f V_f + E_m (1 - V_f)$$
E_f : Modulus of reinforcement fiber
E_m : Modulus of matrix resin
V_f : Fiber volume fraction

In both case of CF/PPS and CF/PA66, as V_{unimp} decreased, E_{exp}/E_{theo} increased at the same ratio to V_{unimp}. This result means that the bending modulus depends on the impregnated states, regardless of matrix resin.

Relationship between the bending strength and V_{unimp} is shown in Figure 9. In both case of CF/PPS and CF/PA66, as V_{unimp} decreased, the bending strength increased at the different ratio to V_{unimp}. When V_{unimp} of both CF/PA66 and CF/PPS were the same value, the bending strength of CF/PA66 was higher than that of CF/PPS. This result means that the decrease in bending strength of CF/PPS caused by unimpregnated region is bigger than CF/PA66. The reason of that could be explained as follows; in the case of CF/PPS, the crack from unimpregnated region is easy to occur, compared to CF/PA66, because of poor adhesion between carbon monofilaments and PPS resin and its low fracture toughness. Actually, the crack from unimpregnated region was observed in the specimens of CF/PPS after fracture, as shown in Figure 10. On the other hand, such crack was hardly observed in CF/PA66.

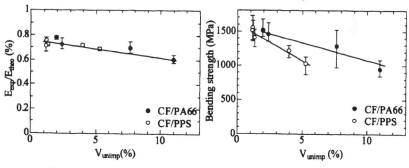

Figure 8 Relationship between E_{exp}/E_{theo} and V_{unimp}

Figure 9 Relationship between bending strength and V_{unimp}

To examine adhesion properties between carbon monofilaments and each matrix

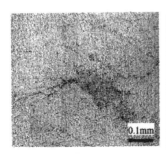
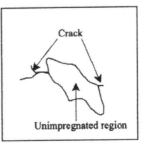

Figure 10 Photograph and schematic drawing the
crack from unimpregnated region of CF/PPS

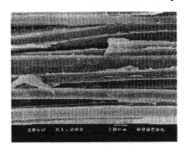
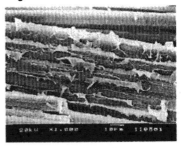

(a) CF/PPS   (b) CF/PA66

Figure 11 SEM images of fracture surface
after 90° direction 3 point bending test

resin, the 3 point bending test in transverse direction to reinforcement fiber (90° direction 3 point bending test) was also performed. E_{exp}/E_{theo} of CF/PPS and CF/PA66 were 0.25 and 0.63, and bending strength were 24.0MPa and 62.3MPa respectively. Fracture surfaces after 90° direction 3 point bending test are shown in Figure 11. In the case of CF/PPS, the matrix resin hardly existed on the carbon fiber. On the other hand, in the case of CF/PA66, existence of the matrix resin and considerable plastic deformation of the matrix resin were observed. These results indicate that PA66 resin have good adhesion to carbon monofilaments compared to PPS resin.

CONCLUSION

· CF/PPS composites is superior to CF/PA66 composites in impregnation property.

· The bending modulus depends on the impregnated states, regardless of matrix resin.

· PA66 resin have good adhesion to carbon monofilaments compared to PPS resin.

· The decrease in bending strength of CF/PPS caused by unimpregnated region is bigger than CF/PA66.

Fabrication and mechanical properties of textile reinforced SMC

A.Nakai*, H. Hamada and T.Ohkawa
Advanced Fibro Science
Kyoto Institute of Technology
Matsugasaki, Sakyo-ku
606-8585, Japan
*nakai@ipc.kit.ac.jp

ABSTRACT

In this study, textile reinforced SMC was fabricated to restrain initial fracture of SMC. We used braided fabric as the reinforcing textile. Braided fabrics have high mechanical properties because of the continuously oriented fiber bundle. Moreover, braided fabric can make a hole, namely, braided hole, whose fiber bundle continues around a hole.

Tensile test was carried out to investigate the mechanical properties of textile reinforced SMC with a circular hole. Four types of specimen were prepared; SMC with braided fabric with a braided hole, or a machined hole, SMC with cloth with a machined hole and SMC with machined hole for the comparison. From the results of tensile test, it was clarified that a braided hole could improve the strength around the hole.

Moreover, textile reinforced SMC was applied to coil cover. Three point bending test was carried out to the element specimen of the coil cover. From these results, braided fabric could improve the strength around the corner of the coil and the usefulness of braided fabric as a reinforcement of SMC was verified.

INTRODUCTION

SMC (Sheet Molding Compound) is one of typical composite materials, which consists of glass fiber mat and paste resin which consists of unsaturated polyester resin, curing catalyst, filler, release agent, low profile additives, pigment, and styrene monomer. SMC products have smooth surface without any particular technique. SMC molding process is simple, namely easy to fabricate complex shape products. In these points, SMC is superior to other FRP and SMC products have been used in structural parts of transportation vehicles, water tank, bathtub and the other variety of products.

Although SMC is widely used, the initial fracture occurs at relatively small strain. When the composite materials are used as structural members, connecting them with other members is one of the important problems. Mechanically fastened joints connecting each member by pins or bolts have been well known as one of the simplest joint methods. In producing the hole by machining, decrease in the mechanical properties of SMC is expected because of the stress concentration around the hole.

As one of the methods to improve the mechanical properties of SMC, we have proposed the textile reinforced SMC. Braided fabric exhibited high strength and high modulus because of continuously oriented fiber bundle. Moreover, braided fabric can keep the strength around a hole by making braided hole whose fiber bundle continues around a hole.

In this study, textile reinforced SMC were fabricated to keep the strength around a hole and improve the mechanical properties of SMC. Tensile test was carried out in order to investigate mechanical properties of textile reinforced SMC with a circular hole. For the actual usage, tensile test with a pin in the circular hole was also performed. Moreover, as an application to actual structure, coil cover was made of textile reinforced SMC. Three point bending test for the element of the coil cover was carried out and the usefulness of textile reinforced SMC to the coil cover was confirmed.

EXPERIMENTAL METHOD

Fabrication of textile reinforced SMC

SMC (1740; Asahi fiber glass co., ltd.) was composed of random glass mats of short glass fiber and paste resin that consists unsaturated polyester, hardener, filler, and so on. SMC plates reinforced with textile were produced by compression molding machine with 300mm × 300mm of the cavity mold. Processing conditions were as follows; 140℃ of molding temperatures, 5.6MPa of molding pressure, 100% of charge ratio.

In molding textile reinforced SMC there is a problem of resin impregnation to fiber bundle of textile fabric. First, only textile fabric was laminated on the SMC sheet and compressed. This product had much uninpregnated region in the textile fabric as shown in Fig.1 (a). Next, textile fabric was impregnated with paste resin in order to improve impregnation state of paste resin into textile fabric. Paste resin was diluted by acetone in order to decrease the viscosity of paste resin because the paste resin with high viscosity was difficult to be impregnated into textile fabric. 10wt% of acetone was added to the paste resin before the impregnation into braided fabric. Laminate of textile with paste resin and SMC sheets was compressed. In this product unimpregnated region was not observed as shown in Fig.1 (b).

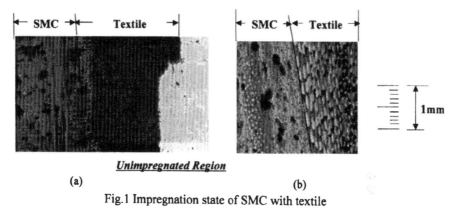

(a)

(b)

Fig.1 Impregnation state of SMC with textile

Tensile test of textile reinforced SMC with a circular hole

In this section, the mechanical properties of textile reinforced SMC with a circular hole was examined. To reinforce SMC, two types of textile fabric were used. One was E-glass cloth (WE18W; Nitto Boseki Co., Ltd.) and the other was flat braided fabric.

Figure 2 shows the geometry of the specimens for static tensile test; 250mm in length, 25mm in width, and SMC without textile was 4mm in thickness, and both SMC with cloth and SMC with flat braided fabric were 4.8mm in thickness. The diameter of a hole was 6 mm. Static tensile test was performed with the gage length of 170 mm. Tensile test was conducted by Instron Universal Testing Machine (Type 4206) by crosshead speed of 1mm/min.

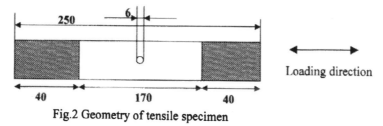

Fig.2 Geometry of tensile specimen

Table 1 shows the types of specimen; SMC with machined hole (SH), SMC with cloth with machined hole (CH), and SMC with braided fabric with Machined hole (MH; Fig.3 (a)), SMC with braided fabric with Braided hole (BH; Fig.3 (b)). In the BH specimen, fiber bundles were continuously oriented around a hole. The specimen to which pin was inserted are called SHP, CHP, MHP, and BHP, respectively.

Table 1 Type of reinforcing textile

Reinforcing textile	With a circular hole	
	Without a pin	With a pin
Nothing	SH	SHP
Cloth	CH	CHP
Flat braided fabric	MH	MHP
	BH	BHP

MH...Machined hole

BH... Braided hole

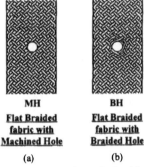

MH	BH
Flat Braided fabric with Machined Hole	**Flat Braided fabric with Braided Hole**
(a)	(b)

Fig.3 Schematic drawings for braided fabric with a circular hole

RESULTS AND DISCUSSION

SMC under tensile load, initial fracture, namely transverse crack, occurred at the interface inside of a fiber bundle oriented in transverse direction. In load-displacement (L-D) curves of SMC products, load increased linearly at the beginning and showed a knee point. It was confirmed that the transverse cracks caused the decrease in slope of the stress-strain curves, that is, knee point[1]. L-D curves of SH, CH, MH and BH are shown in Fig.4 (a). All specimens of SMC with textile showed higher maximum load than SMC without textile. As for SMC with textile, BH showed higher maximum load than CH and

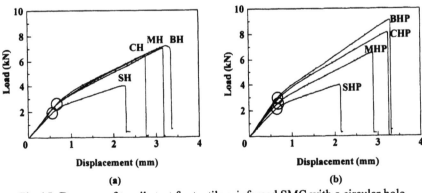

Fig.4 L-D curves of tensile test for textile reinforced SMC with a circular hole

MH. Figure 4 (b) shows L-D curves of SHP, CHP, MHP, and BHP. As shown in Fig.4 (b), all specimens of SMC with textile showed higher maximum load than SMC without textile. The behavior of L-D curves, especially after knee point, depended on the types of textile as reinforcement.

Figure 5 shows photograph of the specimen after tensile test. Fracture of BH and BHP specimen did not pass the hole. This means that the fracture caused by the stress concentrations around a hole were restrained in the BH and BHP specimens.

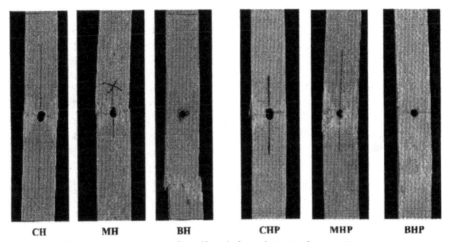

| CH | MH | BH | CHP | MHP | BHP |

Fig.5 Fracture aspect of textile reinforced SMC after tensile test

Tensile properties are summarized in Table 2. Here, 1st modulus and 2nd modulus represents the slope of L-D curves before and after knee point. Normalized load is the load divided by the thickness of the specimen. We defined knee point load and maximum load normalized by those of SMC (SH or SHP) as the improvement ratio.

Table 2 Tensile properties of textile reinforced SMC with a circular hole

	Thickness (mm)	1st modulus (kN/mm^2)	2nd modulus (kN/mm^2)	Normalized load [knee (kN/mm)]	Normalized load [Max (kN/mm)]	Improvement ratio [knee]	Improvement ratio [Max load]
SH	3.96	0.90	0.26	0.39	0.99	1.00	1.00
CH	4.95	0.87	0.29	0.38	1.39	0.99	1.40
MH	4.78	0.85	0.36	0.35	1.46	0.90	1.47
BH	4.78	0.87	0.35	0.36	1.47	0.93	1.48
SHP	3.96	0.83	0.28	0.29	0.88	1.00	1.00
CHP	4.95	0.90	0.39	0.40	1.59	1.38	1.82
MHP	4.82	0.80	0.31	0.30	1.47	1.06	1.67
BHP	4.77	0.95	0.42	0.37	1.76	1.30	2.01

In the case of specimen without a pin, the 1st modulus and knee point load of SMC with textile was almost the same value as SMC without textile. The 2nd modulus and maximum load was improved by textile. However, as for the specimen with a pin, 1st and 2nd modulus, knee point load and maximum load of SMC with textile were higher than those of SMC without textile. Especially, the braided hole contributed to the improvement in the maximum load. The reason of these results is the continuity of fiber bundle around the hole.

Application of textile reinforced SMC to coil cover

Figure 6 shows an appearance of coil cover which is made of SMC. This coil cover is 0.36m square, and the thickness of Coil part (Fig.6 (a)) and Flat part (Fig.6 (b)) is 57mm and 7mm, respectively. Coil was covered with SMC and laminated SMC sheets were put on the coil and molded. In this type of coil cover, it is predicted that stress concentration will occur around the joint part of (a) Coil part and (b) Flat part when bending load is applied. So, coil cover was made using braided fabric covered over this joint part.

Figure 7 shows the photograph during the fabrication of coil cover using SMC with braided fabric. Molding conditions were as follows; 130℃ of molding temperature, 5.6MPa of molding pressure, 30minutes of molding time.

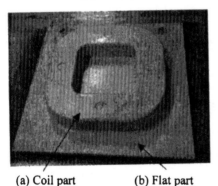

(a) Coil part   (b) Flat part   Fig.7 Fabricating method

Fig.6 Coil cover product

Three point bending test for the element specimen of coil cover

We conducted three point bending test for the element specimen of the coil cover. Two types of specimen were prepared; T01 and T11 (Fig.8). In the specimen T11 the reinforcing braided fabric covered over the joint part of coil part and flat part, whereas the braided fabric did not cover the joint part in specimen T10. As shown in Fig.9, 'a' and 'b' were different loading direction. Testing conditions were as follows; 114mm in span length, 30mm in width, and 1mm/min. of crosshead speed. Three point bending test was conducted by Instron Universal Testing Machine (Type 4206).

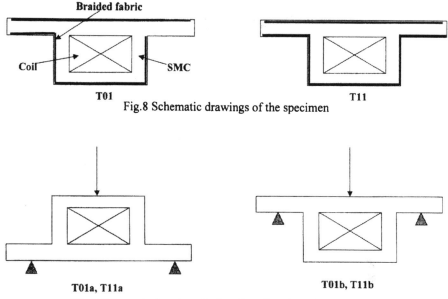

Fig.8 Schematic drawings of the specimen

Fig.9 Schematic drawings for bending direction of the specimen

Figure 10 shows L-D curves of T01 and T11 for bending test. In both 'a' and 'b' direction, T11, that has the braided fabric continuously arranged on the edge of the corner, showed higher maximum load than T01. Figure 11 shows the cross-sectional area of the eT01b and T11b specimen after bending test and schematic drawings of it. The crack occurred in T01b was generated inside of SMC and propagated along the weld line to the edge of the corner. These figures show that braided fabric restrained the propagation of crack to the edge of the corner.

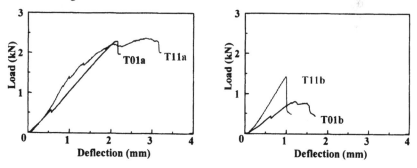

Fig.10 L-D curves for bending test of element of coil cover

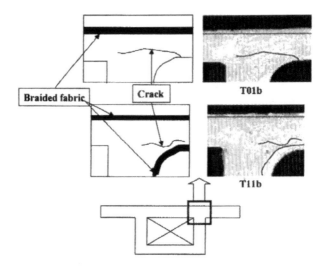

Fig.11 Fracture aspect of element of coil cover after bending test

CONCLUSION

In this study, textile reinforced SMC were developed. From tensile tests of the textile reinforced SMC with a circular hole, SMC with textile showed higher properties than SMC without textile. These differences were due to continuously oriented fiber bundle around the circular hole. It was confirmed that textile was more effective on improvement of maximum load than that of initial fracture load. Especially, the braided hole contributed to the improvement in the maximum load in the case of the hole with a pin. Moreover, textile reinforced SMC was applied to coil cover. From the bending test of element of the coil cover the usefulness of textile reinforced SMC with braided fabric was verified.

REFERENCES

1) A. NAKAI, T. OHKAWA, *Fabrication and Mechanical Behavior of Textile reinforced SMC*, Proceedings of the 7th Japan International SAMPE Symposium pp.749-753 (2001)

Effect of orthotropic electric resistance on delamination detection with electric resistance change method

Akira Todoroki, and Yoshinobu Shimamura, and Hideo Kobayashi
Tokyo Institute of Technology, Department of Mechanical Sciences and Engineering, 2-12-1, O-okayama, Meguro-ku, Tokyo 152-8552, Japan

Miho Tanaka
Graduated student of Tokyo Institute of Technology
E-mail: mtanaka@ginza.mes.titech.ac.jp

ABSTRACT

An electric resistance change method is successfully applied to the delamination detections of CFRP laminates. In the present study, orthotropic electric resistances of CFRP laminates are measured experimentally of three volume fractions of carbon fiber. The experiments in the present paper reveal significant change of the electric conductance values in the transverse direction and thickness direction owing to small change of the volume fraction.

On the basis of the measured orthotropic electric conductance, the effect of the electric conductance is investigated in detail here. The effect of the orthotropic electric resistance change on electric resistance measurements for delamination detection is investigated using FEM analyses. For the analyses, beam type specimens of cross ply laminates are employed. Two types of stacking sequence are [0/90]s and [90/0]s. Electric resistance change owing to a delamination is discussed for both types of the specimens on the basis of the results of the electric current density diagrams. Results obtained are followings. (1) Electric resistances of thickness direction and transverse direction change significantly due to the small change of fiber volume fraction. (2) The mechanism of delamination detection with electric resistance change method was elucidated. (3) Electrodes should be mounted on fiber direction.

237

INTRODUCTION

Composite laminates have low delamination resistance, and that causes delamination by slight out-of-plane impacts. Since the delaminations are usually invisible or difficult to detect by visual inspections, automatic systems for delamination identifications in-service are desired. Recently, an electric resistance change method is employed to identify the internal damages of CFRP laminates [1]-[3]. Since this method adopts reinforcement carbon fiber itself as sensors for damage detections, this method does not cause reduction of static strength or fatigue strength, and applicable to existing structures. Moreover, the electric resistance change method does not cause increase weight.

In the present research, a relationship between fiber volume fraction and the orthotropic electric conductance are experimentally measured. The effects of orthotropic electric conductance are investigated and the effect of electric current charging direction is also investigated using the conductance measured by the experiments.

EXPERIMENTAL MEASUREMENT

In order to investigate a relationship between the fiber volume fraction and electric conductance, unidirectional laminates of three types of carbon fiber volume fraction were fabricated. The material employed here is Q-111 2500: unidirectional carbon fiber/epoxy prepreg produced by Toho Rayon Co., LTD. The conductance of longitudinal, transverse and thickness direction was experimentally measured using the different small size specimens. In order to measure the electric resistance of the specimen, silver paste is painted as electrodes. The results of the measurement are shown in Table.1.

Table 1   Fiber volume fraction Vf and Electric conductance σ

Vf	$\sigma 0$(S/m)	$\sigma 90/\sigma 0$	$\sigma t/\sigma 0$
0.40	3700	1.8×10^{-4}	1.6×10^{-5}
0.47	4600	1.1×10^{-3}	2.2×10^{-4}
0.62	5500	3.7×10^{-2}	3.8×10^{-3}

Measured values of electric conductance of σt and σ90 are significantly smaller than σ0, but they are not zero. Since carbon fiber is not straight but a wavy configuration in practical composites, fiber contact exists in the transverse and thickness direction. Besides, σt is smaller than σ90, due to a resin rich interlamina. Since the fiber contacts depend on the fiber volume fraction, the value of σt and σ90 vary significantly by the slight change of the fiber volume fraction.

Effect of orthotropic electric conductance

Computational method and model

FEM analyses are employed for investigations of the effect of the orthotropic electric conductance using the experimental results of conductance. An adopted five-electrode-beam specimen was analyzed as shown in Figure.1. Two types of stacking sequences are computed here; [0/90]s and [90/0]s. Thickness of a ply is fixed to 0.5mm. Electrodes are mounted on the surface of the laminates, in order to investigate the effect of delamination creation and matrix crack. FEM analyses of two types of damages were performed as shown in Figure.1 and Figure.2 respectively. FEM analyses are performed using a commercially available FEM tool named ANSYS. Four-node-rectangular elements are adopted for the analysis, and the size of the each element is approximately 0.125mm. By using the auto mesh generation system of ANSYS, the specimen model is divided into approximately 19200 elements of the 2-dimentional elements. In order to analyze electric current density and electric resistance change, direct current of 30mA is charged from the electrode A to electrode B and electrode B is fixed to 0V.

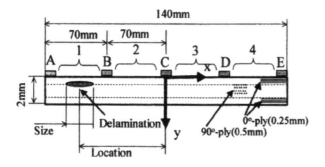

Figure1 Specimen with delamination   [0/90]s

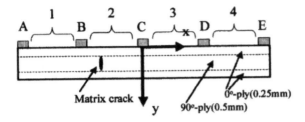

Figure2 Specimen with matrix crack   [0/90]s

Result of internal voltage contour

The internal voltage contour of an orthotropic material is shown in Figure3. The abscissa is the location between the electrode A(-70mm) and B(0mm). The ordinate is the location of the thickness direction of the specimen.

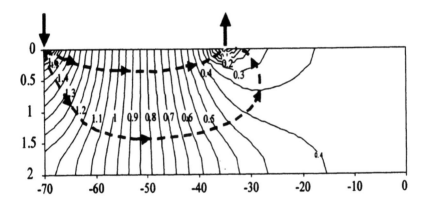

Figure3 Internal voltage contour [0/90]s

In the case of orthotropic material, the current mainly flows into the 0-ply of the top surface of the specimen. The electric current in the thickness direction flows gradually as shown Figure3. When a delamination exists between the top 0-ply and 90-ply, the current into the thickness direction is disturbed by the delamination. That causes electric resistance change observed at the segment.

<u>Electric current density</u>

The electric voltage distribution on the surface of the specimen is computed. Delamination locates between x=-57.5mm and x=-47.5mm at the interlamina between the upper 0°ply and the middle 90°ply (the size is 10mm). Current is injected from electrode A to B. The electric voltage distribution on the surface of the specimen is shown in Figure4.

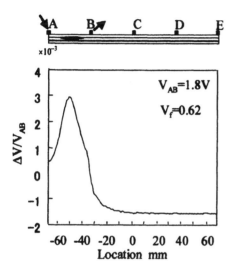

Figure4 Electric potential change

The abscissa is the electric voltage distribution on the surface of the specimen and the ordinate is the location of the thickness direction. Electric voltage distribution exists outside the charged segment as shown in Figure4. This implies that the circular electric current exists outside the charged segment. For this circulation, the electric resistance change in the segment adjacent to the segment where a delamination exists.

<u>Effect of charging direction</u>

In order to investigate the effect of charging direction, two types of stacking sequence, [0/90]s and [90/0]s, are adopted and the electric voltage distribution is computed respectively. Delamination locates between x=-57.5mm and x=-47.5mm at the interlamina between the upper 0°ply and the middle 90°ply (the size is 10mm). Three sets of conductance measured by the experiments are employed. The current is charged from the electrode A to the electrode B. Surface voltage of [0/90]s and [90/0]s between the electrode A and B is showed in Figure5 and Figure6 respectively. The abscissa is the location of the longitudinal direction and the ordinate is the electric voltage distribution. Electrode A corresponds to x=-70mm, and electrode B corresponds to x=-35mm in these Figures.

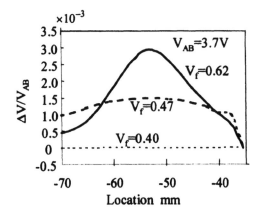

Figure5 $\Delta V/V$ of the [0/90]s

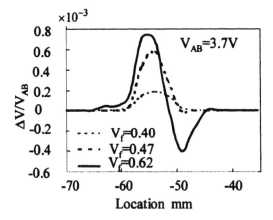

Figure6 Dx of the [90/0]s

The voltage change at the electrode A means an existence of electric resistance change between the segment. As shown in Figure5, electric resistance changes can be observed for the laminates of [0/90]s except for the case of Vf=0.40. On the other hand, electric resistance changes can not be observed for the laminates of [90/0]s because no voltage change is observed at the electrode A as shown in the Figure6. The result reveals that we should charge electric current to the fiber direction.

Results of electric resistance change

Effects of damages of delamination and matrix crack on the electric resistance measurements are investigated. Delamination locates between x=-26mm and x=-18mm and the size is 8mm. Matrix crack locates at x=-22mm. In the case of [0/90]s, matrix crack locates in the middle 90°ply. In the case of [90/0]s, matrix crack locates in the surface 0°ply.

Table 2   Electric resistance change $\Delta R/R$ in the case of [0/90]s

Segment number	$\Delta R/R$ with delamination	$\Delta R/R$ with matrix crack ($\times 10^{-4}$)
1	19.7	0.55
2	9.06	7.36
3	2.83	0
4	0	0

Table 3   Electric resistance change $\Delta R/R$ in the case of [90/0]s

Segment number	$\Delta R/R$ with delamination	$\Delta R/R$ with matrix crack ($\times 10^{-4}$)
1	0	0
2	27.1	3.25
3	0	0
4	0	0

The electric resistance change due to the delamination and the matrix crack are shown in the Table2 and Table3. Table2 is the results of the $\Delta R/R$ of each segment in the case of [0/90]s. Table3 is the results of the $\Delta R/R$ in the case of [90/0]s. The delamination and the matrix crack locate in the segment 2. For the laminate of [0/90]s, the $\Delta R/R$ due to the delamination is observed in several segment and the $\Delta R/R$ due to the matrix crack is observed in the segment 2. This results of the $\Delta R/R$ indicated that the delamination detection is possible in the case of [0/90]s.   For the laminate of [90/0]s, on the other hand, the $\Delta R/R$ due to the delamination is quite small value and the $\Delta R/R$ due to the matrix crack is strongly observed in the segment 2. This indicates that for the laminate of [90/0]s only the matrix crack is detected and the delamination can not be detected.

CONCLUSIONS

(1) The relationship between fiber volume fraction and electric conductance is experimentally measured. $\sigma 90$ and σt are significantly smaller than $\sigma 0$, and they vary depend on the fiber volume fraction.

(2) The strong orthotropic electric resistace makes gradual electric current in the thickness direction, for the laminate of [0/90]s. The current in the thickness direction detects the delamination existence.

(3) The electrode should be mounted on the fiber direction.

REFERENCES

1.Todoroki A., Suzuki H., Kobayashi H., Nakamura H., and Shimamura Y., Transactions of the Japan Society of Mechanical Engineers Series A.; 64(622); 1654(1998), (in Japanese).

2. Todoroki A. and Suzuki H., Applied Mechanics and Engineering; 5(1); 283(2000).

3. Todoroki A. Tanaka Y., and Shimamura Y., Structural Health Monitoring 2000, edited by F.Kchang, Technomic.; 308(1999).

4. Todoroki A. Tanaka Y., and Shimamura Y., Special Technical Publication-2, JSMS, 139(2001).

An experimental device and an inverse method to determine the thermal conductivity of a fiber reinforced composite as a function of temperature and state of cure

H. Menge, F. Trochu and A. Millischer
Department of Mechanical Engineering,
Applied Research Centre on Polymers (CRASP),
Ecole Polytechnique,
C.P. 6079, Station « Centre-Ville »,
Montreal, H3C 3A7, Quebec, Canada
trochu@polymtl.ca

V. Sobotka
Laboratoire de Thermocinétique
Ecole Polytechnique de l'Université de Nantes
Rue C. PAUC
B.P.90604
44306 Nantes Cedex
Tel.: 02.40.68.31.19
Fax.:02.40.68.31.41
e-mail:vincent.sobotka@polytech.univ-nantes.fr

ABSTRACT

The thermal conductivity of a thermosetting composite varies significantly during the injection and curing. These variations play an important role on the heat transfer within a thick part and thus on resin kinetics. In fact, it was observed experimentally that the temperature gradient through the thickness of a composite has a strong influence on the quality of surface finish. In order to perform precise non isothermal mould filling and curing simulations, a good knowledge of thermal conductivity is needed in function of temperature and state of cure. This work is concerned with the measurement of heat conductivity through the thickness of a composite part. For this purpose, an inverse method is developed together with an experimental device designed to measure the transverse thermal conductivity. The heat capacity of the material, which is needed to derive by inverse analysis the thermal conductivity, is obtained following a classical method based on DSC measurements. The data processing procedure based on inverse analysis will first be described. Then the approach will be validated for a set of experimental results obtained for a polyester resin formulation.

INTRODUCTION

Thermal phenomena occuring during RTM injections are complex and play an important role on the surface finish of moulded parts. The chemical reaction induces a change of structure, which results in a variation of thermo-physical properties (1, 2). In particular, the transverse thermal conductivity of the composite is an important parameter that governs the heat exchanged through the thickness of the part. In fact, it was observed experimentally that the transverse temperature gradient is closely connected with the quality of surface finish. The strong variations of transverse heat conductivity with temperature and state of cure must be measured in order to provide a good prediction of temperature and cure in the moulded part. This work will ultimately lead to a better control of the injection process.

Many results have been obtained concerning the thermal conductivity of a composite material (3, 4). Some models have been validated by Bailleul et al. (3). The purpose of this study is to describe the experimental device and the methodology used to measure the transverse thermal conductivity with an experimental apparatus especially designed to minimize experimentation time and cost. The experimental protocols for measuring the heat capacity and thermal conductivity are presented, along with the inverse method used to derive the heat conductivity from temperature recordings through the thickness of the part. The methodology will be validated for experimental results obtained for a polyester resin formulation.

EXPERIMENTAL PROTOCOL

The objective of this study is to determine experimentally the effective thermal transverse conductivity of a thick composite part. The experimental protocol includes three steps :
1. Determination of the heat capacity C_p.
2. Preparation of the raw samples (for a degree of cure $\alpha =0$).
3. Determination of the thermal conductivity $\lambda_\perp$ by an inverse heat conduction analysis.

The experimental device must allow the moulding of a sample by *Resin Transfer Moulding* (RTM) and the measurement of the temperature through the thickness of the composite in the raw and cured states. It is made of a mould, a heating device, a press and a data acquisition system. The injection is performed with a pressure pot that controls the pressure at the injection gate (see Figure 1). The inverse method needs the measurement of at least three temperatures through the thickness of the sample. In order to yield reliable results, the temperature gradient must be sufficiently large compared to measured perturbations. Following preliminary results obtained by Sobotka (4) with a prototype experimental device, a thickness of 15mm was chosen for the moulding cavity. The resin is injected on the upper side of the mold, through an injection hole of diameter between 5 and 10 mm in order to obtain a sufficient flow rate. The injection gate is located along one edge of the mould where no fibers have been placed, so that a straight resin front is obtained. The mould is made out of metal, because of the high temperatures involved and the mechanical properties required to withstand the injection pressure. Moreover, the high thermal conductivity of metals ensures uniform temperatures on the mould plates so as to obtain a perfectly unidirectional heat flux across the thickness of the composite. Finally, aluminium was preferred to iron, because it is much easier to heat thanks to a higher thermal conductivity by an order of magnitude. Finally, the mould is insulated on all faces: top, bottom and lateral sides. An epoxy fibre glass composite was selected as rigid insulating material. A preliminary design of the measurement mould can be found in Jauffres (5).

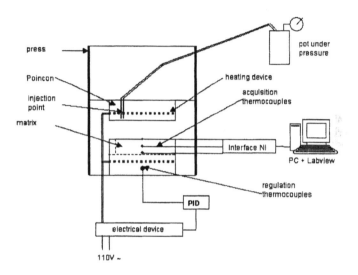

Figure 1 – Schematics of the experimental measurement device.

Thermal simulations have been conducted in order to determine the smallest size of the mould that will minimize cost and experimentation time. For a part of size 120 x 100mm (including the injection zone), the heat flux can be considered as unidirectional in the zone where measurements are recorded. The press is a simple mechanical clamping system, which allows to withstand an internal cavity pressure of 5 bars. Moreover, the experimental device must allow to vary the thickness of the sample in order to characterize the thermal conductivity in function of fibre volume content. In order to obtain an homogeneous temperature in a relatively small mould, tubular heaters of the smallest possible diameter were chosen (5mm). A geometry to allow a uniform heating of the mould cavity was devised. The heating device associated with regulation thermocouples and a cooling device (by air, or another fluid) allows to regulate completely the temperature of the mold.

Determination of the heat capacity Cp

In order to use the method exposed hereafter to measure thermal conductivity, it is necessary to know the heat capacity of the material in function of temperature and state of cure. This is done by a classical DSC method for the raw (C_0) and cured resin (C_1) respectively. The heat capacity for a degree of cure α is obtained afterwards by a simple rule of mixture (see equation (1) below). So the problem consists of measuring C_0 and C_1 before and after the chemical reaction. The heat capacity of the cured material is easy to determine, because no chemical reaction takes place any longer in the material. The heat capacity C_0 is measured for the resin without catalyser, in order to avoid reticulation during the measurement. The results of Figure 2 were obtained for a copolymer polyester / styrene.

$$C_{P_{resin}}(\alpha,T) = \alpha C_1(T) + (1-\alpha)C_0(T)$$
$$with \quad \begin{cases} C_0(T) = C_{P_{resin}}(0,T) \\ C_1(T) = C_{P_{resin}}(1,T) \end{cases} \quad (1)$$

The heat capacity in function of temperature can be readily obtained for the raw and cured resin (with T in °C):
- Raw resin : $Cp(T) = 2.100 * T + 1631.6$ (J/kg.K)
- Cured resin : $Cp(T) = 4.7086 * T + 1157.6$ (J/kg.K)

These results are acceptable. As a matter of fact, the gap between the averaged and measured values and is 6% for the raw resin and 4% for the cured resin.

Preparation of the samples

The samples are cut out of the reinforcing material and piled up in plies to the thickness needed. Approximately 15 mm thick of material is needed in order to obtain a sufficient thermal gradient for this methodology. Chromel-alumel thermocouples are inserted between plies in the plane of the fibres. A maximum of seven thermocouples can be used for measurement and a minimum of three (two for the plies in contact with the top and bottom mould plates, which will be used as boundary conditions, and one for the core temperature. The exact position in the core of the thermocouples is measured with a microscope, after cutting a slice out of the sample.

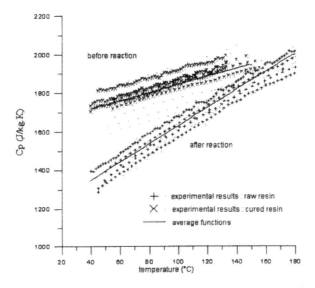

Figure 2 – Specific heat Cp for the raw and cured resin in function of temperature.

Determination of the thermal conductivity

The method used by Sobotka (4) to determine the termal conductivity is based on inverse heat conduction analysis. Knowing the boundary conditions on the top and bottom plates of the cavity, the thermal conductivity can be derived from the temperature recorded in the core of the sample. A specific thermal cycle is required to perform the measurement as described below (see Figure 3) :
   (1) a plateau at ambient temperature in order to get an initial isothermal state;
   (2) a rise in temperature during which the conductivity of the raw material is determined by recording the temperature evolution during this phase;
   (3) a plateau in order to complete the cure of the material;
   (4) a decrease of temperature and a second lower plateau to get a new isothermal state;
   (5) a second rise in temperature in order to obtain the conductivity of the cured composite.

The method is the same for the raw and cured conductivities. With an initial guess for the thermal conductivity, the temperatures through the thickness are computed and compared with the measured ones. The material is assumed to be homogeneous, and its specific heat is already known. A least square criterion is then evaluated and the computed conductivity is modified in order to minimize it.

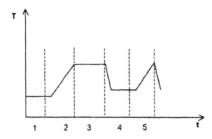

Figure 3 – Typical cure cycle described in the proposed methodology.

MODELS AND NUMERICAL VALIDATION

Effective thermal conductivity model

In order to model the evolution of thermal conductivity as a function of temperature and rate of cure, it is necessary to take into account the non homogeneity of the composite material. When the conductivity is measured, the fibres and the matrix have different contributions. The variations between the raw and cured states can only be due to the matrix, as the reinforcement is not taking part in the chemical reaction and is not structurally modified during the resin transformation. One can find in the literature several models giving the thermal conductivity of a composite material in function of its matrix and fibres, and its principal directions. A simple model was chosen in Equation (3) that allows to decouple the influence of the matrix and the fibers. This model, which is commonly used for composite materials, gives the transverse thermal conductivity $\lambda_\perp$ from the conductivities of the fibres λ_f and of the matrix λ_m.

$$\lambda_\perp = \lambda_m \frac{(1+v_f)\lambda_f + (1-v_f)\lambda_m}{(1-v_f)\lambda_f + (1+v_f)\lambda_m} \tag{3}$$

where v_f is the fibre volume fraction. Knowing the degree of cure and the thermal conductivities of the raw and cured resin, it is possible to find by the rule of mixture the thermal conductivity of the matrix in function of cure:

$$\lambda_m(\alpha,T) = (1-\alpha)\lambda_{mraw}(T) + \alpha.\lambda_{mcured}(T) \tag{4}$$

Then equation (3) can be used to derive the expression of the transverse thermal conductivity of the composite in function of temperature and cure:

$$\lambda_\perp(\alpha,T) = \lambda_m(\alpha,T) \frac{(1+v_f)\lambda_f(T) + (1-v_f)\lambda_m(\alpha,T)}{(1-v_f)\lambda_f(T) + (1+v_f)\lambda_m(\alpha,T)} \tag{5}$$

Validation of the program for the inverse method

Figures 4 shows one of the tests carried out for a heating speed of 5°C/min to validate the software program developed for the inverse heat conduction method. The distance between thermocouples used as boundary conditions is 20mm. The identification is made on a thermocouple placed in the center (10mm). The functions represented are the ones that the code must identify. Several tests were conducted for data files without no source of error, then with the following errors:

- noise of ± 0.7K on temperature,
- error of ± 0.2mm on the position of the thermocouples;
- error of ± 4% on the heat capacity:.

After all these error sources have been cumulated, the maximal gap between the values determined by the program and the exact values is never greater than ± 4%. This completes the validation of the computer code.

CONCLUSION

The present study describes a methodology to measure the transverse thermal conductivity of a fibre-reinforced composite in function of temperature and cure, along with an experimental apparatus and testing protocol especially devised to reduce the cost and time required to perform such measurements. An inverse method was developed for the heat conduction problem to derive the transverse heat conductivity from the measurement of at least three temperatures in time at three different locations through the thickness of a composite. The experimental device described in this article should permit to perform more precise and reliable measurements. The next step will be to develop a method to measure the in-plane conductivity of a composite in function of temperature and cure. Together with a resin kinetics analysis, one would then obtain all the parameters that are required to conduct full non isothermal numerical simulations of mould filling and cure for fibre-reinforced composites made by liquid composite moulding.

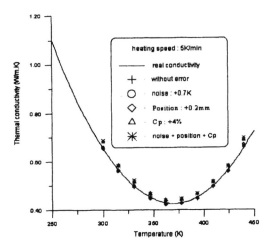

Figure 4 : Determination of transverse conductivity by inverse heat conduction analysis.

REFERENCES

1. J.D. Farmer, E.E. Covert, « Thermal Conductivity of a Thermosetting Advanced Composite During its Cure» , Journal of Thermophysics and Heat Transfer, Vol. 10, N°3, July-September 1996, pp 467-475.

2. H. Tai, «Equivalent Thermal Conductivity of Two- and Three-Dimensional Orthogonnally Fiber-Reinforced Composites in One-Dimensional Heat Flow», Journal of Composites Technology & Research, Vol. 18, N°3, July 1996, pp. 221-227.

3. J.-L. Bailleul, D. Delaunay, Y. Jarny, T. Jurkowski, «Thermal Conductivity of Unidirectional Reinforced Composite Materials – Experimental Measurement as a Function of State of Cure», Journal of Reinforced Plastics and Composites, Vol. 20, No. 01/2001.

4. V. Sobotka, «Détermination des paramètres thermophysiques et cinétiques d'une résine polyester insaturée», Rapport de stage, CRASP, Ecole Polytechnique de Montréal, août 2001.

5. D. Jauffres, «Conception d'une presse expérimentale de moulage de pièces composites par le procédé RTM (Resin Transfer Molding) dédiée à des mesures de conductivité thermique», rapport de stage, INSA de LYON, École Polytechnique de Montréal, janvier 2002.

SIMULATION AND MODELLING

Interpolation by dual kriging of a moving flow front and conservation of the fluid mass

Vincent Gravier, François Trochu, Carl-Éric Aubin
Department of Mechanical Engineering
Applied Research Centre on Polymers (CRASP)
École Polytechnique
C.P. 6079, Station « Centre-Ville »
Montreal, H3C 3A7, Quebec, Canada
trochu@polymtl.ca

Alexandre Plouznikoff and Benoit Ozell
Department of Computer Engineering
École Polytechnique

ABSTRACT

One of the main issues when simulating the filling of a mould cavity is not so much solving the equations governing the flow, but rather to make sure that the calculation is accurate, although the shape of the geometrical domain changes constantly. In order to improve the accuracy of the numerical approximation, it is interesting to refine locally the mesh in the neighbourhood of the flow front and project on the new mesh the different scalar fields involved in the calculation. Very often a significant inaccuracy may result from these operations. The numerical error is cumulative and relatively small errors at each time step can lead to significant discrepancies at the end of the simulation. For all these reasons, errors on the conservation of the fluid mass represent a major challenge in injection moulding simulations. In this paper, a new approach based on dual kriging interpolation is proposed to project the flow front from one mesh to another, while conserving automatically the fluid mass in the cavity. The advantage of this methodology dwells in its generality, as it is valid both in the two and three-dimensional cases. It solves a major problem that occurs commonly in the numerical approximation of moving boundary problems, namely the conservation of the fluid mass. It consists of following precisely the displacement of the flow front by updating at each time t a saturation function $f(x,t)$ defined at all points x in the cavity. At any given time, this function is one in the saturated domain, zero in the empty zone and is bounded between zero and one to reflect partial saturation in the vicinity of the flow front. An interpolation of the saturation function based on dual kriging provides an implicit equation of the moving front. From this equation the moving front can then be interpolated on the new mesh, so as to conserve exactly the fluid mass in the mould. After describing the main equations of dual kriging, the article presents the algorithm devised to conserve the fluid mass at each calculation step. Numerical experiments are carried out to validate these concepts for three-dimensional fluid fronts. The conservation of fluid mass is verified for multiple or merging flow fronts, obstacles, ribbed connections and parts of complex geometry.

INTRODUCTION

Resin transfer moulding (RTM) is an efficient and attractive technique for manufacturing advanced fibre-reinforced composite materials ranging from small and simple pieces to larger parts of complex shape and high structural performance. Compared to hand lay-up or autoclave processing, the RTM process provides the advantages of low labour requirements, relatively short cycle time, low equipment costs, simplification of environmental issues, reduction of styrene emissions and good potential for automation. Over the past decade, an extensive effort has been made to simulate the RTM process. Besides the problems related with the determination of the complex physical phenomena that come into play in liquid composite moulding, a major barrier to further development of these techniques is connected with the high computational time required by three-dimensional simulations compared to two-dimensional calculations. Thus, every method speeding up the computation time will help considerably to implement three-dimensional simulations.

The numerical treatment of moving boundary problems is an important issue in the simulation of injection moulding. In a broad sense, the numerical procedures fall into two categories: moving grid and fixed grid schemes (1, 2). The moving grid approach provides an accurate representation of the flow front position. However, since the saturated domain must be remeshed at each calculation step, it is a computationally intensive process. Moreover, as pointed out in Trochu and Gauvin (2), this approach turns out to be quite difficult to implement for moulds with inserts or in the case of multiple injection gates.

Trochu et al. (1) used non conforming finite elements to calculate the resin pressure in the saturated domain during the filling process, together with a filling algorithm based on conservation of the resin mass to displace the flow front on a fixed mesh. Among the numerous numerical techniques available, the methods based on finite elements and control volume have become the most versatile and computationally efficient to simulate mould filling in a cavity of complex shape including inserts. In these methods, the regeneration of the mesh is not needed at each calculation step, since the resin flow is tracked on a fixed mesh. Filling factors between zero and one are associated to each node or each element of the mesh. The filling factors represent the ratio of the quantity of resin contained inside each cell of the mesh, whether it is a control volume around a node or an element itself, to the total pore volume of the cell. At the beginning of mould filling all the filling factors are zero. At the end their values have become one everywhere in the cavity. The filling factors change in time as the resin progressively fills up the mould.

In order to improve the accuracy of numerical simulations, adaptive mesh refinement techniques have already been implemented for the RTM process. However, conservation of the resin mass is not guaranteed by these algorithms. This has motivated the development of a new approach based on dual kriging to project the flow front position from one mesh to another, while at the same time conserving automatically the fluid mass in the cavity. A description of dual kriging can be found in Trochu (3). After recalling the main equations of kriging, the article presents the algorithm that allows to conserve the fluid mass at each calculation step. Numerical experiments are carried out to validate the concepts for three-dimensional examples. The conservation of the fluid mass is verified for multiple or merging flow fronts, obstacles, ribbed connections and parts of complex geometry.

DUAL KRIGING APPROXIMATION

Kriging is a statistical method which consists of building the best linear unbiased estimator of a random function defined on a geometrical space. A mathematical model of a physical phenomenon or of a geometrical shape can be constructed by kriging from a list of experimental observations or from the known dimensions of the geometric object. Kriging can be used to model curves, surfaces, solid volumes or any function of multiple real variables. In kriging, the model function $u(x)$ that describes the phenomenon is decomposed in the sum of two terms: a mean value and another term representing the fluctuations around the mean:

$$u(x) = a(x) + W(x) \qquad (1)$$

where $a(x)$ is the mean value called the *drift* and $W(x)$ represents the fluctuations, which will be expressed about each observation with a function of arbitrary shape, $K(h)$, called the *generalized covariance*. The expression of $u(x)$ in kriging writes as follows:

$$u(x) = a(x) + \sum_{i=1}^{N} b_i \cdot K\big(|x - x_i|\big) \qquad (2)$$

where b_i for $1 \leq i \leq N$ are coefficients to be determined and $| - |$ denotes the norm of a vector in the geometrical space. The N degrees of freedom b_i will allow function $u(x)$ to fit the observations. Usually, $a(x)$ is a polynomial or trigonometric function and $K(h)$ an arbitrary function of the Euclidean distance between two observations or samples.

The linear system of dual kriging can be written for a linear drift in a three-dimensional geometrical space as follows:

$$
\begin{bmatrix}
 & & & 1 & x_1 & y_1 & z_1 \\
 & K(h) & & \dots & \dots & \dots & \dots \\
 & & & \dots & \dots & \dots & \dots \\
 & & & 1 & x_N & y_N & z_N \\
1 & \dots & \dots & 1 & 0 & 0 & 0 & 0 \\
x_1 & \dots & \dots & x_N & 0 & 0 & 0 & 0 \\
y_1 & \dots & \dots & y_N & 0 & 0 & 0 & 0 \\
z_1 & \dots & \dots & z_N & 0 & 0 & 0 & 0
\end{bmatrix}
\cdot
\begin{bmatrix}
b_1 \\ \dots \\ \dots \\ b_N \\ a_0 \\ a_1 \\ a_2 \\ a_3
\end{bmatrix}
=
\begin{bmatrix}
f_1 \\ \dots \\ \dots \\ f_N \\ 0 \\ 0 \\ 0 \\ 0
\end{bmatrix}
\qquad (3)
$$

where h represents the distance between two observations, x_i, y_i, z_i are the coordinates of the observation i, f_i is the scalar field observed, and a_i, b_j are the coefficients to be determined. The solution of the above system gives in equation (2) the expression of function $u(x)$ in the geometrical domain. As described by Trochu (5), the last three equations mean that $a(x)$ is equal to the mathematical expectation of $u(x)$ and the first N equations simply express the fact that function $u(x)$ fits the observations, i.e., in the example considered here:

$$u(x_i, y_i, z_i) = f_i, \quad 1 \le i \le N \tag{4}$$

PRESENTATION OF THE ALGORITHM

Five main hypotheses are necessary to project the flow front position from one mesh to another while conserving the fluid mass:

✓ A filling field $f(x,t)$ is assumed to exist in the mould cavity at each time t, the values of which correspond to the discrete filling factors at the centre of gravity of each element of the mesh (or at each node surrounded by a control volume).

✓ The actual position of the flow front on the new mesh at a given time is assumed to correspond to a contour line $f(x,t) = \alpha$ of the filling field for a given value of a parameter α, $0 < \alpha < 1$. Obviously, the flow front must be located between the values of the filling scalar field lower than one and greater than zero. This hypothesis amounts to claiming that the contour line of value α is continuously separating the mould cavity in two regions: the empty and the saturated domains. The saturated domain corresponds to the region containing the injection gates, and the empty domain to the region where the vents are located.

✓ The filling factors between any two points close to each other may be calculated by linear interpolation. In the algorithm, this interpolation will be performed only inside the same triangle or tetrahedron.

✓ The region filled with resin at each time step (saturated domain) is determined by assuming that both the resin and the reinforcement are incompressible.

Although the algorithm is valid also in the 2D case, its mathematical foundation will be presented in the sequel for tetrahedrons, i.e., in the 3D case, in order to understand how the projection of the flow front is performed. Three files are required for the calculation:

- A file that contains the description of the first mesh, i.e., the nodal coordinates and the information to create the elements.

- A file that describes the new mesh containing once again the nodal coordinates and the information to create the elements.

- A file that contains the filling field on the first mesh, i.e., specifying at each time step and for each element a filling coefficient between zero and one. This filling coefficient indicates the level of pore saturation of each elementary volume.

1. Centres of gravity of elements

For each tetrahedron of the new mesh, the coordinates of the centre of gravity can be easily calculated as follows:

$$x_{centre} = \frac{1}{4}\sum_{n=1}^{4} x_{corner(n)} \qquad y_{centre} = \frac{1}{4}\sum_{n=1}^{4} y_{corner(n)} \qquad z_{centre} = \frac{1}{4}\sum_{n=1}^{4} z_{corner(n)} \qquad (5)$$

The volume of the tetrahedron is calculated at any given time by the following formula:

$$Volume\ of\ tetrahedron = \pm \frac{1}{6}[a, b, c] = \pm \frac{1}{6}\begin{vmatrix} a_x & a_y & a_z \\ b_x & b_y & b_z \\ c_x & c_y & c_z \end{vmatrix}$$

2. Kriging interpolation

Based on the values of the filling factors at the centres of gravity of each element of the old mesh, a kriging interpolation of the filling field is performed that provides the filling factors at the centres of gravity of the elements of the new mesh. These values will give the amount of resin contained inside each element of the new mesh.

3. Correction of the filling factors

As a result of these calculations, when dual kriging is not used as an interpolator, i.e., in the zones where a new filling factor is calculated by extrapolation, some filling coefficients might come up with a value greater than one or lower than zero. In this case, a simple solution is adopted here for consistency: the filling coefficients are replaced by one if they are greater than one and by zero if they are negative.

4. Position of the flow front on the new mesh

The most probable position of the flow front is assumed to be given by a contour line or surface of the filling field for a certain α, $0 < \alpha < 1$:

$$f(x, y, z) = \alpha \qquad (7)$$

The value of α will be determined such that the total fluid volume $V_\alpha(t)$ (wetted volume) contained in the saturated domain defined by the contour surface (7) is conserved exactly. Therefore this volume $V_\alpha(t)$ must be equal to the total resin volume $V(t)$ injected in the cavity at time t. The parameter α will be chosen by a dichotomy algorithm such that $V(t) = V_\alpha(t)$ at each calculation step.

5. Dichotomy algorithm

First, we initialize the process by choosing $\alpha = 0.5$. The first step consists of calculating $V(t)$ by the formula:

$$V(t) = porosity \times \sum_{n=1}^{nb.tetrahedrons} vol.tetrahedron(n)_{old\_mesh} \times filling\_coeff\_tetra(n)_{old\_mesh} \qquad (8)$$

Once α is fixed, it is necessary in order to calculate $V_\alpha(t)$ to locate each element of the new mesh with respect to the flow front. For each element, three cases are possible:

- The element lies outside $V_a(t)$. The three nodes of each triangle (or four nodes of each tetrahedron) are tested. If the filling coefficients associated with each node are lower than α, the element is not included in the filled area. In this case, the volume of this specific element will not be taken into account in the calculation of the total fluid volume on the new mesh.
- The element lies completely inside $V_a(t)$. This time, the volume of this specific element will be taken into account.
- The contour line or surface cross the element. This is the case if one of the nodes has a filling coefficient greater than α, and another one lower than α. The element will have to be truncated. In order to determine how many intersection points with the contour line or surface exist inside that element, it is necessary to find out if the filling coefficient increases between two nodes from a value lower than α to a value greater than α. In all cases, there is a point on the line joining the two nodes where the filling coefficient is exactly equal to α. The coordinates of this point have to be calculated. The algorithm uses a recursive dichotomy between the two nodes to obtain the coordinates of the point where the filling coefficient is α. The accuracy of the dichotomy can be chosen by the user. Three cases may arise to calculate the fluid volume inside an element lying in the neighbourhood of the flow front:

1. Only one point lies in $V_a(t)$. Two intersection points are found in 2D (three in 3D); the wetted volume may be readily calculated inside the element.
2. Two points lie in $V_a(t)$. There are two points of intersection in 2D (four in 3D). We can calculate the wetted volume by dividing the elementary volume into two elements, which will be clearly identified by the algorithm.
3. In the 3D case, three points may lie in $V_a(t)$. As in the first case, it is easy to calculate the volume outside $V_a(t)$. The wetted volume is obtained readily by comparison with the volume of the whole element.

Once all the elements are treated, $V_a(t)$ is calculated and compared to $V(t)$. If $V_a(t) \le V(t)$, α has to be increased. If $V_a(t) \ge V(t)$, α has to be decreased. The algorithm can find the right value of α by successive iterations. The operation has to be performed at each time step in order to get the final filling field on the new mesh. Finally, a new position of the flow front is obtained that will conserve exactly the fluid quantity, while respecting as much as possible the most probable shape of the flow front at any given time in the cavity.

NUMERICAL EXPERIMENTS

Some experiments will now be presented, each one illustrating a different set of difficulties. In each case, two pictures will be shown: the first one displays the filling field on the old mesh; the second one the result of the algorithm on the new mesh.

First experiment: filling of a basic 3D geometry

In this case, a cube is filled from one of its face. On the old mesh which contains 500 elements approximately, some shadow can be observed due to the different values of the filling field.

On the new mesh that contains nearly 1500 elements, only the flow front is represented. Parameter α is the value of the contour line or surface which allows to conserve the resin quantity. The resin threshold set in this calculation was 0.15%, and the error on the conservation of the fluid volume on the new mesh was only 0.06%.

Figure 2: Basic three-dimensional geometry

Second experiment: obstacle in a 3D geometry

In this case, a thin square plate with a large hole in the centre is filled up, so that the resin encounters an obstacle during the filling process. Once again, the algorithm provides a new shape of the flow front that conserves exactly the fluid mass.

Figure 3: Obstacle in a two-dimensional geometry: filling on the old mesh and result on the new mesh.

Third experiment: multiple flow fronts in a simple three-dimensional geometry

In this case, a double injection is performed in a three-dimensional geometry. Two separate flow fronts are created. Although the visualisation on a black and white picture is not easy, the animation on the computer screen confirms once again that the algorithm respects the exact conservation of the fluid mass.

Figure 4: Multiple flow fronts in a three-dimensional geometry: filling on the old mesh and result on the new mesh.

CONCLUSION

In this paper, a new approach based on dual kriging interpolation is proposed to project the flow front position from one mesh to another, while conserving with precision the fluid mass in the cavity. The algorithm used here was tested on several examples: a basic three-dimensional geometry, a mould with an obstacle and the case of multiple flow fronts. The algorithm is able to project the position of the moving flow front on the new refined mesh while conserving the fluid mass. More numerical experiments are needed to evaluate quantitatively the accuracy of this approach.   Dual kriging interpolation provides a good compromise between minimization of the computational effort and conservation of the fluid mass. This method brings a promising and reliable new solution to a major challenge in the simulation of injection moulding processes, namely the conservation of the fluid mass in the numerical calculations.

REFERENCES

1. F. Trochu, R. Gauvin, D. M. Gao, "Numerical Analysis of the Resin Transfer Molding Process by the Finite Element Method", Advances in Polymer Technology, Vol. 12, 1993, 329-342.

2. F. Trochu, R. Gauvin, "Limitations of a Boundary-Fitted Finite Difference Method for the Simulation of the Resin Transfer Molding Process", Journal of Reinforced Plastics and Composites, Vol. 11, 7, 1992, 772-786.

3. F. Trochu, "A Contouring Program Based on Dual Kriging Interpolation", Engineering with Computers, Vol. 9, 1993, 160-177.

Experimental optimization of process parameters to obtain class A surface finish in resin transfer molding

Mohsan Haider and Larry Lessard
Department of Mechanical Engineering
McGill University, 817 Sherbrooke Street West
Montreal, H3A 2K6, Quebec, Canada

Francois Trochu and Marc-Andre Octeau
Applied Research Centre on Polymers (CRASP)
Ecole Polytechnique, CP 6079, Station Centre-Ville
Montreal, H3C 3A7, Quebec, Canada

ABSTRACT

Resin Transfer Molding (RTM) is becoming more widely used in the fabrication of large and complex structural components. The main objective of this research is first to understand the influence of the parameters that govern the RTM process, especially those concerning the surface finish of final parts. Then it will be possible to optimize these factors in order to obtain a "class A" surface finish. The chemical reactions, manufacturing parameters and quality of tooling have been identified as the main parameters that affect the surface finish of the final part. The type of resin, styrene content, type of low profile additive, types and quantities of ingredients in resin, charge, wetting agent, degassing agent and fiber reinforcement are possible factors that must be optimized in order to get a resin formulation resulting in minimum shrinkage and best possible surface finish. Molding parameters like molding temperature, through thickness temperature gradient, injected flow rate (or injection pressure), rate of cure and injection temperatures must also be optimized. The Taguchi method will be used to study the effect of the different parameters. A flat plate mold with dimensions 17" x 17" has been designed and set up to produce a number of parts in order to study the effect of the above-mentioned parameters on the surface finish. Geometry and surface finish of the mold, positions of vents and injection gate, in-mold coating, gel coat applications and types of reinforcement will also be considered in order to study their influence on surface finish. Taguchi control factors and their levels of influence will be identified. A matrix of experiments will be designed, carried out and analyzed using statistical calculation software. Optimum levels and performance of the above mentioned control factors will be determined. In the end, verification experiments will be carried out. This investigation will demonstrate how surface finish can be improved in a practical way and optimized. Ultimately, it will also provide a better understanding of the factors that govern surface finish in the RTM process.

INTRODUCTION

Resin Transfer Molding (RTM), also known under the generic term (including its various process variants) of liquid composite molding, is increasingly being used in the aerospace and automotive industries.   This family of related processes allows the molding of components with complex shapes and large surface areas. RTM is a process suited for short and medium production runs and is employed in many different applications.   Some of the benefits afforded by RTM include low capital investment, tooling flexibility and the ability to mold large and complex shapes with ribs, cores and inserts. It also allows greater parts integration and a range of resin systems and fibrous reinforcements are available for RTM injection with a controllable fiber volume fraction (see Michael C. Niu (1)).

RTM process

Resin Transfer Molding has the potential of becoming a dominant low cost process for the fabrication of large, integrated, high performance composite products for the consumer segment of the economy and ultimately, for a segment now dominated by higher precision laminated fabrication techniques. In this process (figure 1), a dry reinforcement material that has been cut and shaped into a preformed piece, generally called a preform, is placed in a prepared mold cavity. Once the mold has been closed and clamped shut, resin is injected into the mold cavity, where it flows through the fibrous reinforcement, expelling the air in the cavity and wetting out or impregnating the preform. When excess resin begins to flow from the vent areas of the mold, the resin flow is stopped and the molded component begins to cure. When cure is completed, which can take from several minutes to hours, it is removed from the mold and the process can begin again to form additional parts. A detailed description of the whole procedure can be found in Ludick (5).

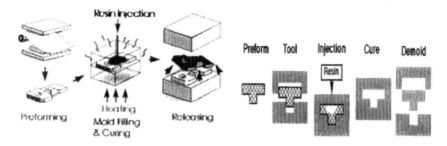

Figure 1: Schematics of the Resin Transfer Molding process

On first inspection, RTM appears to be a three-step process: preforming followed by resin injection and cure. Each step of the process affects the subsequent ones. Every decision has an impact on both processing of the part and the performance of the final product. For example, the preform must be designed in order to take into account not only the thermo-mechanical loading on the part, but also the influence of the fibrous structure on permeability. Therefore it has an effect on time to fill the mold (thus affecting resin selection, viscosity, and processing temperature, which in turn have an influence on the tooling material and demolding time). Similarly, the injection must reflect the delicate balance among the following: a shorter cycle time, the integrity of the fibrous structure (by avoiding processing-induced movements of the reinforcement), the wet-out of fiber bundles and the removal of entrapped air. So resin selection, molding temperature, pressure, and viscosity must be controlled to preserve the integrity of the reinforcement. Similarly, the tooling must be designed to take into account not only the shape and features of the part, but also the specific characteristics of the injection process, such as permeability e.g., resin cure and demolding. These are but a few examples of the challenges associated with RTM, making it essential that processing and performance issues be considered simultaneously to ensure that the final product is of high quality and yet economical (4-6).

Surface finish issues

In the manufacturing of polymer composites, RTM is thought to be an efficient and economical process for fabricating large and complicated structural components. However RTM has not fulfilled its potential for use in high volume applications. Surface finish of the part in these applications is the foremost concern, especially in automotive industry, which is striving to achieve *class A surface finish*. Class A surface finish is thought to exhibit aspects of flatness, smoothness and light reflection similar to that of finished stamped steel sheeting, typically with a DORRI (Distinctness of Retro Reflective Image) value of 60-90, as measured by Mathews et al. (7) using the D-sight optical enhancement technique.

Class 'A' finish is a perfectly polished high luster surface, free of porosity and scratches of any kind. The term originated in the marine and automotive industries. Examples of such a finish can be found on high quality boat hulls and automobiles. However, this quality of surface finish is achieved through two different procedures. Cars have primers and paint systems sprayed over medium quality metal surfaces. The paint flows into a self-leveling thin film and requires polishing to achieve a true Class 'A' surface. The boat hull, however, receives its finish directly from the mold itself. If the mold has a Class 'A' finish, all the parts produced from it will also have the same high quality surface. Construction of quality molds can decrease final finishing time and increase overall part quality. Special gloss (brightness, shine) can be measured easily using a gloss meter (BS 3900/ASTM D523-89) with reference to a standard surface, a viewing angle and a material color as described by Mathews et al. (7).

Several different techniques can be employed to achieve class A surface finish such as:

1. Finding out optimum chemical composition of the resin system. This will reduce shrinkage and ultimately, decrease the defects on the surface of the part. The types and quantities of resin, styrene, catalyst, inhibitor, accelerator, internal de-molding

agent, external de-molding agent, charge, wetting agent, degassing agent and fiber reinforcement have to be optimized. In recent years, resins with low profile additives have been introduced to reduce shrinkage and enhance surface finish (8-10).

2. Part and tool design. The quality and surface finish of a part depends highly on its design and on the surface finish of the mold. The mold should be designed in such a way to eliminate sharp edges and corners where it is difficult for the resin to reach. Gates and vents locations must be optimized in order to remove entrapped air and allow a smooth flow of the resin (11,12).

3. Use of in-mold coating (IMC). In-mold coating and gel coat applications are being used to achieve optimum surface finish, which in most cases increase manufacturing time and cost (13).

4. Selection of molding process parameters. For example, molding temperature (upper and lower mold wall temperatures), injection rate or pressure, rate of cure and injection temperature as shown by Gauvin et al. (14) will also have an influence on part quality and surface finish.

5. Secondary and finishing operations on the manufactured part (7,11). This is very often necessary in RTM and increases significantly the final manufacturing cost.

The purpose of this research is to design a methodology based on the above-mentioned techniques in order to obtain a class A surface finish. The chemical composition referred by Gordon et al. (8) is the starting point of this research. The mold for making test components has been designed, manufactured and adapted for surface finish investigation by Ford Motor Company Inc., Detroit, MI in collaboration with Ecole Polytechnique and McGill University. A series of temperature and pressure sensors will be set up in the mold together with a data acquisition system. The emphasis will be on the modification of molding process parameters. Experiments will be designed using the Taguchi method of design of experiments (2,3). Taguchi control factors and their levels of influence will be identified. A matrix of experiments will be designed, experiments carried out and their results analyzed using statistical calculation software. Optimum levels and performance of the above mentioned control factors will be determined. In the end, verification experiments will be carried out. Ultimately, this investigation will provide a better understanding of surface finish aspects of the RTM process.

The Taguchi method

The Taguchi method refers to techniques of quality engineering that includes both statistical process control and innovative quality related management techniques. The Taguchi method leads to the best selection and setting of product/process design parameters and tolerances. The Taguchi method has become well known because it has led to major reductions in product/process development lead-time and has improved the manufacturability of complex products. The method analyzes, in a systematic way, the complex cause-effect relationship between design parameters and performance. This technique indicates how to design and perform experiments in order to investigate processes where the output depends on many factors (variables, inputs) without having to tediously and uneconomically run the process using all possible combinations of values of those variables. By systematically choosing certain combinations of variables, it is possible to separate their individual effects (2,3).

In order to determine and subsequently minimize the effect of factors that cause variation, the design cycle is divided into three phases: system design, parameter design, and tolerance design. In the first stage, the designer uses knowledge of the process being investigated to produce an initial design of a product or process. The objective of the second stage is to choose suitable values for the parameters of a product or process. In the third stage (tolerance design), higher quality parts replace less reliable components, improving the quality of the product or process (2,3). Parameter design is the focal point of this discussion. An overview of the Taguchi method (3) is shown in figure 2.

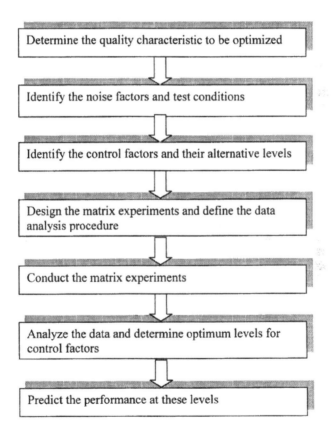

Figure 2: Flow chart of the Taguchi method

Surface finish optimization procedure

The chemical reactions, manufacturing/molding parameters and tooling have been identified as the main parameters that affect the surface finish of the final part. The main objective of this research is to focus on the optimization of manufacturing parameters, which include molding temperature, upper and lower mold temperatures, injection rate, rate of cure and injection temperature. Three levels of performance for each of the above-mentioned factors would be considered for this research. An L_{18} Taguchi experimental array has been selected for carrying out experiments, which means that 18 experiments are necessary to find out the impact of each factor and each level. The factors and their corresponding levels are:

Factors	Level 1	Level 2	Level 3
Molding temperature (A)	70 C	85 C	100 C
Upper and lower mold temperature (B)	5 C	10 C	20 C
Injection rate (C) (depends on resin)	N-10%	Nominal (N)	N+10%
Rate of cure (D) (depends on DSC tests)	N-10%	Nominal (N)	N+10%
Injection temperature (E)	N-5 C	Nominal (N)	N+5 C

In other words, 18 flat plates will be manufactured using different quantitative levels of the above mentioned factors and the surface finish of each plate will be measured using a *QMS oblique angle optical diffusion device*, which gives surface finish on a percentage basis from 0 to 99.9. The Taguchi quality characteristics which will be used for interpretation of the results is *bigger is better* (3). Statistical numbers such as capability indices (C_p, C_{pk}), mean squared deviation, signal to noise ratio, standard deviation, variance and the percent contribution of each factor level will be calculated (2,3).

Experimental planning

Many parameters are to be studied in order to master the RTM process. In this project, the surface finish of the final part is the key element to optimize. In order to do so, it is crucial to define a proper experimental plan to focus on the parameters that have the most influence on the end results.

Some mathematical models evaluate with statistical equations the different parameters involved in a process and permit the elaboration of an experimental plan to minimize the number of experiments and measurements. Using the Taguchi method 18 different experiments will be carried out with the factors and their corresponding levels listed in the following table:

Experiment #	Factor A	Factor B	Factor C	Factor D	Factor E
1	1	1	1	1	1
2	1	2	2	2	2
3	1	3	3	3	3

4	2	1	1	2	2
5	2	2	2	3	3
6	2	3	3	1	1
7	3	1	2	1	3
8	3	2	3	2	1
9	3	3	1	3	2
10	1	1	3	3	2
11	1	2	1	1	3
12	1	3	2	2	1
13	2	1	2	3	1
14	2	2	3	1	2
15	2	3	1	2	3
16	3	1	3	2	3
17	3	2	1	3	1
18	3	3	2	1	2

CONCLUSION

An experimental procedure based on the Taguchi method has been proposed to study the process parameters that influence surface finish in RTM process. The plan of experiments, which has been described in this paper, is in progress. The authors are working on the results and their interpretation. More experiments will be designed and carried out for the validation of results.

ACKNOWLEDGEMENTS

The authors of this paper acknowledge the financial and technical support by Ford motor company of Detroit, MI and the Networks of Centers of Excellence Auto21, Theme C (Materials & Manufacturing), Project C3 (Polymer Composites). Authors would like to thank M. DeBolt, D. Steenkamer, P. Blanchard from Ford and K. Kendall from Aston Martin for their technical advice and valuable support.

REFERENCES
Book References

1. Michael C. Niu, Composites airframe structures practical design information and data, Conmilit Press Ltd., Hong Kong, 1993.

2. Ranjit K. Roy, Design of experiments using Taguchi approach: 16 steps to product and process improvement, John Wiley and Sons, New York, NY, 2001.

3. Jiju Anthony and Mike Kaye, Experimental Quality: A strategic approach to achieve and improve quality, Kluwer Academic Publishers, Norwell, MA, 2000.

Edited Conference Proceedings

4. W. Krenkel, and V. Dollhopf, "RTM processing of high performance composites", Proceedings of ICCM-11, Gold Coast, Australia, 14th- 18th July 1997.

5. J. Ludick, "Resin Transfer Molding for missile shroud production", Proceedings of ICCM-11, Gold Coast, Australia, 14th- 18th July 1997.

Journal References

6. J.C.I. Chang, "Aerospace Materials and structural research into the next millennium", Composites applications and design, Volume 1, 134-158.

7. F.L. Matthews, C. Dutiro, and R.N. Alaka, "Factors controlling surface finish in resin transfer molding", SAMPE Journal, Volume 33 No. 5, 1997.

8. S. Gordon, R. Boukhili, and F. Trochu, "The effect of Low Profile Additives on the surface aspects of RTM molded composites" Proceedings of third Canadian international composite conference, Montreal, Canada, 21st –24th August, 2001.

9. Y.J. Huang, and C.C. Sue, "Effect of polyvinyl acetate and polymethyl methacrylate low profile additives on the curing of unsaturated polyester resins. Curing kinetics by DSC and FTIR", Journal of applied polymer science, 1995, Volume 55, 305-322.

10. C.J. Liu, M.S. Kiasat, A.H. Nijhof, H. Blokland, and R. Marissen, "The Effect of the Addition of a Low Profile Additive on the Curing Shrinkage of an Unsaturated Polyester Resin", Polymer Engineering and Science, Volume 39, 1999, 18-25.

11. M.Y. Lin, M.J. Murphy, and H.T. Hajn, "Resin transfer molding process optimization", Composites Part A: Applied science and manufacturing, 2000, 361-371.

12. N.R.L. Pearce, F.J. Guild, and J. Summerscales, "An investigation into the effects of fabric architecture on the processing and properties of fiber reinforced composites produced by resin transfer molding", Composites Part A, 1998, 19-27.

13. J.W. Rayle, and D.W. Cassil, "Advancements in injection in mold coating technology", Metal Finishing, 1995, 40-44.

14. G. Lebrun, R. Gauvin, and K.N. Kendall, "Experimental investigation of resin temperature and pressure during filling and curing in a flat steel RTM mold", Composite Part A, Volume 27A, 1996, 347-355.

Simulation of Process-Induced Residual Stresses in Thick Filament Wound Tubes

Chun Li, and Michael R. Wisnom
University of Bristol, Department of Aerospace Engineering, Bristol, BS8 1TR, UK

L. Graeme Stringer
Structures and Materials Centre, QinetiQ, Farnborough, Hants, GU14 0LX, UK.

ABSTRACT

The development of residual stresses throughout the cure process in thick composite components is strongly influenced by the cure history. Residual stresses in thick filament wound tubes can lead to delamination and intralaminar cracking during the cure. A major challenge in the manufacturing of filament wound structures is to have a fundamental understanding of the development of residual stresses.

A quasi three-dimensional finite element model has been developed to predict the process-induced residual stresses in thick composite cylinders. The focus of the process modelling of the residual stresses in the composite tubes is on the early stage of the cure, namely the cure cycle prior to cool-down, during which the residual stresses could cause delamination since the composite has low strength. The main identified factors affecting residual stresses such as heat transfer, resin cure kinetics, gelation, resin thermal expansion, chemical shrinkage and mandrel interaction have been integrated into the main model. The simulation identifies the important influence of the thickness of the tubes, the chemical shrinkage of the resin, and the mandrel interaction on the development of residual stresses. A good match between the model prediction and experimental results obtained at QinetiQ was obtained in most cases. A heated mandrel was proposed for reducing the residual stresses during the cure of the composite tube, which has been proven effective in the experimental work at QinetiQ.

271

INTRODUCTION

Residual stresses occurring in a composite tube during cure are generally undesirable. The tensile residual stresses occurring during manufacturing could have a significant effect on the mechanics and performance of composite structures by inducing warpage, interlaminar cracks, delamination, distortion and wrinkling of the fibres. These defects of the composite result in the degradation of its properties and even the failure of the whole structure. Previous attempts to manufacture 50 mm thick E-glass/MY750 filament wound tubes resulted in severe delamination.

Much work has been devoted to modelling the process-induced stresses generated in composites during the filament winding process [1-11]. Theoretical models for thermosetting composites have progressed from treating the filament wound material as a thin, linear elastic shell material to a thick, viscoelastic material of arbitrary cross-section with anisotropic properties. Although numerous studies have been performed on the effect of residual stresses on failure, damage tolerance, mechanical performance, and dimensional stability, the origins and development of residual stresses in composite is still not well understood. In this paper, the main reasons for residual stresses -- anisotropy, temperature and cure gradients, and tooling interaction were taken into account. An integrated model is proposed for predicting the process-induced residual stresses in thick filament wound tubes and investigating the causes of cracking. It is achieved by incorporating codes into a commercial finite element software ABAQUS and correlating the simulation results with the measured temperature profiles and strain history during the cure of the filament wound tubes.

SIMULATION OF RESIDUAL STERSSES

Delamination Prior to Cool-down

Previous studies on residual stresses in filament wound tubes suggested that delamination occurred at the early stages of cure. A tube experiment proposed by Plepys [12] was undertaken to measure the residual stresses generated in epoxy resin with three-dimensional constraints. MY750 resin was cured in a long thin-walled glass tube, which induced 3-D constraints on the resin. The state of the resin could be easily observed in the glass tube and the strains on the glass tube were recorded. The strains are directly related to the stresses developed in the resin. Cracks were observed before cool-down for a cure at 100°C. Meanwhile, drops in strain were recorded in the hoop direction at the middle and quarter of the tube when the temperature of the resin was still rising. No obvious voids or debonding was seen before the development of the cracks. The measurements and observations indicated that the stresses developed in the early stages of the cure can lead to cracks in the resin. In this study, the stress and strain development in the filament wound tubes before the cool-down stage is investigated.

Introduction to Simulation

A 50 mm thick 400mm long tube wound onto a 540mm long steel mandrel with solid steel end domes is modelled. The outer diameter of the mandrel is 100 mm, which is the same as the inner diameter of the composite. The wall thickness of the mandrel is 5mm. The composite tube on the mandrel cures at elevated temperatures with dwells at 90, 130 and 150°C, and then cools down to room temperature.

A quasi-three dimensional elastic model is proposed for the simulation. A radial slice of the model is taken in the simulation due to the symmetry in the circumferential direction. The mandrel dome ends are simplified as cylindrical of the same mass because the ends only affect the heat transfer. The nodes of each element of the axisymmetric body are actually nodal "circles" and the volume of material associated with the element is that of a body of revolution. The value of a prescribed nodal load or reaction force is the total value on the ring, which is the value integrated around the circumference. Regular axisymmetric solid elements are used for the analysis, representing the whole axially symmetric body.

The simulation includes two models, namely, a thermochemical model and a stress-strain model. The thermochemical model is not coupled with the stress-strain analysis. The thermochemical model is responsible for predicting the temperature and degree of cure history of the tube, and the temperature of the mandrel. The predicted results are then input into the stress-strain model. In the process-induced stress analysis, cure-dependent strains and stresses are predicted. The important properties used for the simulation are listed in Appendix.

Thermochemical Model

The temperature and cure history of the composite is calculated in the thermochemical model, which is essential for prediction of the residual stresses developed during the cure of composite structures. A cure kinetics model [13] provides the exothermal heat released by chemical reactions and predicts the degree of cure history. The heat released by the curing resin system is assumed to be directly proportional to the extent of cure, β, as indicated in Eqn. (1). Modulated DSC measurements at QinetiQ give an average value of total exothermic heat for MY750 epoxy of 278 kJ/kg.

$$\frac{dq}{dt} = \frac{H_t}{\beta_m} \frac{d\beta}{dt} \qquad (1)$$

Where β is the extent of cure, q is the heat flow, H_t is the total heat of reaction of the resin. The maximum degree of cure for MY750 for the cure cycle, β_m, is defined as 1.0.

Stress-Strain Model

Introduction to Stress-Strain Analysis

The stress-strain analysis is responsible for prediction of process-induced strains and residual stresses throughout the structure of interest. An ABAQUS [14] process simulation was developed to calculate residual stresses developed during the cure of the filament wound composite tube by integrating the output from the thermochemical model. The geometry and the mesh used for the stress-strain analysis are the same as for the thermochemical. There are three main submodels in the stress-strain simulation model, submodel UMAT for determining the cure-dependent material properties, submodel UEL for interface interaction between the tooling and the composite tube and submodel UEXPAN providing temperature and cure dependent strains.

Submodel of Volume Change

Volume changes of the material are temperature- and cure-dependent. The output from the heat transfer analysis is required to calculate the volume change of the composite due to thermal expansion and chemical shrinkage. The linear strain ΔL in direction i is calculated with Eqn(2). It is updated as a variable used for the calculation of the strain and stress in subroutine UMAT.

$$\Delta L_i = \alpha_i \times \Delta T - \frac{dV}{d\beta}_i \times \Delta \beta \tag{2}$$

In which, α_i and ΔT are the thermal expansion coefficient in direction i and the temperature change of the composite respectively; $dV/d\beta_i$ is the gradient of the chemical shrinkage of the composite in the direction i, which changes with degree of cure; $\Delta \beta$ is the degree of cure increment.

A coefficient of thermal expansion of $100.26*10^{-6}$ /K is used in the axial and radial directions and $5.16*10^{-6}$ /K in the hoop direction for the 90° filament wound E-glass/MY750 composite tube. Total chemical shrinkage of 6.1% is used for the matrix [15]. There is no chemical shrinkage assumed in the hoop direction because of the constraint of the fibres. Half of the total chemical shrinkage occurs in the radial direction and the other half in the axial direction.

Submodel of Composite Properties

The development of stresses and strains during the cure involves a complicated physical change and chemical reaction of the matrix, giving rise to phenomena such as reaction heat, cure shrinkage and the evolution of mechanical and thermophysical properties of the matrix. The submodel UMAT is responsible for the calculation of cure-dependent properties and prediction of the strains and stresses in the composite. A simplified approach is adopted to try to capture the basic mechanisms without requiring a large amount of experimental data. Rather than a continuous variation, two step changes in properties are incorporated at the gel point and modulus development point. The stresses prior to gelation are considered negligible because the resin is effectively liquid at this stage and any changes in volume of the resin are likely to lead to flow rather than development of real stresses in the composite. Once the resin gels, hydrostatic stresses are allowed to occur. Significant stresses can only arise when 3-D constraints exist. The shear moduli remain low until the modulus development point. This is defined as the point at which appreciable shear moduli develop. Fully relaxed elastic properties are used after this point. No viscoelastic behaviour of the resin is considered in the modelling because the resin is above its T_g before cooldown, and relaxation times are very short. A degree of cure of 0.25 was obtained for the gel point in washer experiments [15]. The modulus development point at a degree of cure of 0.45 was also determined experimentally based on readings during the cure from strain gauges embedded in the material [16].

Submodel of Mandrel Interaction

The mandrel has an important effect on the development of residual stresses during the cure of a composite tube. As observed in the cure of 50 mm thick tubes carried out at QinetiQ, sliding and sticking at the interface occurred during the cure, followed by separation during cool-down. This indicates that the contact between the composite and the mandrel is neither perfectly smooth nor fully locked. The contact of the mandrel is closely related to the thermal expansion, cure-related material properties, interface friction, and the development of stresses. A user defined interface with friction between the mandrel and the tube has been developed, which describes the separation, sticking and sliding occurring at the interface during cure. A friction coefficient of 0.5 is used for the interface based on measurements for gelled epoxy [17].

RESULTS AND PARAMETRIC STUDY

Correlation with Measurement Results

The temperature history of the 50mm thick 90° filament wound tube at the mid-plane is predicted as shown in Fig. 1a. The simulation shows that there is a significant lag in temperature between the inner and the outer region of the thick composite tube during the cure. There is an overshoot in temperature at the 130 °C dwell due to the exothermal heat from the chemical reaction of the resin during the cure. The simulation results of the temperature history are in reasonably good agreement with the experimental data as shown in Fig. 1b. The same trend of the strain development at the mid-length of the tube is predicted as the strain measurements obtained at QinetiQ, as shown in Fig. 2a and 2b. The simulation indicates that the chemical shrinkage is dominant after gelation, which induces radial tension in the tube. The mechanism of residual stress development in composite tubes is different from laminated composites. For cylinders the geometric constraint of the cylinder itself plays an important role. The stiffness of the fibres in the hoop direction effectively prevents free expansion and contraction during cure. This causes radial and hoop stresses. With the tube gelling from outside to inside, from both ends to the mid-length, the gelled outer layer and ends prevent the inside part contracting freely, which imposes tensile radial stresses at the inside of the tube. As indicated in the simulation, see Fig. 3, the maximum tensile stresses in the radial direction during cure occur around the quarter and central regions of the tube, suggesting that the 90°C filament wound tube would tend to initiate delamination in these regions. This is confirmed by the experimental observations by ultrasonic C-scanning of the 50 mm thick tube carried out at QinetiQ as shown in Fig. 4a. The tube contains extensive delamination concentrating in the quarter and central regions. The delamination covers virtually the entire length and circumference. The maximum radial stress of 1.56 MPa as calculated in the stress-strain simulation is not high but it may be enough to cause delamination in the partially cured composite with low strength.

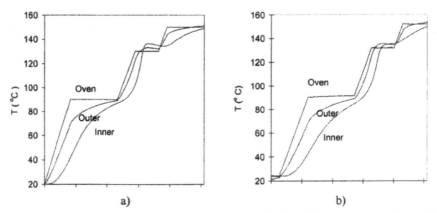

Figure 1 a) Predicted temperature history of 50mm thick 90° filament wound tube; b) Measured temperature history of 50mm thick 90° filament wound tube at QinetiQ

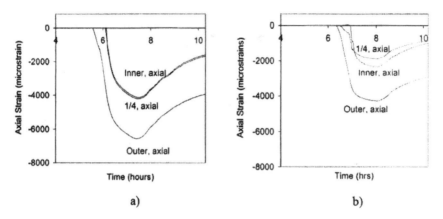

Figure 2 a) Predicted strain history in axial direction; b) Measured strain history in axial direction of 50mm thick 90° filament wound tube

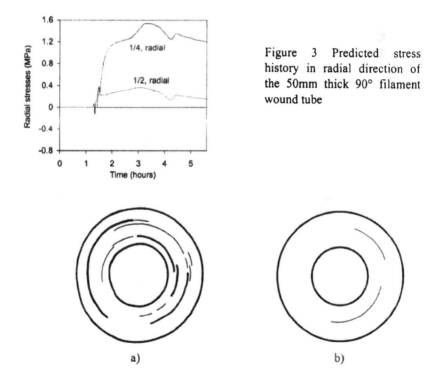

Figure 3 Predicted stress history in radial direction of the 50mm thick 90° filament wound tube

Figure 4 a) Delamination in 50mm thick FW tubes cured on a unheated mandrel; b) Delamination in 50mm thick FW tubes cured on a heated mandrel.

Parametric Study

The sensitivity of the model to the parameters and properties used for the simulation of the residual stresses during the cure of the composite tube is investigated. The cure of the 50mm thick E-glass/MY750 90° tube was employed for the parametric studies. The main parameters investigated in this part include gel point, modulus development point, thermal expansion coefficient of the composite and the mandrel, chemical shrinkage of the matrix, friction coefficient and separation stresses at the interface. The tensile stress in the radial direction is responsible for delamination of the filament wound tube, therefore the maximum radial stress is used as the criterion to investigate the influence of the parameters and properties. All the parameters apart from the modulus development point were changed by +/-50% in the simulation. The modulus development point in the parametric studies was decreased by 22% instead of 50% because the point should be higher than the gel point.

The parametric study indicates that the chemical shrinkage of the resin is the key to determine the residual stresses of the composite. The simulation indicates that the modulus development point and the thermal expansion coefficient also have a great influence on the development of the stresses. The gel point, thermal expansion of the mandrel, friction coefficient and separation stress at the interface have a comparatively small effect in the simulation for this particular composite tube following the standard cure cycle. The gel point of the composite has a low effect on the strains and stresses in the tube in that only the hydrostatic stresses are assumed to evolve after that point, which are low for the tube without full constraints. The thermal expansion of the mandrel has a small effect on the thick composite tube with low ratio of inside diameter versus outer diameter (ID/OD), which will be discussed in the following case study. The parameters to define the interface such as the friction coefficient and the separation stress have negligible influence on the model for this case. The parameters provide criteria at which the interface between a pair of nodes fails and therefore slip occurs. The magnitude of the parameters does not change the separation-stick-slipping nature of the interface. That means that a higher separation stress for instance would delay the failure at a pair of nodes but the built-up stresses would still lead to slip at the interface. The marginal sensitivity of the parameters indicates that it is not necessary to measure those data precisely.

CASE STUDY

Two typical cases are investigated to study the characteristics of the cure history and the development of stresses and strains evolving in the composite tubes, including cure of a 3mm thick hoop wound tube to show the effect of ID/OD, and a thick tube cured on a heated mandrel. The same material properties and the cure cycle are used for the following simulation.

Effect of ID/OD

The simulation shows that radial stresses are mainly compressive in the 3mm thick tube as shown in Fig. 5, indicating that cracking is not likely to occur. In comparison, there was mainly tension in the radial direction in the 50mm thick tube. This indicates tubes with high ratios ID/OD are less likely to initiate delamination. This is because of higher constraints and less contribution of the thermal expansion of the mandrel to the relaxation of the radial tension for the tubes with high ID/OD. Besides, thinner tubes cure more evenly with little temperature and cure gradient. This correlates with experimental observations that delamination occurred in the 50 mm thick tube but not in the 3mm thick tube. The prediction of the stain history also matches the observations that there is neither separation nor slip at the interface between the mandrel and the composite tube.

Cure on a Heated Mandrel

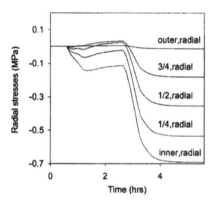
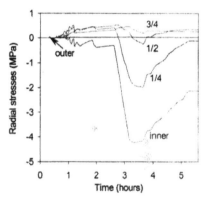

Figure 5 Predicted radial stress history of 3mm thick 90° filament wound tube

Figure 6 Predicted radial stress history of 50mm thick 90° filament wound tube, with the mandrel temperature following the standard cure cycle

One of the reasons for residual stresses during the cure of a composite tube is differential cure within the composite. Therefore, a way of reducing residual stresses was proposed by QinetiQ to achieve more uniform cure within the composite by using a heated mandrel. The simulation indicates that as shown in Fig. 6, the radial stresses are quite constant after gelation, with a maximum tension of 0.5 MPa, much lower than the stresses for the cure of the tube on an unheated mandrel. Compared with the cure of the tube on a unheated tube, less delamination occurs in the tube cured on a heated mandrel as shown in Fig. 4b. The delamination develops mainly in the middle of the tube as it cures from both the mandrel side and the outside surfaces. The simulation matches the observation of the experiments on the cure of the tubes at QinetiQ.

CONCLUSIONS

A quasi three-dimensional linear elastic model has been proposed for predicting the process-induced residual stresses in filament wound tubes. The simulation includes a heat transfer analysis and stress-strain analysis of the cure of the tubes for a standard cure cycle. Integration of cure kinetics, composite material properties and the interface model with the main model allows the main sources of residual stresses to be taken into account. The simulation matches the experimental results of temperature and strain histories obtained by QinetiQ in most cases. Parametric studies are carried out to identify the crucial parameters for the simulation, which are found to be the thermal expansion, chemical shrinkage and the modulus development point. The case study shows correctly the relatively high interlaminar tension in a 50 mm thick E-glass/Epoxy tube and the low stresses in a thin 3mm thick tube. This correlates with experimental observations that delamination occurred in the 50 mm thick tube but not in the 3mm thick tube. The prediction of cure on a heated mandrel indicates that lower residual stresses are developed during cure and thus there is less possibility to initiate delamination. This has been verified by experimental observations.

ACKNOWLEDGEMENTS

We would like to thank the Structures and Materials Centre, QinetiQ, Farnborough, Hants, UK for funding the work. Special thanks are given to Mr. Richard. J. Hayman and Mr. Lawrence Greaves for their contributions to the work.

REFERENCE

1   Calius, E. P., G. S. Springer, "Model of Filament-Wound Thin Cylinders", International Journal of Solids and Structures, Volume 26, No. 3, 1990, 271-297.

2.   Kim, Y. K.; S. R. White, "Cure-Dependent Viscoelastic Residual Stress Analysis of Filament-Wound Composite Cylinders", Mechanics of Composite Materials and Structures, No. 5, 1998, 327-354.

3.   Lee, S. Y. and G. S. Springer, "Filament Winding Cylinder: I. Process Model", Journal of Composite Materials, Volume 24, 1990, 1270-1298.

4.   Olofsson, K. S., "Stress Development in Wet Filament Wound Pipes", Journal of Reinforced Plastics and Composites, Volume 16, No.4, 1997, 372-390.

5.   White, S. R., Z. Zhang., "The Effect of Mandrel Material on the Processing-Induced Residual Stresses in Thick Filament Composite Cylinders", Journal of Reinforced Plastics and Composites, No. 12, 1993, 698-710.

6.   Gabrys, C. W. and C. E. Bakis, " Simplified Analysis of Residual Stresses in In-Situ Cured Hoop-Wound Rings", Journal of Composite Materials, Vol.32, No. 13, 1998, 1325-1343.

7.   Springer, G. S, "Modeling the Cure Process of Composites", SAMPE Journal, Volume 22., 1986, 22-27

8.   Calius, E, G. S. Springer, "Selecting The Process Variables for Filament Winding", National SAMPE Symposium and Exhibition (Proceedings), 1986, 891-899.

9.   Mantel S. C. and G. S. Springer, "Filament winding process models, Composite Structures", No. 27, 1994, 141-147.

10.   Cai Z. T. G and S. Allen, "Winding and Consolidation Analysis for Cylindrical Composite Structures", Journal of Composite Materials, No. 26, 1992, 1374-1399.

11.   Korovtkov V. N., Y.A. Chekanov and B. A. Rozenberg, "The Simultaneous Process of Filament Winding and Curing for Polymer Composites", Composites Science and Technology, No. 47, 1993, 383-388.

12.   Plepys, A. R., R.J. Farris, Evolution of residual stresses in three-dimensionally constrained epoxy resins, polymer, Volume 31, 1990, 1932-1936.

13.   Crosby. P. A. and G. F. Fernando, "Report to DERA Farnborough - Characterisation of the Cure of Ciba MY750 + HY917 + DY073", Sensors and Composites Group, Engineering Systems Department, Cranfield University (RMCS), 1999.

14.   ABAQUS Theory Manual, Version 6.2, Hibbitt, Karlsson & Sorensen, Inc.

15.   Chun Li, Kevin Potter, Michael R. Wisnom, Graeme L. Stringer, In-situ Measurement of Chemical Shrinkage of MY750 Epoxy Resin by a novel gravimetric method", submitted to Composites Science and Technology, April 2002.

16.   C. Li, M. R. Wisnom, G. L. Stringer, K. Potter, "Relationship between Chemical Shrinkage and Gelation of MY750 Epoxy Resin", 10th European Conference on Composite Materials, Brugge, Belgium, June 3-7, 2002.

17.   Chun. Li, Michael R. Wisnom, Graeme L. Stringer, Richard Hayman, Effect of Mandrel Contact on Residual Stresses during Cure of Filament Wound Tubes,

Eighth International Conference on Fibre Reinforced Composite, pp. 105-112, Newcastle upon Tyne, Sept. 2000.

APPENDIX: MATERIAL PROPERTIES

Table 1 Material properties for E-glass/MY750 composite

Properties	composite
Density at 20 °C, ρ (kg m^{-3})	2.135×10^3
Fibre Fraction, V_f (%)	69
Convective Coefficient, (Wm^{-2}K^{-1})	30
Specific heat capacity, C_p (Jkg^{-1} K^{-1})	931 ~1076
Through-thickness thermal conductivity, λ_2 and λ_2 (Wm^{-1} K^{-1})	0.464
Fibre direction thermal conductivity , λ_1 (Wm^{-1} K^{-1})	0.727
Fibre direction modulus , E_1 (GPa)	54.03
Transverse modulus , E_2 (MPa)	105.4
Through thickness modulus , E_3 (MPa)	105.4
Poisson's ratio , ν_{12}	0.32
Poisson's ratio , ν_{21}	0.000624
Poisson's ratio , ν_{13}	0.32
Poisson's ratio , ν_{31}	0.000624
Poisson's ratio , ν_{23}	0.99104
Poisson's ratio , ν_{32}	0.99104
Shear modulus ,G_{12} (MPa)	26.47
Shear modulus ,G_{13} (MPa)	26.47
Shear modulus ,G_{23} (MPa)	26.47

Direction 1 is the fibre direction, 2 and 3 are the transverse direction.

Table 2 Material properties for steel mandrel material

Properties	Mild Steel
Young's modulus (GPa)	207
Poisson's ratio	0.3
Coefficient of thermal expansion (K^{-1})	11×10^{-6}

Development of TTMSP to Predict the Performance of Discontinuous Natural Fibre-Polymer Composites

W.S. Lin, A. Pramanik and M. Sain
Wood Composite Group, Earth Science Center & Forestry
University of Toronto
Ontario, Canada

ABSTRACT

Natural fibre-polymer composites are becoming increasingly important for load-bearing application due to their unique benefit of lightweight and good mechanical performance. Polymers alone exhibit time dependent mechanical properties influenced by many environmental factors. More significantly, when natural fibres are present in the polymer matrices, the thermomechanical properties of such composites are changing by the environmental conditions and exhibit an internal and a global time dependent behavior. While durability analysis of the natural fibre-polymer composites is the main concern in real-life application, prediction methods are essential to be developed in evaluating their mechanical performance.

Both temperature and moisture diffusion are known to influence the time dependent mechanical properties of polymer based composites; however, conventional analysis method only consider the temperature factor only, instead of a concurrent analysis for both factors. Therefore, an evaluation technique is presented in this paper to describe the nonlinear viscoelastic behavior of natural fibre-polymer composites under concurrent temperature and moisture diffusion effects. Using an analogous approach to the TTSP, time-moisture superposition will be used to identify the moisture-dependent shift factors under an assumption that the effects of temperature and moisture can be decoupled and separately determined, and reassembled later on.

INTRODUCTION

The use of natural fibres as filler in polymer-based composites is rapidly advanced over the past years because they are abundant, recyclable, biodegradable, and economically sound. Due to their economical advantage, the use of natural fibres-polymer composites in load-bearing application is widely investigated, and various researchers (1,2) indicated that tensile strength and flexural modulus of the composites are increased with the percentage of natural fibre filler. Due to this nature, natural fibre fillers can also be used to tailor the performance of polymeric resins (3). However, although it is obvious that by incorporating natural fiber in polymeric resin, the hydroscopic properties of the composites will be altered, which in turn alters the mechanical performance of the composites, a quantitative analysis of such mechanism is not available; therefore, to quantify the hydroscopic effects on the mechanical performance of composites by formal mathematic modeling will be the primary motivation in this research.

The approach led to the time-temperature-moisture-superposition-principle (TTMSP) based on an assumption that the temperature and moisture effects can be evaluated separately. Using an analogous approach to the time-temperature-superposition-principle (TTSP) (4), the temperature shift factor (a_T) and moisture shift factor (a_M) are separately determined for various environmental conditions, a combination of different temperature and relative humidity levels. Then the temperature and moisture effects are coupled by reestablishing their relationship through addition of their induced strain values.

All previous investigations are based on the well knows Schapery's theory on nonlinear viscoelastic materials (5,6) such as Hoyle et al. (7) who developed a relationship of primary creep for structure applications of wood-base materials. Popelar et al. (8), and Strganac et al. (9) demonstrated the validity of above mentioned Schapery's theory and its modified version as a nonlinear constitutive model of thermoplastic polymers, and proved them to be accurately characterizing the material nonlinear response. In terms of experimental analysis, Dutta and Hui (10) used the TTSP to predict creep of fibre-reinforced plastics (FRP) using nonlinear viscoelastic parameters (based on Schapery's theory) for their constitutive model. For hygroscopic aging effects, Sain (11) demonstrated time-dependent mechanical properties of polymers influencing by temperature and stress with measurement of viscoelastic properties using time reflectometry. Dai (12) indicated that flake stress relaxation rates vary with the loading strain levels that are most likely related to characteristics of free volume change in cell walls of wood flake. However, summing up from all available investigations, the effect of viscoelasticity due to hygrothermal changes is not known.

ANALYTICAL APPROACH

Nonlinear Viscoelastic Theory

The nonlinear viscoelastic theory, also known as the Schapery's theory (5), presents a constitutive behavior, a stress-strain relation, of polymeric materials. The theory starts with a constitutive equation for uniaxial loading. According to the underlaying thermodynamic theory (13), an increment of work per unit initial volume equals to the follow:

$$Work\ per\ Unit\ Initial\ Volume = \sigma\delta\varepsilon \qquad [1]$$

where σ and ε are the corresponding stress and strain of the loaded composite slabs, and all uniaxial derived theories can be applied to other stress and strain pairs as long as the virtual work condition is satisfied. Due to the load bearing nature of the composite slabs, only equations relating to their creep behaviors will be presented in this review.

The creep compliance is defined as follow:

$$D(t) = \frac{\varepsilon(t)}{\sigma} \qquad [2]$$

where σ represents a constant applied load. The thermodynamic theory permits us to express the nonlinear material properties in strain (13) as follow:

$$\varepsilon(t) = g_0 D_0 \sigma + g_1 \int_0^t \Delta D(\psi - \psi') \frac{dg_2 \sigma}{d\tau} d\tau \qquad [3]$$

where
$$D_0 \equiv D(0) \qquad [4]$$
$$\Delta D(\psi - \psi') \equiv D(\psi - \psi') - D_0 \qquad [5]$$

In Equations [4] and [5], D_0 is the initial value of the creep compliance, $\Delta D(\psi - \psi')$ is the transient component of the creep compliance, and ψ is the reduced-time calculated as follow:

$$\psi = \int dt / a_\sigma \qquad \text{for} \qquad (a_\sigma > 0) \qquad [6]$$
$$\psi' = \psi(\tau) = \int dt / a_\sigma \qquad [7]$$

, and g_0, g_1, g_2, and a_σ are the material properties as a function of stress. Conventionally, if the applied load is sufficiently small, g_0, g_1, g_2, and a_σ can be assumed to be unity. However, in general these stress-dependent properties have specific thermodynamic significance and the changes in g_0, g_1, and g_2 reflect third and higher order dependence of the Gibb's free energy on the applied stress and can be analytically determined as follow (5):

$$g_0 = \frac{\sinh \sigma / \sigma_e}{\sigma / \sigma_e} \qquad [8]$$

$$g_1 g_2 = a_\sigma^n \frac{\sinh \sigma / \sigma_m}{\sigma / \sigma_m} \qquad [9]$$

where σ_e, σ_m, and n are constants which have values that depend on the particular material, e.g. for polyethylene, they are 400psi, 185psi, and 0.0890 (14), respectively.

Equation [3] can be simplified since only a single step load is applied to the specimen in this study:

$$\varepsilon(t) = g_0 D_0 \sigma + g_1 g_2 \Delta D(\psi) \sigma \qquad [10]$$

By substituting a constant stress into equation [3]:

$$\frac{dg_2 \sigma}{d\tau} = 0$$

except when τ equals to zero, it yields:

$$\varepsilon(t) = g_0 D_0 \sigma + g_1 g_2 \Delta D\left(\frac{t}{a_\sigma}\right) \sigma \qquad [11]$$

Equation [11] for nonlinear creep shows that initial elastic response is particularly linear even though the creep is strongly nonlinear, and the transient component of the creep $\Delta D(\psi)$ can be modeled by the power law as (5):

$$\Delta D\left(\frac{t}{a_\sigma}\right) = D_1\left(\frac{t}{a_\sigma}\right)^n \qquad [12]$$

where the constant D_l has values that depend on the particular material. Findley and Khosla (14) had reported the creep behavior of various unfilled thermoplastics which follows very closely to the creep equation:

$$\varepsilon = \varepsilon_o' \sinh \sigma / \sigma_\varepsilon + m't'' \sinh \sigma / \sigma_m \qquad [13]$$

where ε_o' and m' are material constants having the value of 1.530% and 0.397%, respectively, for polyethylene (16). By comparing Equations [10] and [11] with the solution of g_o, g_l, and g_2 from Equations [8] and [9], D_0 and D_1 in Equations [11] and [12] respectively can be determined as follow:

$$D_o = \frac{\varepsilon_o'}{\sigma_\varepsilon} \qquad \text{and} \qquad D_l = \frac{m'}{\sigma_m} \qquad [14 \text{ \& } 15]$$

TTSP Method

Temperature-Time-Superposition-Principle (TTSP) method is widely used for extrapolating experimental measurements made at short times to the mechanical response of a polymeric material at an extended times. It is called the method of *reduced variables*, a tool for predicting the material behavior at one certain time (or frequency) and temperature from experiments performed at some other time (or frequency) and some other temperature (17), and thus it serves to expand the time scale.

Ferry's method of reduced variables in viscoelasticity (18) provides an analytical method by a temperature-dependent shift factor, a_T, through a shift function. The shift function is a very sensitive function of the temperature, and the explicit form of a_T will depend upon the reference temperature T_o selected. Since this T_o is completely arbitrary, it is insightful to select some temperature which is characteristic of a polymer. Therefore, the glass transition temperature, T_g, is often taken as the reference temperature, and thus, a_T becomes a universal function of temperature (19).

$$\log(a_T) = \frac{-C_1(T - T_g)}{C_2 + (T - T_g)} \qquad [16]$$

Above relation holds from T_g to (T_g + 100°C), and the constants C_1 and C_2 are the approximate universal constants with values 17.4 and 51.6 (20).

Free Volume Theory

The study of Free Volume Theory (FVT) refers to the influence of small changes of volume, which associates with modifications of molecular mobility, a "structural" contribution, for the viscoelastic materials. Since accommodations of molecule chains need sufficient inter-molecular space for differential motion that occurs with time such as to produce the viscoelastic mechanical response (21), this time dependent characteristic is a very sensitive reaction. Thus, even if there is only a small volume change, a very large change in the time scale of macroscopic deformations can be produced. Accordingly, this volume change must directly affect the interstitial space that molecules need to respond to macroscopically imposed stress and deformations, effects of material dilatation due to solvents, temperature changes, or mechanical stresses.

Free volume is defined as the region that is accessible to the center of a molecule through a drifting motion that does not require interaction with its neighbor molecules (22). For the purpose of deriving the FVT model, an assumption is made that all molecular motion mechanisms are affected by volume changes equally; therefore, no change in the phenomenological distribution functions of relaxation or retardation times (21). Moreover, the viscoelastic equations are identical to those of linear temperature-dependent viscoelasticity except the shift function, a_M, which will be an instantaneous functional of the free volume in terms of solvent concentration, c:

$$a_M = a_M\{c\}$$ [17]

Dolittle had (22,23) empirically derived the shift function in terms of the free volume based on the phenomenological time-temperature superposition principle:

$$\log(a_M) = B\left(\frac{1}{f} - \frac{1}{f_o}\right)$$ [18]

where f_o is the reference (fractional) free volume at some reference temperature, in which the time-dependent (linear) material properties are measured, and f is the fractional free volume assuming a linear functional of c at any instant time:

$$f = \frac{V_f}{V}$$ [19]

where V_f is the free volume, and V is the total volume.

For a better understanding of the FVT, it may also be visualized through a mechanical analog for viscoelastic behavior as shown in Figure 1:

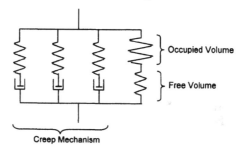

Figure 1: Mechanical Analogue of the Constitutive Behavior of Volumetric Influence. (23)

Seefried and Koleske (24) indicated that if it is assumed that the fractional free volume of the polymer increases linearly above the T_g, then the relation of T_g with f can be written as follow:

$$f = f_g + \alpha_f(T - T_g)$$ [20]

where α_f is the thermal coefficient of expansion of the fractional free volume above T_g ($\alpha_f = 2.10 \times 10^{-4}\ K^{-1}$ for polyethylene at 298 K). The combination of Equations [18] and [20] yields:

$$\log(a_M) = \frac{-B}{f_g}\left[\frac{T - T_g}{(f_g/\alpha_f) + (T - T_g)}\right]$$ [21]

Comparison shows that Equation [21] is identical with Equation [16] where:

$$C_1 = \frac{B}{f_s} \qquad \text{and} \qquad C_2 = \frac{f_s}{\alpha_f} \qquad [22 \ \& \ 23]$$

EXPERIMENTAL

A flexural creep test is designed based on the standard test method of ASTM D 6112. The four-points loading configuration will be used because the plastic lumbers are expected to be relatively ductile and thus do not normally fail by the maximum strain (3%) under the three-points loading. The plastic lumber specimens tested are commercial railing products in square shape with sides of 1.5 inch and thickness of 1/4 inch, which consist a resin matrix of recycled HDPE and filled with natural fiber. The cellulose fiber content of the tested plastic lumbers is about 50%. All test specimens are employed in the "as-manufactured" form in all cases.

The test rack is used to provide support of the test specimen at both ends with a span equal to sixteen times the depth of the specimen with tolerant of plus four and minus 2, and consequently the support span of the test rack in this case will be:

$$L = 16 \times 1.5 \ inch = 24 \ inch$$

According to the test standard, the span for the loading beam will be one-third the length of the support span and locates mid-span of the test specimen. The noses of both the support and loading beam are configured with cylindrical surfaces with a radius of 0.5 inch in order to avoid excessive indentation of the specimen. In order to allow for overhanging, at least 10% of the support span should be maintained at each test specimen ends. Therefore, the total specimen span must be at least 30 inch in all time. Finally, the deflection of the specimen is measured at the midpoint of the load span at the bottom face of the specimen with a set of electrical displacement transducers (Figure 2) connected to an on-line computer controlled data acquisition system.

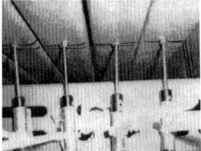

Figure 2: Displacement Transducers.

The specimens are preconditioned in the test environment for forty-eight hours to ensure an established equilibrium condition prior to be being tested. Two extreme cases of environment conditions are being tested, which are normal room condition (20°C and 34%RH) and maximum humidity (40°C and 92%RH).

The flexural creep is started by properly mounting the specimens on the creep fixture of flexural creep rack. A pre-selected load of 70 lbf is applied rapidly and smoothly to the specimen, preferably in 1 to 5 s, and start the timing at the onset of loading. The flexural deflection of the specimen is measured at the bottom of the test specimen at the midpoint of the load span in accordance with the approximate time schedule of 1, 6, 12, and 30 min, and 1, 2, 5, 20, 50, 100, 200, 500, 700, and 1000 hr.

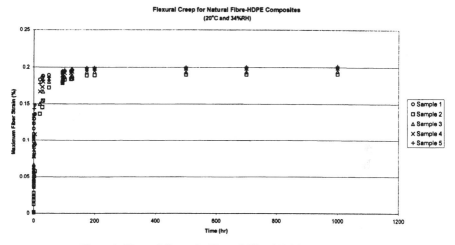

Figure 3: Flexural Creep for Natural Fibre-HDPE Composites.

CONCLUSION

The model of TTMSP had been successfully developed based on the Schapery's nonlinear viscoelastic theory. The most significantly contribution of the TTMSP was that it provided an analytical approach to incorporate both the time-temperature and time-moisture effects in the long-term performance of the natural fibre-thermoplastic composites. The model of the time-temperature characteristic was adapted from the well-established TTSP method, and that for the time-moisture was developed based on the free volume theory of viscoelastic materials as a function of moisture content in different relative humidity levels. One the shift factors for both the temperature and moisture effects were determined, the mechanical properties over time of the composites could be predicted by determining the coupling strain effects due to the combined condition.

REFERENCE

1. Woodhams, R.T., Law, S., and Balatinecz, J.J., Proceedings of Wood Adhesives Symposium, May 16-18, Madison, Wisconsin, 1990, pp. 177-182.
2. Meyers, G., and Clemens, C., *Wastepaper in Plastic Composites Made by Melt Blending: Demonstration of Commercial Feasibility*, USDA Forest Service, Madison, WI, 1993.
3. Balatinecz, J.J., and Woodhams, R.T., Journal of Forestry, 91, pp. 22-26, 1993.

4. Ferry, J.D., *Viscoelastic Properties of Polymers – Chapter 11: Dependence of Viscoelastic Behavior on Temperature*, John Wiley & Sons, Inc., 1985.
5.
6. Schapery, R.A., Polymer Engineering and Science, July, Vol. 9, No. 4, pp. 295-310, 1969.
7. Hoyle, R.J., Griffth, M.C., Itani, R.Y., Primary Creep in Douglas-Fir Beams of Commerical Size and Quality, Wood and Fibre Sci., 17(3): 300-314, 1985.
8. Popelar, C.F., Popelar, C.H., and Kenner, V.H., Polym. Eng. Sci, 30, p.577, 1990.
9. Strganac, T.W., Letton, A., Payne, D.F., and Biskup, B.A., (1995) AIAA J. 5, 904-910.
10. Dutta, P.K. and Hui, D., Computers and Structures, 76(2000), 153-161.
11. Sain, M., Design, Manufacturing and Application Composites – CANCOM 2001, pp. 335-343.
12. Dai, C., Wood and Fiber Science, 33(3), 2001, pp. 353-363.
13. Passaglia, E., and Knox, J.R., *Engineering Design for Plastics: Chapter 3 – Viscoelastic Behavior and Time-Temperature Relationships*, ed. Baer, E., Van Nostrand Reinhold Company, N.Y., 1964.
14. Ferry, J.D., J. Am. Chem. Soc., 72, 3746 (1950).
15. Catsiff, E., Offenbach, J., and Tobolsky, A.V., J. Colloid Sci., 11, 48 (1956).
16. Williams, M.L., Landel, R.F., and Ferry, J.D., J. Am. Chem. Soc., 77, 3701 (1955).
17. Knauss, W.G., and Emri, I., Polymer Engineering and Science, Mid-January, 1987, Vol. 27, No.1, pp. 86-100.
18. Losi, G.U., and Knauss, W.G., Polymer Engineering and Science, April 1992, Vol. 32, No. 8, pp. 542-557.
19. Doolittle, A.K., J. Appl. Phys., 22, 1471 (1951).
20. Seefired, C.G., and Koleske, J.V., *Plastics Polymer Science and Technology: Chapter 7 - Mechanical Behavior*, ed. Baijal, M.D., Wiley-Interscience, 1982.
21. Simpson, W., Wood and Fiber, 12(3), 1980, pp. 183-195.
22. Brunnauer, S., Emmett, P.H., and Teller, E., J. Am. Chem. Soc., 60, 309-319 (1938).
23. Hartley, I.D., and Schneider, M.H., Wood Sci. Technol. 27: 421-427 (1993).
24. Meyer, J.A., Wood polymer materials. In: Rowell, R.M. (ed.) The chemistry of solid wood. Advances in chemistry series 207. Washingon, D.C.: American Chemical society, pp. 257-189, 1984.
25. Schneider, M.H, Brebner, K.I., and Hartley, I.D., Wood Fiber Sci. 23(2): 165-172, 1991.
26. Knauss, W.G., and Kenner, V.H., J. Appl. Phys. 51(10), October 1980, pp. 5131-5136.

27. Maskavs, M., Kalnins, M., Reihmane, S., Laka, M., and Chernyavskaya, S., Mech. of Comp. Mat., Vol. 35, No. 1, 1999, pp. 55-62.
28. Royers, C.E., *Engineering Design for Plastics: Chapter 9 – Permeability and Chemical Resistance*, ed. Baer, E., Van Nostrand Reinhold Company, N.Y., 1964.
29. Plushchik, O.A., and Aniskevich, A.N., Mech. of Comp. Mat., Vol. 36, No. 3, 2000, pp. 233-240.
30. Balachandar, M.A., Iyer, C.V., and Raghavan, J., Design, Manufacturing and Application Composites – CANCOM 2001, pp. 319-326.

Scaling Phenomenon in the Dynamic Analysis of Composite Beams

D.-H. Xu, X.-X. Jiang and D. Nikanpour

Advanced Material & Thermal Group
Space Technology/Engineering
Canadian Space Agency
6767 Route De L'Aeroport
St-Hubert, QC, J3Y 8Y9
Canada

ABSTRACT

Shape Memory Alloy (SMA) is a type of smart material which is currently being investigated for the application in control of the dynamic mechanical response of various types of flexible structures. In this paper, composite cantilever beams were fabricated by attaching SMA (Nitinol) strips on the top and the bottom surfaces of an aluminum base beam. Their dynamic responses were subsequently studied both theoretically and experimentally. After the finite element model had been proved by the experiment to be valid and accurate, it was applied to study the influence of the length of SMA strips on the natural frequencies of the composite beams. The results clearly shows that there exists a scaling effect in such composite beams in terms of natural frequency and the length ratio of SMA strips to the aluminum base beam. This implies that the natural frequencies of big such composite cantilever beam can be obtained in practice by testing in a laboratory a small beam with the same length ratio of SMA strips to base beam.

Keywords: Natural Frequency, Scaling, Composite Beam, Finite Element Method

INTRODUCTION

Although shape memory alloy (SMA) has a slower response than piezoelectric materials, it has been seen growing use in mechanical and aerospace industries because of its high strain/stress output. Because of this high strain/stress high output, SMA has been widely used in shape and vibration control of structures, especially the flexible structures, since the late 1980s [1-4]. Our research work that leads to the study of the natural frequencies of composite beams began with the vibration control of flexible beam with SMA reinforced. Besides these works on the shape and vibration control of composite beams, there are some works on the natural frequencies of the composite beams in recent years. J.R. Banerjee [5] presented exact expressions for the frequency equation and mode shapes of composite Timoshenko beams with cantilever boundary condition. The method of power series expansion was developed by H. Matsunaga [6] to estimate the natural frequencies of laminated composite beams. It is noticed that the mechanical properties of the beams discussed above are anisotropic on the cross section but these properties along the length of the beams keep unchanged.

In this paper, the beam consists of a homogeneous aluminum base beam and two pieces of nitinol strips. The nitinol strips cover only a portion of the length of the whole composite beam. Thus, the dynamical properties of the composite beam will change with the length of the nitinol strips. The objective of this paper is to investigate the influence of the strip length on the natural frequencies of this composite beam. By comparing this influence among the composite beams with different lengths, it is found that there exists a frequency scaling effect between short and long beams.

ANALYSIS MODEL

Model

The composite beam studied here is a cantilever beam and formed by an aluminum base beam and two pieces of nitinol strips. These two strips are respectively attached by glue to the top and bottom surfaces of the base beam as shown in figure 1 (a). Figure 1 (b) is the cross section of the composite part that consists of the aluminum beam, the glue layers and the nitinol strips. The cross section is rectangular and assumed to be symmetric about the axis z. Axes x and z form the symmetric plane of the beam. Thus, we have the following equations if we take t_{al}, t_{gl}, and t_{ni} to present the thickness of the aluminum beam, the glue layer and the nitinol strip.

$$\begin{aligned}
h_1 &= t_{al} \\
h_2 &= t_{al} + 2t_{gl} \\
h_3 &= t_{al} + 2t_{gl} + 2t_{ni}
\end{aligned} \tag{1}$$

According to elastic theory [7], the expression for the bending stiffness of the aluminum beam k_{al} is written as follows.

$$k_{al} = \int E_{al} y^2 dA_{al} = \frac{1}{12} bt_{al}^3 E_{al} \tag{2}$$

where k_{al} is the bending stiffness of aluminum beam, E_{al} is the elastic modulus of aluminum, A_{al} is the area of the cross section of aluminum beam and b is the width of the beam.

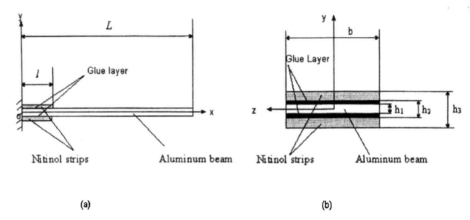

(a)   (b)

Figure 1: Figure 1 (a) is the sketch of the composite beam and figure 1 (b) is the sketch of the cross section of the composite part of the beam.

From the equation (2), the bending stiffness of the composite part of the beam can be expressed like the following equation.

$$k_{com} = \int E_{al} y^2 dA_{al} + \int E_{gl} y^2 dA_{gl} + \int E_{ni} y^2 dA_{ni} \tag{3}$$

where k_{com} is the bending stiffness of the composite part of the beam. E_{gl} and E_{ni} stand for the elastic modulus of glue and nitinol respectively. A_{gl} is the cross section area of glue layers and A_{ni} is for that of nitinol strips.

Because of the symmetry of the cross section, equation (3) can be re-written like equation (4).

$$
\begin{aligned}
k_{com} &= 2b\left(\int_0^{\frac{h_1}{2}} E_{al} y^2 dy + \int_{\frac{h_1}{2}}^{\frac{h_2}{2}} E_{gl} y^2 dy + \int_{\frac{h_2}{2}}^{\frac{h_3}{2}} E_{ni} y^2 dy\right) \\
&= \frac{b}{12}[E_{al} h_1^3 + E_{gl}(h_2^3 - h_1^3) + E_{ni}(h_3^3 - h_2^3)]
\end{aligned}
\tag{4}
$$

where b is the width of the beam, h_1, h_2 and h_3 are as shown in figure 1 (b).

The mass per unit length of the composite part is defined as m_{com} and its expression is shown below.

$$
\begin{aligned}
m_{com} &= \int \rho_{al} dA_{al} + \int \rho_{gl} dA_{gl} + \int \rho_{ni} dA_{ni} \\
&= b[\rho_{al} h_1 + \rho_{gl}(h_2 - h_1) + \rho_{ni}(h_3 - h_2)]
\end{aligned}
\tag{5}
$$

and

$$
m_{al} = \rho_{al} A_{al} \tag{6}
$$

where ρ_{al}, ρ_{gl} and ρ_{ni} are the mass density of aluminum, glue and nitinol respectively and m_{al} is the mass per unit length of the aluminum beam. The subscription al, gl, and ni represent aluminum, glue and nitinol respectively.

Geometry Parameters and Material Parameters

Some of the geometry and material parameters which will be used in calculation are listed in table 1. The beam is 13mm wide and at least 180mm long so that the beam can be taken as a Bernoulli-Euler beam considering its width and thickness.

Table 1. Geometry and material parameters for calculation

Material	Thickness (mm)	Elastic modulus (GPa)	Density (kg/m^3)
Al (beam)	0.79	70	2700
SMA (strip)	0.2	28.27	6450
Glue	0.1	4.6	1450

CALCULATION RESULTS

Verifying the Analysis Results

Nastran, a commercial finite element analysis software, is employed to perform the analysis task. There are two kinds of beam elements used in calculation. One is the aluminum beam element and another is the composite beam element. The expressions for their bending stiffness are shown in equation (2) and equation (3).

Before performing analysis, we tested the natural frequencies of two specimens to verify the finite element results. The parameters of the specimens and the results of the first natural frequencies obtained by test and analysis are shown in table 2.

Table 2. Comparison of results between test and analysis

	L (mm)	l (mm)	t_{gl} (mm)	E_{ni} (GPa)	f_{exp} (Hz)	f_{ana} (Hz)
Specimen 1	195.0	85.0	0.175	32.0	27.1	27.31
Specimen 2	180.0	40.0	0.1	28.27	25.25	25.45

where the letter l is used to represent the length of the nitinol strips and the capital letter L stands for the length of the whole beam. f_{exp} is the natural frequency of the composite beam obtained by experiment and f_{ana} is that obtained by finite element analysis.

The results in table 1 shows obviously that the difference between the natural frequencies obtained by experiment and analysis is very small. The maximum of the relative error is less than 0.8%. Therefore, the finite element method used here is effective.

The Influence of the Length of Nitinol Strips on the Natural Frequency of the Composite Beams

The whole length of the beam is L and the length of nitinol strips l is theoretically from 0 to L. Figure 2 (a) illustrates the first natural frequency of the composite beam whose whole length is $L=180$mm changes with the length of the nitinol strips. We also calculated other beams with different total length of 300mm and 400mm. The results are shown in figure 2 (b) and figure 2 (c) respectively. From figure 2, it is seen that the natural frequency of the composite beam increases with the length of the strips until it reaches its maximum. Then, the natural frequency decreases with the strip length increasing.

Scaling Effect

From figure 2, it is also seen that the pattern of the curves reflecting the relation between the natural frequency and the length of the nitinol strips are almost same even if the whole length of the beam changes from 180mm to 400mm. Here, we define two ratios as follows.

$$RL = \frac{l}{L} \qquad (7)$$

$$RF = \frac{f}{f_0} \qquad (8)$$

where RL is the ratio of the length of nitinol strips to the whole length of the beam and RF is the ratio of the natural frequency of the composite beam f to that of the aluminum beam without nitinol strips f_0 ($l = 0$). According to equation (7) and (8), the natural

frequencies of the aluminum beams with length of L=180mm, 300mm and 400mm are
20.078Hz, 7.228Hz and 4.066Hz respectively.

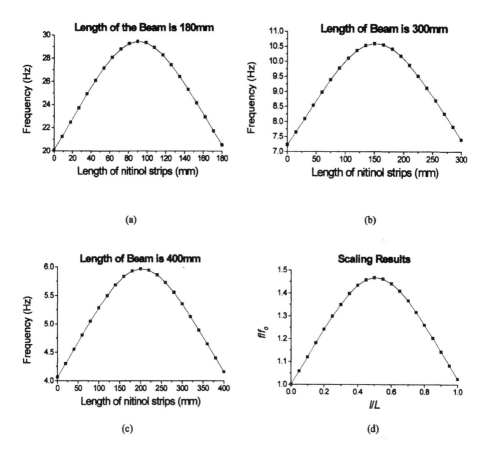

Figure 2: The influence of the length of the nitinol strips on the natural frequency of the
beams with different length. By comparing the curve pattern of figure 2 (a), (b)
and (c), the scaling results as shown in figure 2 (d).

If we take RL as the abscissa and RF as the ordinate and re-plot figure 2 (a) (b) (c), we will find that the three curves are actually the same one as shown in figure 2 (d). This discovery means that there is a scaling effect existing in natural frequencies between short and long composite beams if their material and geometry parameters are the same except the length. If we have

$$\left(\frac{l}{L}\right)_{sb} = \left(\frac{l}{L}\right)_{lb} \tag{9}$$

, then we will have

$$\left(\frac{f}{f_0}\right)_{sb} = \left(\frac{f}{f_0}\right)_{lb} \tag{10}$$

where the subscription sb stands for short beam and lb stands for long beam.

DISCUSSION

The curves representing the relation of the length of the nitinol strips and the natural frequency of the composite beams are "increasing-decreasing" ones. These "increasing-decreasing" curves show that there are two effects of the nitinol strips on the natural frequency of the composite beam. One is the stiffness effect and another one is the mass effect. When the strips are put on the beam, they do not only give the beam an extra stiffness but also an extra mass. For this cantilever beam, the stiffness effect dominates when the value for the length of strips is small. In the region that the stiffness effect dominates, the natural frequency increases with the length of the strips. In the region dominated by mass effect, the natural frequency of the composite beam decreases with the length of the strips increasing.

CONCLUSION

The analysis results show that there exists a scaling phenomenon in the natural frequency of composite beams. With this phenomenon, the natural frequency of long composite beams in practice can be evaluated by short beams in a laboratory if the short and long beams have the same length ratio of the nitinol strips to the base beam.

ACKNOWLEDGEMENTS

The authors would like to thank for the helpful discussion at the preparation stage of this paper with Professor S. Shrivastava who is working at Department of Civil

Engineering of McGill University. The authors also appreciate highly the help with the experiment from Mr. M. Mullavey who is a student at McGill University. The first author would like to thank NSERC for its Visiting Fellowship.

REFERENCES

1. D. Inman, Vibration with Control, Measurement and Stability, Prentice-Hall, Englewood Cliffs, NJ, 1989
2. G. Song, B. Kelly and B.N. Agrawal, "Active Position Control of a Shape Memory Alloy Wire Actuated Composite beam", Smart Mater. Struct., 9, 2000, 711-716.
3. L. Leng, A. Asundi and Y. Liu, "Vibration Control of Smart Composite Beams with Embedded Optical Fiber Sensor and ER Fluid", Journal of Vibration & Acoustics, 121, 1999, 508-509.
4. J.R. Banerjee, "Frequency Equation and Mode Shape Formulae for Composite Timoshenko Beams", Composite Structures, 51, 2001, 381-388.
5. A. Baz, K. Iman and J. McCoy, "Active Vibration Control of Flexible Beams Using Shape Memory Actuators", Journal of Sound and Vibration, 140, 1990, 437-456.
6. H. Matsunaga, "Vibration and Buckling of Multilayered Composite Beams According to Higher Order Deformation Theory", Journal of Sound and Vibration, 246, 2001, 47-62.
7. S. Timoshenko, and J.N. Goodier, Theory of Elasticity, 2nd ed., McGraw-Hill Book Company, New York, 1951

FINITE ELEMENT MODELING OF A TRIAX COMPOSITE MEMBRANE SECTOR

QI ZHAO
Concordia Center For Composites
Department of Mechanical and Industrial Engineering,
Concordia University
Montreal, Quebec, Canada H3G 1M8

*S. V. HOA
Concordia Center For Composites
Department of Mechanical and Industrial Engineering,
Concordia University
Montreal, Quebec, Canada H3G 1M8

hoasuon@vax2.concordia.ca

R. MOUDRIK
EMS Technology Ltd.
Montreal, Quebec, Canada

ABSTRACT

Some reflectors for communication purposes are made of the triaxial woven fabric (TWF) composites. The stress analysis of a membrane sector of the reflector subjected to pressure is presented here using Superelement 2 developed previously in [2,3]. Since the full-scale membrane sector is too large to model by finite element method using a personal computer, an investigation was made on the effect of the size of the sector of similar shape on the deflection and stresses. A similitude law was proposed. This paper presents these results.

---

* Author to whom correspondence should be addressed.

INTRODUCTION

Satellite dishes incorporate a parabolic or circular curve into the design of their bowl-shaped reflectors. The parabolic curve has the property of reflecting all incident rays arriving along the reflector's axis of symmetry to a common focus located to the front and center. The parabolic antenna's ability to amplify signals is primarily governed by the accuracy of this parabolic curve. The spherical antenna with circular curve creates multiple focal points located to the front and center of the reflector, one for each available satellite. The curvature of the reflector is such that if extended outward far enough along both axes it would become a sphere. Poor antenna performance often results from construction errors. For mesh antennas, they also are highly susceptible to environmental effects. Heavy wind storms, for example, can loosen the clips holding the mesh to the frame and distort the curve from its original shape or even blow out one or more of the mesh panels. The ART EM reflector is the mesh spherical antenna of the communication satellite. It is constructed of ribs and membrane surface, shown in Figure 1. The membrane surface is a part of a spherical surface between two inner and outer circles. The sphere radius is 111 inch. Its outer perimeter diameter is D=57 inch and inner perimeter diameter is d=20 inch. The entire membrane is divided uniformly into 12 sectors by ribs. Each membrane sector is a 30^0 degrees angle part of the spherical surface [1]. Compared with ribs, the membrane sector is a very weak load-bearing structural part. It is necessary to investigate its mechanical behavior. The ribs and membrane surface of the ART EM reflector is made of the triaxial woven fabric (TWF) composites. The reason for this construction is due to lightweight, and also due to the facility to allow air to escape through the holes during launching. This air escape reduces significantly the air impact, which significantly reduces the loading on the structure.

The TWF composite is composed of three sets of yarns, which intersect and interlace with each other at 60^0 angles. As shown in Figure 2, the entire fabric can actually be obtained by assembling many unit cells. There are many ways to identify a unit cell in the structure of Figure 3. One type of unit cell is shown in Figure 3. This was modeled as Superelement 2 in the finite element development [2]. Superelement 2 is a 15-node superelement constructed of six identical 16-node 3D isoparametric elements and three identical 4-node 2D isoparametric laminate elements. There are 12 DOFs at each node. This element takes into account the geometric and material properties of the twisted yarns. The assembly is done by the pseudo element technique and the static condensation procedure [2]. In this paper, the stress analysis of a membrane sector of ART EM reflector subjected to pressure is performed using Superelement 2. This paper presents the finite element model and the deformed configuration of the membrane sector.

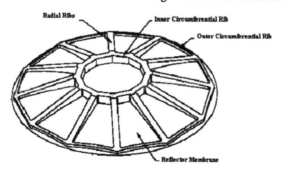

Figure 1 EM reflector bottom

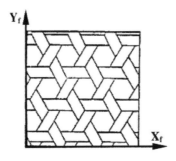

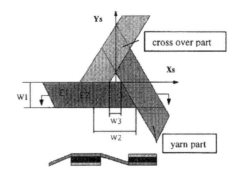

Figure 2: Triaxial woven fabric   Figure 3: A unit cell (superelement)

FINITE ELEMENT MODEL

Consider a finite element model of the membrane sector subjected to pressure. A pressure P = 200,000 Pa is applied on its inner surface. Since ribs surround the membrane sector, ribs' effect on the sector must be taken into account. Compared with the ribs, the membrane sector is much weaker in stiffness than the ribs. So the membrane sector can be assumed being clamped on its boundaries.

Through the automatic meshing program, the membrane sector is discretized into 10,603 elements and 96,269 nodes. Since there are 12 DOFs at each node, 96269*12 unknowns will be solved for this problem. This will require the computer to have ability to handle 96269*12 by 96269*12 matrix. The computer should have at least 96269*12 * 96269*12*32 bits memory because calculation must be performed in double precision condition. Even if some memory-saving skills are considered, it still is a large-scale problem that needs much more memory and can hardly be processed in a personal computer. Therefore, this problem will have to be simplified before a solution can be obtained. As we know, for continuum and homogeneous problem, one may reduce the size of the object with similitude law and solve the problem of the size-reduced object to obtain the solution of the full-scale problem. But for the TWF composite with open holes, this strategy is not suitable. If the structure of the TWF composite is reduced in size, the unit cells built for it also are reduced in the same proportion. The number of the unit cells of size-reduced structure is still the same as one of full-scale structures. Since the element is set up on modeling a single unit cell, that means that the number of elements is not reduced in finite element model of the size-reduced structure. So the similitude law is not suitable for accurate solution of the TWF composite with open holes. We have taken a different approach, that of studying different sizes of similar shape sectors and extracted from the results a similitude rule.

NUMERICAL RESULTS AND DISCUSSIONS

In this paper, we will make an approach by solutions of several size-reduced membrane sectors. Membrane sectors of sizes 1/12, 1/ 10 and 1/8, 1/7 and 1/6 of full scale are studied. The pressure still is P = 200,000 Pa applied to these sectors. Since the sector thickness is reduced for these examples, Superelement 2 also is reduced correspondingly in thickness, i.e., yarn thickness is reduced in proportion, but its plane dimensions are kept the same. We will investigate the mechanical behavior of these different-scale membrane sectors subjected to pressure.

Table 1 gives the geometric parameters of the unit cell (Superelement 2) measured from the microscopic photographs of the TWF composites. Table 2 lists the fiber and matrix properties for the fiber volume fraction $V_f = 0.695$. Table 3 shows the effective elastic properties of the impregnated yarn obtained by using the composite cylinder assemblage (CCA) model. These data are used for finite element models.

Table 1: Geometric parameters

Geometric parameters	Measured values
Yarn thickness (mm)	0.14
Yarn width w1 (mm)	0.85
w2 (mm)	1.10

Table 2: Elastic properties of fibers and matrix

Material	E_L (GPa)	E_T (GPa)	G_{LT} (GPa)	G_{TT} (GPa)	v_{LT}	ρ (gm/cm^3)
Carbon Fiber	500.0	40.0	24.0	14.3	0.26	2.10
Epoxy Resin	3.5	3.5	1.3	1.3	0.35	1.17

Table 3 Elastic properties of the impregnated yarn of TWF composites

Material	E_L (GPa)	E_T (GPa)	G_{LT} (Gpa)	v_{LT}	v_{TT}	V_f	ρ (gm/cm^3)
Carbon/epoxy	338.57	12.40	5.61	0.287	0.437	0.695	1.8164

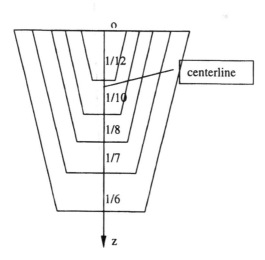

Figure 4 Different-size sectors

1. Finite element Models

Investigation on the mechanical behavior of sectors of different sizes but similar shape was carried out. Figure 4 shows the configuration of the sectors of different sizes. Sectors of 1/12, 1/10, 1/8, 1/7 and 1/6 of the full size sector are examined. Coordinate z has the same common origin at 0 for all sectors. Figure 5 shows the finite element mesh for a sector of 1/12 scale. There are 88 elements and 868 nodes. The boundary conditions are assumed to clamped all around. Pressure is applied on the concave side of the sector.

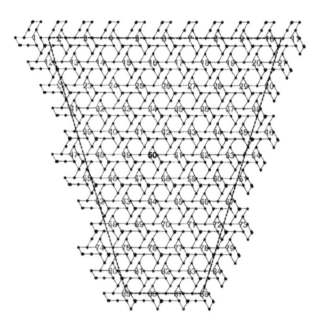

Figure 5 Mesh of 1/12 sector (Front)

2. Sector deflection along the centerline Oz

The deflection in the x direction at the nodes (along the Oz line) for the 1/12 scaled sector are shown in Figure 6. The deflection is 0 at two ends due to the clamped boundary conditions. The deflection is increasing as one moves away from the larger side of the sector to the smaller side of the sector. If the sector were flat, this contradicts common sense since larger side would be more flexible than the smaller side. However, the reflector is a shell with curvature.

The deflection for other scaled sectors also follow similar shaped curves. The superposition of the deflections for all sectors is shown in Figure 7. The origin O in Figure 7 corresponds to the origin O in Figure 4. It is interesting to note that apart from end conditions and scatter, the deflection for all five scaled sectors fall along the same line. The amplitude for the scatter is about $\pm 0.6 \times 10^{-4}$ m. It is important to remember that the thickness of the yarn for each scaled sector is also reduced by the same proportion. This means for the 1/12 scaled sector, the thickness of the yarn is t/12 where t is the thickness of the full scale sector. For the 1/10 scaled sector, the thickness of the yarn is t/10 etc. The deformed configuration for the 1/12 scaled sector is shown in Figure 8.

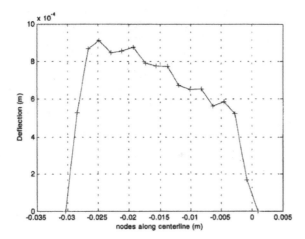

Figure 6: Defection of 1/12 sector along its centerline

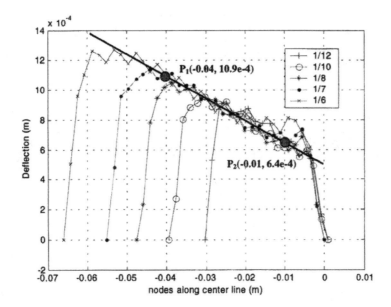

Figure 7: Defection of different-scale sectors on their centerlines

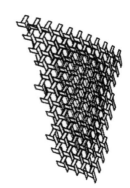

Figure 9: Deformed and undeformed configurations

3. Similitude observation for deflection

The appearance of the common straight line begs for a similitude observation. An equation for the straight line can be obtained by using the coordinates of two points. Point P_1(-0.04 m, 10.9 x 10^{-4} m) and point P_2(-0.01 m, 6.4 x 10^{-4} m) on the curve are used (note that the first argument in the parenthesis represents the position along the Oz line and the second argument in the parenthesis represents the deflection at that point). The following equation is obtained:

$$v = -0.0150z + 4.9e - 4 + \delta \quad (m) \tag{1}$$

where v is the deflection (m).

z is the coordinate distance from O as shown Figure 4 (m).

δ represents the magnitude of the scatter, which in this case is ±0.6 x 10^{-4}.

Equation (1) is plotted as shown in Figure 10. This equation is not valid close to the clamped boundaries. The range of validity of this equation for the full scale model can be -0.38 m < z < -0.03 m.

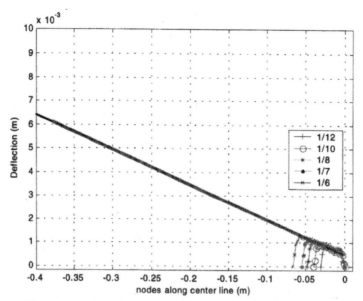

Figure 9: Predicted defection of full-scale sector on their centerlines

CONCLUSIONS

In this paper, the deformation behavior of a sector of the reflector subjected to uniform pressure is approached by modeling several different-scale sectors using Superelement 2. In this approach, the centerline deflections of the sector and a formula for them are given. A similitude rule was derived for the deformation of the ART reflector.

From numerical results, we can find that yarns of the sector are in complex couple deformation of extension, bending and twisting, even though the sector is loaded simply with pressure.

ACKNOWLEDGEMENT

The financial support from the Natural Sciences and Engineering Research Council Canada, EMS Technologies and Canadian Space Agency are appreciated. The technical guidance of personnel at EMS Technologies is also appreciated.

REFERENCES

1. M. Long "Satellite Receiving Antennas", Middle East Satellite Today magazine, January 1998.

2. Q. Zhao and S.V. Hoa. "Finite Element Models for Analysis of Triaxial Woven Fabric (TWF) Composites with Open Holes", submitted J. Composite Materials.

3. Q. Zhao, S.V. Hoa and P. Ouellette. "Numerical and Experimental Study of Triaxial Woven Fabric Composites with Open Holes", submitted J. Composite Materials.

Some applications of neural network and genetic algorithm for FRP composites

Prof. G. Ben(O. Byon)
Department of Mechanical Engineering
College of Industrial Technology, Nihon University
1-2-1, Izumicho, Narashino
275-8575, Japan
ben@cit.nihon-u.ac.jp

ABSTRACT

In FRP composite structures, not only their dimensions but also material properties composed of the structures can be considered as design parameters. Since these design parameters are closely connected each others, it is difficult to get an optimal design of FRP composite structures. Finite element methods and some nonlinear programming techniques have been used for getting solutions of optimal problems. Recent progresses of computer hardware technologies are remarkable. Software technologies are also made advance and "soft" algorithms, namely neural networks and genetic algorithms are used for getting solutions of inverse problems and optimal problems. After numerical procedures of neural network and genetic algorithm were briefly explained, their effectiveness for getting solutions in the inverse and optimal problems was demonstrated here. The inverse problem which was a damage detection from natural frequencies and/or their modes, the estimation of strength from damping capacity in CFRP laminates were executed by the neural networks. Furthermore, the genetic algorithms obtained the optimal lamination of FRP cylindrical thick shells under an external pressure and the optimal fiber orientation angles for obtaining the maximum axial buckling load.

INTRODUCTION

Recent progresses of computer hardware technologies are remarkable and solution of complicated problems obtained by a large-scaled computer some years ago can be now obtained by a personal computer. Software technologies are also made advance and "soft" algorithms, namely neural networks and genetic algorithms are used for getting solutions of inverse problems and optimal problems This "soft" algorithm needs not "strict" procedures of conventional numerical methods, like finite element methods.

In FRP composite structures, not only their dimensions but also material properties composed of the structures can be considered as design parameters. Since these design parameters are closely connected each other, it is difficult to get an optimal design parameters and inverse solutions in FRP composite structures. Finite element methods and some nonlinear programming techniques have been used for getting solutions of optimal problems. The inverse problems for obtaining fiber orientation angles in the symmetric angle-ply CFRP laminate having some values of material constants [1] and having some buckling stresses [2] were already solved by the neural networks.

In this paper, after numerical procedures of neural network and genetic algorithm were briefly explained, their effectiveness for getting solutions in the inverse and optimal problems was demonstrated. The inverse problem of damage detecting from natural frequencies and/or their modes in the CFRP laminated beam having some of damage area was executed by the neural network [3]. The relations among the damaged area, the damping capacity and the static or the impact bending strength of the CFRP cross-ply laminated beams were experimentally obtained. By using some of the experimental data, neural networks was developed for estimating the static or impact bending strengths of CFRP cross-ply laminated beams from their damping capacity [4].

Furthermore, the genetic algorithm obtained the optimal lamination of FRP cylindrical thick shells under an external pressure [5] and the optimal fiber orientation angles for obtaining the maximum axial buckling load [6].

DAMAGE DETECTION AND ESTIMATION OF STRENGTH IN CFRP LAMINATE BY NEURAL NETWORK

Neural network

In a neuron model (hereafter called as a unit), the lower subscripts i and j (from 1 to ℓ) denote a unit number and the upper superscript N (from 1 to F) denotes a layer of the hierarchical neural network. Assuming that h_i^N is a threshold of the unit at the layer N, x_j^{N-1} is an input value from the unit j at the layer $N-1$ and that $W_{ij}^{N,N-1}$ is a connection strength between the unit i at the layer N and the unit j at the layer $N-1$, a state value

$\bar{x}_i^N$ and an output value x_i^N of the unit i at the layer N are expressed as follows

$$\bar{x}_i^N = \sum_{j=1}^{\ell} W^{N,N-1} x_j^{N-1} - h_i^N \quad (1) \qquad x_i^N = f\left(\bar{x}_i^N\right) = \frac{1}{2}\left\{1 + \tanh\left(\frac{\bar{x}_i^N}{\mu_0}\right)\right\} \quad (2)$$

In which, $f\left(\bar{x}_i^N\right)$ and μ_0 are the quasi-linear response function and the gradient of this function. The value of x_i^N is changed from 0 to 1. As shown in Fig.1, the hierarchical neural network is composed of the input layer $(N - 1)$, the hidden layers $(N =$ from 2 to $F - 1)$ and the output layer $(N - F)$. When input tutor data x_i^1 are given to the input layer, the output values of x_i^N at the hidden layers are calculated by Eqs.(1) and (2) in order from the input side to the output and the values x_i^F at the output layer can be finally obtained. Using an error back propagation method and the difference $x_i^F - d_i$ (d_i: output tutor data corresponding to input tutor data x_i^1) at the output layer, a learning signal δ_i^F of the unit i at the output layer is calculated by the following equation.

$$\delta_i^F = \left(\frac{2}{\mu_0}\right)\left(d_i - \bar{x}_i^F\right)\bar{x}_i^F\left(1 - \bar{x}_i^F\right) \quad (3)$$

When this learning signal δ_i^F is used as the initial value, the learning signal δ_i^{N-1} ($N =$ from F to 2) at the hidden layers could be successively calculated from the output side to the input by the following equation.

$$\delta_i^{N-1} = \left(\frac{2}{\mu_0}\right)\sum_{j=1}^{\ell}\left(\delta_j^N W_{ij}^{N+1,N}\right)\bar{x}_i^{N-1}\left(1 - \bar{x}_i^{N-1}\right) \quad (4).$$

The modified connection strengths $W_{ij}^{N,N-1}(k+1)$ at $k+1$ iteration is calculated by the following equation

$$W_{ij}^{N,N-1}(k+1) = \eta\delta_j^{N-1}x_j^{N-1} + \alpha W_{ij}^{N,N-1}(k) \quad (5)$$

In which, $W_{ij}^{N,N-1}(k)$ are the modified ones at k iteration, η and α are a learning and a stability constants. After adding the modified connection strengths to the previous ones, the iterative calculation using Eqs (1) and (2) is executed with $\eta = 0.25$ and $\alpha = 0.9$ up to the scheduled

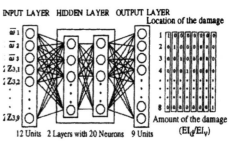

Fig.1 Neural Network

iteration number. That is, after the input tutor data are given at the input layers, the output values at each hidden and the output layers are calculated by Eqs (1) and (2). By using the learning signals of Eqs. (3)   and (4), the connection strengths are modified through the iterative calculation. This process is called the learning and the neural network is developed through this process.

In the next step, other data are given at the input of this developed network and the values at the output layer can be calculated. This process is called the estimating and the inverse problem can be solved through this process.

Damage detection by numerical data

The laminated beam (length=105mm, thickness=2.7mm, width=15mm) made by 16 prepregs of unidirectional carbon fiber reinforced polyimide (CF/PIX, E_L=139GPa, E_T=9.2GPa, G_{LT}=7.08GPa, v_{LT} =0.33, density=1.59g/cm^3) was fixed at one end here. In the FEM analysis, this laminated beam was divided to 10 elements in the axial direction. The damage area was equivalent to the element area and the existence of damage was limited to only one in the CFRP laminated beam. The quantity of damage was simulated by a ratio of EI_d / EI_y , in which EI_d was the bending stiffness of damaged element and EI_y was undamaged one. The 12 input data were three natural frequencies from the first to the third and 9 amplitudes in the third mode. The 9 output data were one value of EI_d / EI_y and 8 row vectors in which one component corresponding to the damage position took the value of 1 and other 7 components corresponding to undamaged positions were 0. If we gave the 24 sets of tutor data which were formed of three cases of 0.2, 0.35 and 0.5 of EI_d / EI_y and 8 damage cases having one damaged element among the 10 elements except the two elements near free edge, their results were calculated by the FEM. Then, these 24 tutor data sets were used for the learning process. This neural network is shown in Fig. 1 and is hereafter called as FM type. Another neural network composed of only three natural frequencies in the input layer was also developed (F type). By using the developed neural network, the damage detection of CF/PIX beams in the cases of EI_d / EI_y =0.30 and 0.45 is shown in Fig.2 for   the F type. The result of FM type was omitted owing to space limitation. Both results showed that 0.30 and 0.45 cases were close to the target lines. The quantities and locations of damage were fairly detected nevertheless the values of 0.45 and 0.30 were not used in the learning process. Although the number of input units in the F type was fewer than those of the FM type,

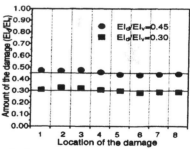

Fig.2 Comparison of estimating values to target values(F type)

the F type showed the same performance.

Damage detection by experimental data

The experimental three natural frequencies of CF/PIX laminated beam were obtained by an impulse exciting method and its EI_d / EI_y was changed to 0.57, 0.46 or 0.42. The experimental data for EI_d / EI_y =0.57 and 0.42 were used to develop the F type neural network and the value of 0.46 was used to estimate the damage location in order to demonstrate the performance of the developed neural network. The symbols of opened square and circle show the learning data of 0.57 and 0.42 in Fig.3, respectively and they were fairly accordant with the target of horizontal

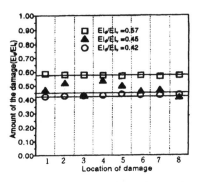

Fig.3 Comparison of learning and estimating values obtained by experiment (F type)

solid lines. The data for EI_d / EI_y =0.46 were not used in the learning process and the exact value of EI_d / EI_y were not obtained but the output data (solid triangular symbol) took between 0.57 and 0.42. Especially the locations of damage were perfectly detected by the present neural network. Therefore, the neural network seems to be sufficient in the first approach of damage detection.

Estimation of bending strength from damping capacity

In order to get non-destructively the bending strength, namely to estimate bending strength from the damping capacity, the neural network was developed. The input layer had the damping capacity and the CFRP thickness, and the output layer had the static bending strength. The neural network was also developed in which all of the results of 2mm and 3mm thickness were used as the tutor data and the results of 2.5mm thickness were not used at all. In Fig.4, the estimating value (opened square symbol) obtained by inputting the damping capacity of 2.5mm thickness to the neural network are compared with the experimental static bending strength (solid square symbol). The maximum error between these two strengths was about

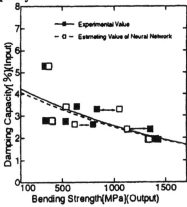

Fig. 4 Relation of damping capacity to bending stress for both results

20 %. However, the solid line obtained by a least square method was closed to the dotted line approximating the estimating values. So, the neural network is able to estimate the static bending strength from the damping capacity even in a case of the different thickness.

OPTIMIZATION OF COMPOSITE STRUCTURES BY GENETIC ALGORITHM

Genetic algorithm

In order to reduce the possibility of delamination failure, it is most important to reduce the maximum tensile strain in through the thickness direction for thick-walled FRP cylindrical shell design under external pressure. This maximum tensile strain was selected as the objective value (Fitness function), and the fiber orientation angle, the stacking sequence, and/or the material system were considered as design parameters in the genetic algorithm. The genetic algorithm was composed of four stages. The numerical procedures of genetic algorithm were simply explained by using an example of the optimal stacking in the thick walled FRP cylindrical shell as follows.

In the first stage, the population, namely a number of individuals had to be decided and a number of genes in one individual should also be decided. The population was 20 individuals here and this meant simultaneously adopting 20 lamination patterns in the search process of optimum. The individual was considered as the combination of design parameters in one lamination pattern. One individual had 24 genes corresponding to the lamination sequence through the thickness direction in the thick shell. The gene codes of 0 , 1, 2 and 3 were assigned to 0° and 90° layers of E-glass/Epoxy and 0° and 90° T300/SP-286, respectively. At first, the gene codes in all of individuals were decided randomly and the tensile strains $\varepsilon_{z\,max}$ in the all of individuals and in through the thickness direction were calculated by the finite cylindrical element method [7].

The second stage was the selection: namely, the individual having the lowest value of $\varepsilon_{z\,max}$ among the 20 individuals was selected as the best individual at the present generation. The best individual always remained at the next generation, which was called the elite preservation strategy. If the stresses corresponding to the lowest value of $\varepsilon_{z\,max}$ were less than the failure criterion of Tsai-Wu, then the algorithm went to the third stage. Otherwise, went back to the selection process for choosing other individual.

The third stage was crossover. The 19 individuals were newly and randomly decided and together with the best individual in the second stage, the number of individuals was also 20 in the third stage. Among the 20 individuals, two individuals were selected randomly, and this ratio of selection was called the crossover probability (one of GA parameter). Using the selected two individuals, some gene codes of one individual were swapped for another gene codes. This operation meant to make other design variables.

The final stage was mutation. Only one gene randomly selected from 20 individuals in the third stage was changed to other value of code depending on the mutation probability (one of the GA parameter). This operation was useful to avoid a local minimum in a search of optimum. After these four stages were executed, the generation proceeded to new one. The values of cross over probability and mutation probability used here were 0.30 and 0.03.

Optimization of hybrid cross-ply thick walled cylindrical shell

The optimum was obtained by the 300th generation (iteration). The optimum stacking sequence and the maximum tensile strain $\varepsilon_{z\,max}$ at the 300th are shown in Table 1, and this optimum value was compared with the optimum results of $[90°_5/90°_5/0°_5]$ only T300/ SP-286. This optimum stacking sequence by the GA was better than other cases and it could reduce the tensile strain by 10 % in comparison with the former optimum result [9].

Table 1 – Results of stacking sequence in hybrid cylindrical thick walled shell

	Stacking Sequence		ε_{zmax} (μ)
	Inner Surface	Outer Surface	
Optimum by GA	(1/1/3/3/3/3/3/3/3/3/3/3/3/3/2/2/2/2/2/2/2/2/2/2)		3256
Stacking sequence in Ref. [9]	(3/3/2/3/2/3/3/2/3/3/2/3/3/2/3/3/2/3/2/3/3/2)		3599

0 = E-glass/epoxy 0° layer, 1 = E-glass/epoxy 90° layer, 2 = T300/SP-286 0° layer, 3 = T300/SP-286 90° layer.

Optimization of axial buckling strength of laminated cylindrical shell

The fitness function adopted here was maximum axial bucking load and design parameters were considered to be the ply angle and the stacking sequence of the laminated cylindrical shell. The fiber orientation angle θ_i was represented by a unit of 3 or 6 degrees. In the case of the 6 degree unit, the fiber orientation angle of each lamina was represented as binary codes of 4 digits and 3 degree unit used 5 digits. The example of 6 ply laminates using 6 degree unit was expressed as follows.

$[0°/60°/30°/48°/78°/90°]=$
$[0000/1010/0101/1000/1101/1111]$ (6)

Table - 2 Optimum configurations of Boron/ Epoxy laminated cylindrical shell

Optimum stacking sequences	(N_x/t) [MPa]	Buckling Type
$[\pm30/\pm84/\pm60/\pm6/\pm48/\pm54]$	755.7	Asymmetry
$[\pm66/\pm18/\pm30/\pm78/\pm60/\pm30]$	755.6	Asymmetry
$[\pm24/\pm72/\pm60/\pm12/\pm30/\pm60]$	755.6	Symmetry
$\#[\pm28/\pm65/\pm86/\pm5/\pm27/\pm63]$	754.9	
$[\pm24/\pm84/\pm42/\pm42/\pm6/\pm66]$	755.4	Asymmetry
$[\pm66/\pm6/\pm48/\pm48/\pm84/\pm24]$	755.4	Asymmetry
$[\pm36/\pm84/\pm0/\pm60/\pm42/\pm48]$	755.4	Asymmetry
$[\pm54/\pm6/\pm90/\pm30/\pm48/\pm42]$	755.4	Asymmetry
$\#[\pm50/\pm9/\pm87/\pm7/\pm54/\pm40]$	750.5	
$[\pm36/\pm90/\pm0/\pm54/\pm48/\pm42]$	755.2	Asymmetry
$[\pm54/\pm0/\pm90/\pm36/\pm42/\pm48]$	755.2	Asymmetry
$\#[\pm55/\pm2/\pm82/\pm45/\pm35/\pm48]$	756.7	

The population size of 100 individuals, the probabilities of crossover (=0.6) and mutation (=0.02) were used here and the material properties of Boron/Epoxy and the dimension of shell were E_L=207GPa, E_T=20.7GPa, G_{LT}=6.89GPa, v_{LT} =0.3, Length=0.3m, Radius=0.1 m, and thickness=1.524$\times 10^{-3}$m. In Table2, the nine kinds of stacking sequences representing the optimum and close to the optimum are shown in order of lager value. This method gave many kinds of stacking sequence having almost the same buckling load under the axial compression. Some buckling loads (# mark) almost agreed with the three kinds of results obtained by nonlinear programming [8].

CONCLUSIONS

In order to demonstrate a validity of neural network, the damage detection from natural frequencies and/or their third mode and the estimation of bending stress from damping capacity in the CFRP laminates were discussed here. As a result, the neural network was very useful as the first approximation of the damage detection and as the estimation of strength by using the experimental data. After the neural network is developed, the network is able to detect the damage and to estimation of strength by comparatively easy summation and multiplication. Therefore, this network is considered one of methods fitted for the field work.

Furthermore, the genetic algorithm was applied to obtain the optimum configurations for the smallest tensile strain in through the thickness direction in the thick walled hybrid cylindrical shell subjected to an external pressure and for the maximum axial buckling load in the Boron/Epoxy cylindrical laminated shell. Then, the genetic algorithm was useful for optimal problems having various design parameters.

REFFERENCES

[1] G. Ben (O.Byon) et al, Transactions of Japan Society for Mechanical Engineers, Vol.58-A,1992, 539-548
[2] G. Ben (O. Byon) et al, ibid, Vol.60-A, 1994, 569-573
[3] G. Ben (O.Byon) et al, ibid, Vol.62-A, 1996, 152-157
[4] G. Ben (O. Byon) et al, Key Engineering Materials, Vols.145-149, 1998, 427-432
[5] G. Ben (O. Byon) et al, Journal of Thermoplastic Composite Materials, Vol.11, 1998, 418-428
[6] Y. Aoki and G. Ben (O.Byon), Proceeding pf ASME-JSMA PVP Conference, 1998,
[7] G. Ben (O. Byon) and J.R.Vinson, AIAA, Vol.29, 1992, 2192-2196
[8] Y. Hirano, Jour. Japan Society for Aeronautical & Space, Science, Vol.32, 1994, 46-51
[9]G. Ben (O. Byon) and J.R.Vinson, "Proceedings of 5th Japan-US Conference on composite materials, 1990, 257-264

Interfacial Reinforcement in Composites with PPT Aramid Fiber Modified by Network-Intercalation Method

Nobuo Ikuta, Akihiro Ohnishi
Department of Materials Science and Engineering
Shonan Institute of Technology
1-1-25, Tsujido-Nishikaigan, Fujisawa, Kanagawa 251-8511, Japan

Eiji Fujioka, and Kazuhiko Kosuge
New Product Development Group
'Kevlar' Product & Application Technology Department
Dupont-Toray Co., Ltd.

ABSTRACT

A novel surface treatment for *p*-phenylene telephthalamide (PPTA) fiber was performed with silanes and urethane binder that are usually used as sizes for glass fiber treatment. The PPTA used for the surface treatment was modified in spinning process to make the gap between PPTA crystallite open. In this treatment, supercritical carbon dioxide fluid method was carried out to impregnate the molecules of sizes into open gap in PPTA fiber. After the impregnation, the fiber was heated at 100 °C – 170 °C. to make the gap close from open-gapped fiber to normal type of PPTA modified with sizes. The interfacial shear strength of fiber to epoxy resin was measured by microdroplet method. As a result, the modified PPTA improved the interfacial shear strength by ca. 65 % to the interfacial shear strength given by normal PPTA without treatment. Those improvement were 33 % without heating, 18 % with only silanes and 12 % with only urethane instead of the mixture of silane and urethane. In addition, the fiber strength showed no remarkable decrease after the treatment.

INTRODUCTION

Poly (p-phenylene telephthalamide) (PPTA) is a famous kind of reinforcing fiber for composite materials. However, the interfacial strength is too weak to raise total performance of the composite, though some surface modification methods have been proposed; plasma treatment [1, 2] and chemical modification [3, 4]. These sorts of modification cause the fibrillation of fiber and results in the reduction of fiber strength.

As well known, general reinforcing fibers such as carbon fiber and glass fiber are respectively treated by cathode oxidation and with silane coupling agents and/or film formers. On the other hand, organic fiber such as aramid fiber and high molecular weight polyethylene has had no industrial method for the surface treatment yet, because of the reason, as described above.

In this paper, we propose a novel treatment of PPTA with normal sizes including silane coupling agents and film formers, which are usually used for surface treatment of glass fiber. In the surface treatment, PPTA used is not normal type of fiber such as Kevlar® such as Kevlar® 29 and 49. This type of PPTA is modified during fiber spinning to make larger gaps between crystallites in fiber structure. We call it open-gapped PPTA fiber.

In the idea that the sizes are fixed in the fiber without chemical bonding between them, there are three important factors; impregnation of sizes into the open gaps, production of macromolecular sizes in the fiber, and closing of the open gap to fasten the sizes. In the first factor, supercritical fluid method has been employed. In the second and the third factors, heating has been accomplished to enhance the reaction between silane and film former to make network formation, and to close the gap between crystallite in the fiber.

EXPERIMENTAL

Surface Treatment by Supercritical Fluid Method

In the treatment, PPTA fiber (12 μm in filament diameter) used was not normal type such as Kevlar® 29 and 49, but specialized with a larger gap between crystallites in fiber structure (open-gapped PPTA). Size emulsion used was a mixture of aminopropyltriethoxysilane (APS) and urethane film former (UB). APS and UB were respectively supplied as a commercial grade: KBE 903, Sin-Etsu Chemical Co., Ltd. and Vondic 1310F, Dainippon Ink and Chemicals Inc. The mixture (APS:UB = 1:4 wt/wt) were be solidified by freeze-dry to be applied in the following procedure.

Supercritical carbon dioxide fluid was used to make molecules of sizes permeated into open gap in the fiber. An instrumental apparatus for supercritical liquid (SFE 2200, ISCO, Inc.) was employed under 250 bars and 150°C. Open-gapped PPTA fiber was separated from freeze-dried sizes in a bottle, which was put into the apparatus. After the permeation, the fiber was heated at 100 C for 20 min and at 170°C for 30 min to close the gap, that is, to change into normal type of PPTA fiber.

Measurement of Interfacial Shear Strength by Microdroplet Method

Epoxy resin was used as a matrix to measure the interfacial shear strength between PPTA fiber and the resin by microdroplet method. Liquid epoxy resin (Epicoat 828, Yuka Shell Epoxy Co. Ltd.) was mixed with 11 parts of a hardening agent, triethylenetetramine (Kohei Chemical Industry Co. Ltd), and smeared on the fiber. The hardening process was performed at 50 °C for 80 min and at 100 °C for 60 min. Microdroplet test was performed by a commercial testing machine (HM 410, Tohei Industrial Co. Ltd.)   Test speed was 0.03 mm/min and more than 10 microdroplet of resin with 70 – 90 μm in diameter were always selected in every measurement to depress a scattering of data.

Fabrication of Unidirectional Composites

Unidirectional composites were produced to evaluate the mechanical property of PPTA fiber composites.   Each type of PPT fiber was impregnated into liquid epoxy resin with 11 parts of triethylenetetramine, and was molded by pultrusion.   Hardening condition was the same as in the preparation of microdroplet samples.   Moreover, the composites were annealed at 170 C for 180 min.   Finally, the unidirectional composites with 46 volume percent of fiber have a rod-shape with 2.5 mm of diameter and 30 mm of length to be supplied to bending test.

Bending Test

Bending test was done by an universal testing machine (Autograph AGS1000-B, Shimadzu Co.) equipped with a three points bending jig.   Span length was 25 mm. Test speed was 0.5 mm/min.   Four coupons were tested to get the average and the standard deviation of bending strength and modulus.

RESULTS AND DISCUSSION

Open Gap between Crystalline

X ray diffraction was measured to evaluate the structure of crystalline in normal

Table 1.   Change of Crystalline Size with different types of
p-Aramid Fiber measured by X-ray Diffraction

Sample	Sheet / hkl	Crystallite Size / Å
Normal PPT (Kevlar® 29)	110	60
	200	50
	004	66
Open Gapped PPT	110	36
	200	46
	004	67
Open Gapped PPT after Heating	110	68
	200	52
	004	69

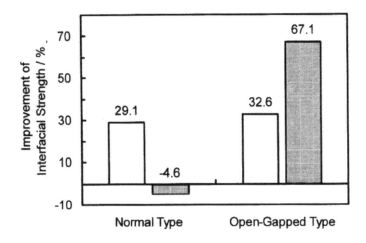

Fig. 1.   Interfacial reinforcement between epoxy resin and normal type and open-gapped type of PPTA fiber; permeated with a mixture of APS and UB (1:4) by supercritical liquid method (□) and heated at 100 °C for 20 C and 170 C for 30 C (■).

PPTA fiber (Kevlar® 29), open-gapped PPTA fiber before and after heating. In all measurements, the degree of crystallization was not almost changed in these fibers. As shown in Table 1, all of fibers were almost the same size of crystalline along z-axis, but thickness of crystalline was changed in those fibers. That is, the crystalline structure in open-gapped fiber was thinner than that in normal fiber, Kevlar 29. This meant that the modified PPTA fiber had an open gap slit between the thin crystalline along fiber. In addition, the heating at 170°C made the modified fiber normal crystalline structure. This suggested that open-gapped fiber became the normal PPTA fiber.

The above results support the procedure in which the surface treatment should be separated into two steps; the penetration of sizes into open-gapped fiber at the first step and the fixation of sizes between closed gap in fiber by heating at the second step.

Effectiveness of Open Gap on Interfacial Reinforcement

Before and after heating in the treatment with silane and film former (APS:UB=1:4), the interfacial shear strength was measured by microdroplet method. As shown in Fig. 1, open-gapped structure in fiber and heating in the process significantly affected the interfacial shear strength. Without heating, normal and open-gapped types of fiber were almost the same value in interfacial shear strength. However, the heating process had open-gapped fiber effective in interfacial reinforcement, while it had normal fiber negative in interfacial reinforcement. Eventually, open-gapped fiber permeated with size and then heated showed the highest

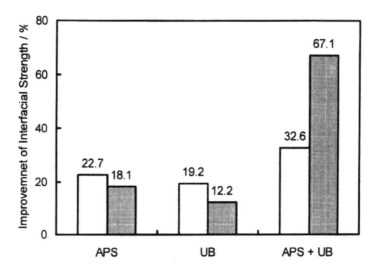

Fig. 2.   Interfacial reinforcement between epoxy resin and open-gapped type of
PPTA fiber; permeated with APS, UB and a mixture of APS and UB (1:4)
by supercritical liquid method (□) and heated at 100°C for 20 C and
170 C for 30 C (■).

improvement in interfacial shear strength.

This result suggests that the open-gapped structure is suitable to be permeated
with silane and film former.   However, only the open-gapped structure is not important.
Considering that the resulting interfacial strength before heating in Fig. 1 is almost the
same, supercritical fluid seems to gave the same effect in the permeation of sizes into
both type of fibers, that is, irrespective of width of gap between crystallites in the fiber.

Effectiveness of Mixture of Silane and Film Former

Another factor is suggested to be the composition of size: APS and UB.   The
interfacial shear strength was evaluated for open-gapped PPT fiber permeated with only
APS or only UB instead of the mixture of APS and UB.   Without heating, the resulting
values gradually increased in the order of UB, APS, and the mixture of APS and UB.
However, the heating effect on the interfacial strength was different in only APS or only
UB from in the mixture of APS and UB.   In particular, the mixture gave the highest
improvement among them.

This result means that the mixture was valuable for interfacial reinforcement.
In the previous paper [5], APS and UB on glass fiber have been found to make the strong
interaction by infrared spectroscopy.   The same interaction is considered to occurs even
in the gap between crystallites in PPTA fiber.   Such an interaction makes APS and UB

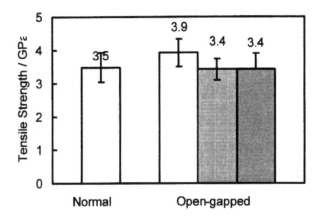

Fig. 3   Tensile strength of normal PPTA fiber and of open-gapped PPTA fiber untreated (□), permeated with with a mixture of APS and UB (1:4) by supercritical liquid method (▤) and heated at 100 °C for 20 C and 170 C for 30 C (■).

networks, which is remaining even in narrow gap after heating.   On the other hand, silane and film former themselves are impossible to make networks.

Eventually, the mechanism of interfacial reinforcement by the treatment with mixture of silane and film former is presumed as follows: First, each component, that is, silane and film former, permeates into open gap between PPTA crystallites.   Second the heating process makes the gap narrow and both components network with strong interaction.   In the final stage, network of size is squeezed in each narrow gap.   As there are a lot of gaps with network size between crystallites, it seems to be difficult to slip the size from the fiber.

If the amount of the mixture is enough on the fiber, this networks structure exists not only in the gap, but also at the surface.   Such a size at the surface is also capable of strong interaction with epoxy resin, as known in industrial technology.

Consequently, the mixture of silane and film former is possible to transfer stress of resin to fiber.

Fiber Strength of Size-Treated PPTA

The previous treatments for aramid fiber such as plasma treatment and chemical modification decreases the tensile strength of the fiber, as mentioned above.   Therefore, it is important to confirm the tensile strength of open-gapped fiber before and after the treatment studied here, compared with normal PPTA fiber (commercial name Kevlar®

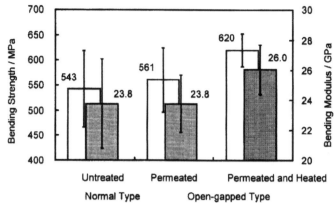

Fig. 4   Bending strength (□) and modulus (■) of unidirectional composites with epoxy resin matrix and normal and open-gapped types of

29).

As shown in Fig. 3, open-gapped fiber without treatment was a little bit of higher than normal fiber.   After the treatment, tensile strength of open-gapped fiber gave the same value as of normal fiber.   That is, the fiber strength was kept irrespective of the treatment.   This result suggested that the treatment gave neither breakage of crystallite nor hindrance to crystallization.

Bending Properties of Composites with Size-Treated PPTA Fiber

The bending strength and bending modulus were measured for unidirectional composite with open-gapped fiber before and after the treatment to know the effect of interfacial reinforcement on mechanical properties.   Figure 4 shows the bending strength and modulus of rod type of unidirectional composite with normal PPTA fiber without treatment, with open-gapped fiber treated with the mixture of APS and UB. Compare with normal fiber without treatment, even the permeation of the mixture raise the bending properties.   The heating process after the permeation significantly elevates the properties.   It should be noted that the bending modulus increases after the treatment. This means that the permeation with silane and film former leads to raise the property of the fiber.

CONCLUSION

This study is concluded as follows:
1. The treatment used in this study is effective in interfacial strength and bending property.
2. The treatment has two important control factors; gap between crystallites in p-aramid fiber and permeation of size consisting of silane and film former.   The process is

like intercalation into clay layer.
3. The adhesion seems to depend on the slipping stress between crystallite of PPTA and thin networks structure of size.
4. From items 2 and 3, this novel treatment for PPTA fiber is specialized in terms of 'intercalation' and 'networks'.   So, we call it 'networks-intercalation method.'
5. The mechanism of the interfacial reinforcement resembles the interaction in glass fiber / resin, except the interaction between fiber and size.
6. In addition, this technique is possible to give the second property to Kevlar; including water-resistance, fibrillation-resistance.
7. Eventually, A lot of previous knowledge on sizes including film former for glass fiber treatment is helpful on the selection of sizes for Kevlar-resin interaction.

ACKNOWLEGDEMENTS

This study was partially supported by a Grant-in-Aid for Scientific Research (B) (Subject No. 14350361) from Japan Society fro the Promotion of Science.

REFERENCES

(1) R. E. Allred, E. W. Merrill, D. K. Roylance, Polym. Preprints, **24**, 223 (1983).
(2) J. R. Brown, P. J. C. Chappell, Z. Mathys, *J. Mater. Sci.*, **26**, 4127 (1991).
(3) Y. Wu, G. Tesoro, *J. Appl. Polym. Sci.*, **31**, 1041 (1986).
(4) M. Takayanagi, T. Kajiyama, T. Katayose, *J. Appl. Polym. Sci.*, **27**, 3903 (1982).
(5) N. Ikuta, K. Tomari, H. Kawada, H. Hamada, *Key Eng. Mater.*, **137**, 207 (1998).

The study of analytical models for damping of laminated composite thick plates

Dr. Y. Ohta
Department of Mechanical Systems Engineering
Hokkaido Institute of Technology
7-15-4-1, Maeda, Teine-ku, Sapporo
006-8585, Japan
ohta@hit.ac.jp

ABSTRACT

This paper studies the analytical models for damping analysis of fiber reinforced plastic laminated composite plates. For this purpose, the maximum strain and kinetic energies of a cross-ply laminated plate are evaluated analytically based on the three-dimensional theory of elasticity. The displacements of the simply-supported rectangular plates are expanded into the polynomial forms with respect to a thickness coordinate, and then governing equations are formulated by using the Ritz's method. In the numerical calculations, natural frequencies and modal damping ratios are calculated for the plates with different stacking sequence and thickness ratios. The validity of the assumption of deformations and the applicability of the other plate theories (e.g. Classical Lamination Theory, First-order Shear Deformation Theory and Higher-order Shear Deformation Theory) to the laminated thick plates are discussed by comparing the numerical results obtained by the present method with the CLT and the FSDT solutions.

INTRODUCTION

Recently Fiber Reinforced Plastics (FRP) are being increasingly used in the structural applications, because the technical merits of high strength to weight ratio and high stiffness to weight ratio of FRP composite materials are outstanding. It is also known that the anisotropic property in FRP materials considerably affects the dynamic characteristics (e.g. vibration and buckling) of structural elements such as plates and shells. Therefore a large number of theoretical and experimental studies have been published on the dynamic problems of FRP laminated composite plates and shells.

In most of the analyses found in the literatures, the laminated plate is macroscopically modeled as a thin plate of general anisotropy, and is analyzed by using the Classical Lamination Theory (CLT) (e.g. Vinson et al.(1)) based on the Kirchhoff's hypotheses, where transverse shear deformations and extension in the thickness direction of the plate are ignored. However the CLT cannot give accurate results for laminated thick plates, because the transverse shear deformations cannot be ignored due to the relatively smaller shear elastic moduli of FRP composite materials than the other elastic moduli. Thus the analyses including the effect of shear deformations are required for the laminated thick plates. Yang et al.(2), Whitney et al.(3) and Dong et al.(4) analyzed laminated thick plates and shells based on the First-order Shear Deformation Theory (FSDT), in which in-plane displacements are assumed to vary lineally through the thickness. However, the FSDT cannot satisfy the free surface conditions on transverse shear stresses on the top and bottom surfaces of the plate and thus requires a shear correction factor multiplied to the transverse shear stiffness of the plate. Reddy(5) proposed the Higher-order Shear Deformation Theory (HSDT), wherein in-plane displacements are expressed in the polynomial forms of third order. This theory satisfies the free surface conditions exactly and does not require the shear correction to the transverse shear stiffness. There are few papers discussing the displacement functions in the evaluation of frequency and damping ratio of FRP laminated composite plates although these theories have been proposed.

The purpose of this paper is to study the analytical models for damping of FRP laminated composite thick plates. For this purpose, the maximum strain and kinetic energies of an FRP cross-ply laminated plate are evaluated analytically based on the three-dimensional theory of elasticity, and the displacements of the rectangular plate, which is simply-supported at all edges, are expanded into the polynomial forms of arbitrary order with respect to a thickness coordinate. A governing equation is derived by using the energy approach of minimizing the Lagrangian defined by the two energies (Ritz's method), and the present problem is reduced to a complex eigenvalue problem by introducing the complex elastic moduli. On the other hand, two governing equations for the laminated plate are also formulated by using the CLT and the FSDT, respectively. In numerical calculations, natural frequencies and modal damping ratios are obtained for the laminated plates with different stacking sequence and thickness ratios by using different sets of displacements. Finally not only the validity of the assumption of deformations but also the applicability of the plate theories (CLT, FSDT and HSDT) applied to the FRP laminated thick plate are discussed by comparing with the numerical results obtained from the present method, the CLT and the FSDT analyses.

THERETICAL ANALYSIS

We consider a N-layer laminated composite rectangular plate $(a \times b \times H)$, which is simply-supported at all edges. In the present analysis, the Cartesian coordinate system (x, y, z) is taken at the middle surface of the laminated plate as shown in Figure 1. The maximum displacements (amplitude) of the plate in the x, y and z directions are denoted by u, v and w, respectively.

Based on the three-dimensional theory of elasticity, the maximum strain and kinetic energies of the plate are expressed as follows :

$$U_{max} = \frac{1}{2} \int_V \left(\sigma_x \varepsilon_x + \sigma_y \varepsilon_y + \sigma_z \varepsilon_z + \tau_{yz} \gamma_{yz} + \tau_{zx} \gamma_{zx} + \tau_{xy} \gamma_{xy} \right) dV$$

$$T_{max} = \frac{1}{2} \int_V \rho_m \omega^2 \left(u^2 + v^2 + w^2 \right) dV$$

(1)

where ρ is a mean mass density of FRP composite material and ω means a circular frequency. For the cross-ply laminated plate, the stress-strain relation in the k-th layer is given by

$$\begin{Bmatrix} \sigma_x \\ \sigma_y \\ \sigma_z \\ \tau_{yz} \\ \tau_{zx} \\ \tau_{xy} \end{Bmatrix}^{(k)} = \begin{bmatrix} Q_{11} & Q_{12} & Q_{13} & 0 & 0 & 0 \\ & Q_{22} & Q_{23} & 0 & 0 & 0 \\ & & Q_{33} & 0 & 0 & 0 \\ & & & Q_{44} & 0 & 0 \\ & \text{Sym.} & & & Q_{55} & 0 \\ & & & & & Q_{66} \end{bmatrix}^{(k)} \begin{Bmatrix} \varepsilon_x \\ \varepsilon_y \\ \varepsilon_z \\ \gamma_{yz} \\ \gamma_{zx} \\ \gamma_{xy} \end{Bmatrix}$$

(2)

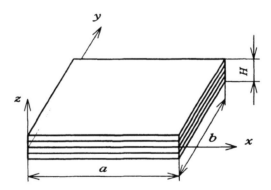

Figure 1   Geometry and coordinates of a laminated composite plate

where the elastic coefficients $Q_{ij}^{(k)}$ are calculated(Vinson et al.(1)) by using Young's moduli, shearing elastic moduli, Poisson's ratios of FRP material and the direction of fiber in the k-th layer. Assuming small amplitude vibration, the strains are related to the displacements by

$$\varepsilon_x = \frac{\partial u}{\partial x}, \quad \varepsilon_y = \frac{\partial v}{\partial y}, \quad \varepsilon_z = \frac{\partial w}{\partial z}$$

$$\gamma_{yz} = \frac{\partial w}{\partial y} + \frac{\partial v}{\partial z}, \quad \gamma_{zx} = \frac{\partial u}{\partial z} + \frac{\partial w}{\partial x}, \quad \gamma_{xy} = \frac{\partial v}{\partial x} + \frac{\partial u}{\partial y}$$

(3)

For the generality of the analysis, the following nondimensional quantities are introduced :

$$\xi = \frac{x}{a}, \quad \varsigma = \frac{y}{b}, \quad \eta = \frac{z}{H/2}$$

$$h = \frac{H}{a}, \quad l = \frac{b}{a}$$

$$\bar{u} = \frac{u}{H}, \quad \bar{v} = \frac{v}{H}, \quad \bar{w} = \frac{w}{H}$$

(4)

$$\lambda^2 = \frac{\rho_m H a^4 \omega^2}{D_0} \quad \text{(Frequency parameter)}$$

where D_0 means a representative bending stiffness $D_0 = E_2 H^3 / 12(1 - v_{12} v_{21})$ in the analysis. (Suffix 1 and 2 denote the major material-symmetry direction and the in-plane transverse directions, respectively.)

When the laminated plate is simply-supported at all edges:

$$\bar{v} = \bar{w} = 0 \quad \text{at} \quad \xi = 0, 1 \ (x = 0, a)$$

$$\bar{u} = \bar{w} = 0 \quad \text{at} \quad \varsigma = 0, 1 \ (y = 0, b)$$

(5)

displacements can be expressed as follows :

$$\bar{u}(\xi, \varsigma, \eta) = \cos m\pi\xi \ \sin n\pi\varsigma \sum_{i=1}^{I} U_i \eta^{i-1}$$

$$\bar{v}(\xi, \varsigma, \eta) = \sin m\pi\xi \ \cos n\pi\varsigma \sum_{j=1}^{J} V_j \eta^{j-1}$$

(6)

$$\bar{w}(\xi, \varsigma, \eta) = \sin m\pi\xi \ \sin n\pi\varsigma \sum_{k=1}^{K} W_k \eta^{k-1}$$

where m and n are the half wave numbers in the x, y direction, respectively, and U_i, V_j and W_k are unknown coefficients.

By substituting eqns (2), (3), (4) and (6) into eqn (1), and minimizing the Lagrangian $T_{max} - U_{max}$ with respect to the unknown coefficients U_i, V_j and W_k for a stationary value :

$$\frac{\partial \left(T_{max} - U_{max}\right)}{\partial U_a} = \frac{\partial \left(T_{max} - U_{max}\right)}{\partial V_\beta} = \frac{\partial \left(T_{max} - U_{max}\right)}{\partial W_\gamma} = 0 \tag{7}$$

$$(\alpha = 1,2,\cdot\ \cdot,I\ ;\ \beta = 1,2,\cdot\ \cdot,J\ ;\ \gamma = 1,2,\cdot\ \cdot,K)$$

a following governing equation is derived.

$$\left[\begin{bmatrix} A_{i\alpha} & B_{j\beta} & C_{k\gamma} \\ & D_{j\beta} & E_{k\beta} \\ \text{Sym.} & & F_{k\gamma} \end{bmatrix} - \lambda^2 \begin{bmatrix} G_{i\alpha} & 0 & 0 \\ & G_{j\beta} & 0 \\ \text{Sym.} & & G_{k\gamma} \end{bmatrix}\right] \begin{Bmatrix} U_i \\ V_j \\ W_k \end{Bmatrix} = 0 \tag{8}$$

The elements in the coefficient matrix of eqn (8) are given by

$$A_{i\alpha} = \frac{m^2\pi^2}{2} q_{11i\alpha}^{00} + \frac{2}{h^2} q_{55i\alpha}^{11} + \frac{n^2\pi^2}{2l^2} q_{66i\alpha}^{00}$$

$$B_{j\alpha} = \frac{mn\pi^2}{2l}\left(q_{12j\alpha}^{00} + q_{66j\alpha}^{00}\right)$$

$$C_{k\alpha} = \frac{m\pi}{h}\left(q_{55k\alpha}^{01} + q_{13k\alpha}^{10}\right)$$

$$D_{j\beta} = \frac{m^2\pi^2}{2} q_{66j\beta}^{00} + \frac{2}{h^2} q_{44j\beta}^{11} + \frac{n^2\pi^2}{2l^2} q_{22j\beta}^{00}$$

$$E_{k\beta} = \frac{n\pi}{lh}\left(q_{44k\beta}^{01} - q_{23k\beta}^{10}\right)$$

$$F_{k\gamma} = \frac{m^2\pi^2}{2} q_{55k\gamma}^{00} + \frac{2}{h^2} q_{33k\gamma}^{11} + \frac{n^2\pi^2}{2l^2} q_{44k\gamma}^{00}$$

$$G_{mn} = \frac{h^2}{2}\int_{-1}^{1} \eta^{m-1}\ \eta^{n-1}\ d\eta \tag{9}$$

$$q_{klmn}^{ij} = \int Q_{kl} \frac{\partial^{(i)}\eta^{m-1}}{\partial \eta^{(i)}} \frac{\partial^{(j)}\eta^{n-1}}{\partial \eta^{(j)}}\ d\eta$$

In the present analysis, complex elastic moduli with loss factors

$$E_1^* = E_1(1 + j\eta_1),\quad E_2^* = E_2(1 + j\eta_2),\quad G_{12}^* = G_{12}(1 + j\eta_{12}),\quad G_{23}^* = G_{23}(1 + j\eta_{23}) \tag{10}$$

are introduced to evaluate not only natural frequencies but also modal damping ratios, and thus the present problem is reduced to a complex eigenvalue problem. In usual manner the real part of each eigenvalue yields a frequency parameter (natural frequency), and the ratio of imaginary part to real part of each eigenvalue gives a modal damping ratio of the laminated plates.

For discussion on the assumption of deformations in vibration analysis, the laminated plate is also modeled based on the CLT and the FSDT, respectively, and frequency parameters and modal damping ratios of the plate are also calculated numerically from the two kinds of analyses. However, the formulation of each analysis is not written here due to the limitation of the space of pages.

NUMERICAL RESULTS AND DISCUSSIONS

By using the analytical schemes developed in the previous section, numerical studies are carried out for cross-ply laminated square plates with different stacking sequence and thickness ratios. In numerical calculations, natural frequencies and modal damping ratios of the plates are obtained by using many sets of I, $J(=I)$ and K terms in the displacement functions [eqns (6)] and by using the CLT and FSDT solutions. This means that frequencies and damping ratios are obtained by employing the displacement functions which have polynomial forms of $(I-1)$-th and $(K-1)$-th order for in-plane and transverse displacements, respectively. A graphite/epoxy, which is highly orthotropic fiber reinforced plastics material, is chosen and the material properties used in numerical calculations are

$$E_1/E_2=20,\ G_{12}/E_2=0.65,\ G_{23}/E_2=0.5,\ v_{12}=0.25,\ v_{23}=0.25$$
$$\eta_1=0.0015,\ \eta_2=0.01,\ \eta_{12}=\eta_{23}=0.016$$

The stacking sequence is expressed by using the notation $[\alpha_1/\alpha_2/\alpha_3/\alpha_4....]$. For instance, [0/90/0] denotes a 3-layered plate which has fiber angles 0°, 90° and 0° from lower to upper layer in the plate.

Table I shows the comparisons of numerical results obtained by the present Ritz's method and by Ohta et al.(6). As shown in these tables the comparisons show an excellent agreement, and thus it demonstrates the validity of the present method.

Figure 2 indicates the nondimensional frequency parameters obtained from present method ($I=J=4$, $K=3$), HSDT($I=J=4$, $K=1$), FSDT and CLT for four symmetrically 8-layered cross-ply thick plates ($h=0.1$) with different stacking sequence,

Table - I   Comparisons of frequency parameters of laminated composites plates
($I=1.0$, [0/90/0], $I=J=4$, $K=3$)

h	Mode	Present	Ohta et al.(6)
0.2	(1, 1)	32.42	32.43
	(1, 2)	52.10	52.32
	(2, 2)	83.87	83.13
0.05	(1, 1)	46.54	46.56
	(1, 2)	72.93	72.99
	(2, 2)	168.2	168.4

and each frequency in the figure is normalized by each CLT solution. As seen in the results not only the present method but also FSDT and CLT give similar, reasonable solutions although there is a little difference in frequencies depending on the stacking sequence.

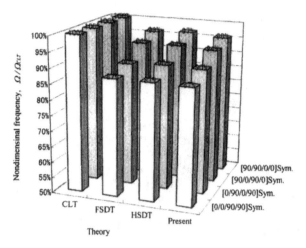

Figure 2   Nondimensional frequencies of laminated composite plates
(l=2.0, h=0.1, Mode (1, 1), I=J, κ^2=5/6)

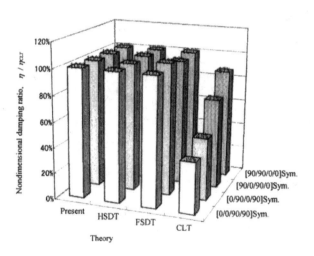

Figure 3   Nondimensional damping ratios of laminated composite plates
(l=2.0, h=0.1, Mode (1, 1), I=J, κ^2=5/6)

Figure 3 indicates the nondimensional modal damping ratios for the case of Figure 2, but each damping ratio in the figure is normalized by each present method's result ($I=J=4$, $K=3$). It is found from these results that there is a greater difference between CLT solutions and the other solutions compared with the case of frequency shown in Figure 2, and that the difference in damping ratio depends on the stacking sequence of the plate.

CONCLUSIONS

The main purpose of this work is to study the assumptions in deformation employed in the vibration analysis of laminated composite plates. For that purpose, the three-dimensional analysis was developed for simply-supported, cross-ply laminated rectangular plates, and three displacements were assumed in the polynomials of arbitrary order with respect to a thickness coordinate. A frequency equation was derived by substituting the displacements into the functional and minimizing it. For comparison, the plate theories (FSDT and CLT) were also applied to this problem. Frequency parameters and modal damping ratios were calculated numerically by the present method and by two plate theories, respectively, and the effect of difference in employed displacements on the accuracy of the solutions are discussed from the comparison of the results obtained. The following conclusions were reached from the numerical experiments.

(1) The accuracy of modal damping ratio is not more sensitive to the displacement functions employed in the analysis than the case of natural frequency.

(2) The difference in damping ratio between CLT solutions and the other solutions is larger than the one in natural frequency, and it depends on the stacking sequence of the plate greatly.

REFERENCES

1. Vinson, J. R. and Sierakowski, R. L., The Behavior of Structures Composed of Composite Materials, Martinus Nijhoff Publishers, Dordrecht, 1986.
2. Yang, P. C., Norris, C. H. and Stavsky, Y., "Elastic Wave Propagation in Heterogeneous Plates", Int. J. Solids Struc., 2, 1966, 665.
3. Whitney, J. M. and Pagano, N. J., "Shear Deformation in Heterogeneous Anisotropic Plates", ASME J. Appl. Mech., 37, 4, 1970, 1031.
4. Dong, S. B. and Tso F. K. W., "On a Laminated Orthotropic Shell Theory Including Transverse Shear Deformation", ASME J. Appl. Mech., 39, 4, 1972, 1091.
5. Reddy, J. N., "A Simple Higher-order Theory for Laminated Composite Plates", ASME J. Appl. Mech., 51, 4, 1984, 745.
6. Ohta, Y. and Narita, Y., "A Comparison of the Multi-layer and Lamination Theories in Vibration Analysis of Laminated Composite Plates", ASME PVP-Vol.279 Developments in a Progressing Technology, 1994, 21.

Simulation Model for Deformation Mechanism of Textile under Bi-axial Loading

Kazuhiro SAKAKIBARA
Division of Advanced Fibro-Science, Graduate School, Kyoto Institute of Technology
Matsugasaki, Sakyo-ku, Kyoto, 606-8585, JAPAN
E-mail: kazu00@ipc.kit.ac.jp

Dr. Atsushi YOKOYAMA
Division of Advanced Fibro-Science, Graduate School, Kyoto Institute of Technology
Matsugasaki, Sakyo-ku, Kyoto, 606-8585, JAPAN
E-mail: yokoyama@ipc.kit.ac.jp

ABSTRACT

The textile is utilized as not only garment material but also industrial material. The technique that predicts the deformation behavior of the textile with diversity becomes a useful tool. However, it is difficult that the deformation behavior of the textile is predicted for complicated woven structure and mechanical properties of self. Then, this study proposes numerical analysis using the computer as the technique. This study aimed at the construction of the numerical model of the textile for using for the analysis. To begin with, the effect of the woven structure on the deformation behavior was examined by the bi-axial tensile test and observation of internal structure. Next, the compressive property of fiber bundle was measured by the experiment and analysis. Woven structure and compressive property were introduced into the model of textile. The analysis was carried out using built model, and its result was compared with actual result. The importance of woven structure and compressive properties of fiber bundle in the analysis was shown.

INTRODUCTION

The textile has been used as not only the garment material but also the industrial material (ex: Composite material). It is necessary to predict the deformation behavior of the textile as tailored materials. However, there are some difficulties to predict the deformation behavior of the textile under various applied loading. These reasons are structure and mechanical properties of the textile. The textile has complicated structure. The geometrical structure of textile affects its mechanical properties. Consequentially, the textile shows a anisotropy behavior and strong non-linear behavior. In addition, the textile is usually used under the bi-axial load. These situation requests to develop the design methodology of the textile under multi-axial loading in actual use. The deformation mechanism of the textile is affected from not only mechanical properties of fiber material of textile but also woven structure Therefore, it is difficult to predict the deformation behavior of the textile. The numerical analysis is proposed as the technique that predicts deformation behavior of the textile. Until now, the trial that analyzes of the textile using the finite element method considering the characteristic of yarn was studied [1][2]. However, the effective analysis technique is not established yet. That is because there is no fundamental model of the textiles expressing the relation between woven structure and the mechanical property. This study aims the construction of the fundamental numerical model of the textile. The deformation behavior of the textile is receiving the effect of the woven structure [3]. Therefore, it is necessary to faithfully express the woven structure in the model of the textile for the numerical analysis. The actual deformation behavior of textile was get from tensile test. The deformation mechanism of the textile is investigated by using experimental data of bi-axial tensile test. The three-dimensional numerical model of textile is proposed by considering the deformation mechanism of textile and evaluate its effectiveness was examined in the comparison between analytical result and deformation behavior of actual textile. In this research expresses yarn as a fiber bundle.

EXPERIMENTAL APPROACH

Experimental procedure

The bi-axial plain weave cloth of the acryl was prepared to be a specimen. The outline of the textile used in this study is listed in **Table.1**. Dimensions of the specimen are length 120 mm, width 120 mm respectively for considering the cramp part (10 mm). In tensile test, the tensile load applied on side of specimen in the weft direction. In this test, the side edge on warp direction is not fixed and moves freely in all deformation process. The reason is for clarifying the change of the internal structure of the textile more. In addition, the deformation of the woven structure by the increase in the load was observed for measurement the change of the internal structure of textile with the extension. As a method of the observation, the textile observed the cross-section by the microscope, after it was fixed under each tensile load.

Experimental result and discussion

Load-Displacement curve of the textile is shown in **Fig.1**. The displacement is the migration length of the clamp, and it means the elongation of the textile. This curve shows that the deformation process of textile can be divided into two stages. The gradient of the curve changed, at the displacement 7.5mm. From this result, this study focuses the mechanism of two stages in the mechanical properties of the textile. Then, the change of the woven structure by the load was measured for examine the relationship between extension and woven structure. The photographs of cross-section of the textile are shown in **Fig.2**. The internal structure of the textile changes with the increase in the load. Aspect ratio of cross-section of fiber bundle and crimp angle was noticed as a parameter of the change of the internal structure of the textile. The change of crimp angle measured from the width-direction fiber bundle is shown in **Fig.3**. The change of the aspect ratio of cross-sectional shape of width-direction fiber bundle is shown in **Fig.4**. The tendency of two curves decreases rapidly at first and becomes constant after 7.5 mm of displacement. This position (7.5mm of displacement) agrees with the position where the gradient of the Load-Displacement curve changes. By these results, this study was considered as following. First stage in **Fig.1** is what the change of the inside of woven structure. Second stage is the behavior of the fiber bundle and filament shows. In short, the deformation behavior of the textile is affected by the change of woven structure and fiber bundle inside of textile.

Table.1 Outline of Textile.

Material of specimen	Acrylic fibers	
	Warp	**Weft**
Fiber bundle orientation number per 100 mm^2	74	71
Initial crimp angle (degree)	20.3	25.5
Aspect ratio of initial cross-section	0.72	0.51

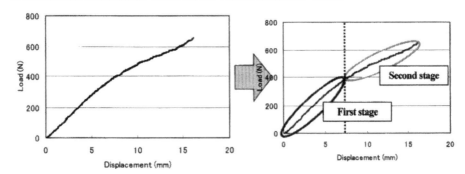

Fig.1 Load-Displacement Curve: Experiment.

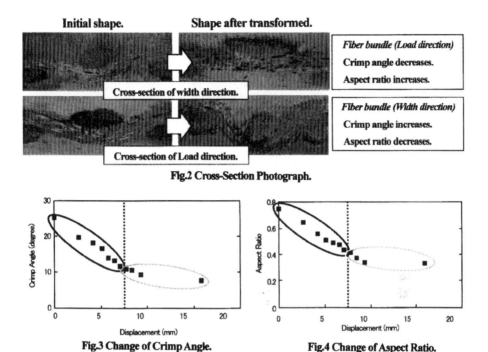

Initial shape.   Shape after transformed.

Fiber bundle (Load direction)
Crimp angle decreases.
Aspect ratio increases.

Cross-section of width direction.

Fiber bundle (Width direction)
Crimp angle increases.
Aspect ratio decreases.

Cross-section of Load direction.

Fig.2 Cross-Section Photograph.

Fig.3 Change of Crimp Angle.   **Fig.4 Change of Aspect Ratio.**

NUMERICAL MODELING AND ANALYTICAL PROCEDURE

By the experiment, it was clarified that the complicated structure of textile affected mechanical properties of textile. Then, it is necessary to consider the influence of woven structure and fiber bundle. This study began to build the model in consideration of them. However, it is impossible to create the numerical model of textile as a single model. Then, by dividing the whole textile into some stages, this study considered this problem. To begin with, the local model was built in each stage. Finally, those combining local model made the model of the textile.

Model of fiber bundle

The fiber bundle was modeled to express the mechanical behavior of the textile. The outline of the fiber bundle used for the build of the model is shown in **Table.2**. The shape of the model was decided from the cross-sectional of the actual fiber bundle. The three-dimensional element was used for the build of the model. This reason is to express the three-dimensional woven structure. The appearance of built numerical model of fiber bundle is shown in **Fig.5**. The orthotropic material was given to the material setting of the model.

The setting of the material property that differed in longitudinal direction and cross-sectional direction of fiber bundle was enabled in order to consider the anisotropy of the fiber bundle. Longitudinal material property (E_1) was deduced by the tensile test of the single fiber bundle. E_{11} and E_{12} were decided by approximating the stress-strain curve to bi-liner like the **Fig.6**. By this, the nonlinear material property was set. The material property of cross sectional direction (E_2, E_3) was deduced from compression test of the single fiber bundle. The method of compression test of fiber bundle is described in the following. To begin with, it puts on the fixed tensile load in fiber bundle. Afterwards, compressive load was given so that the cross-section of the fiber bundle might be crushed. Moreover, the fiber bundle of this condition was fixed by resin. Then, the microscope observed the cross section. From the result, the relationship between compression displacement and compressive load was summarized in the graph. The compressive displacement is the deformation volume of fiber bundle for the compression direction. The Compressive load-Compression displacement curve is shown in **Fig.7**. In this curve, the compression direction is described as "+". This study investigate that it was introduced on the model of this compressive property. However, it was impossible to adopt this result as it is. It is because the contact area changes with the increase in the compression displacement. Then, the numerical analysis on the assumption of compression test was carried out. The numerical model of fiber bundle cross section on the assumption of compression test is shown in **Fig.8**. In the analysis, the material properties of cross sectional direction were changed. Material property of cross sectional direction makes longitudinal material property change as a standard. Changing range is $1/2 \sim 1/20$ for the longitudinal material property. The material property in which analytical result was most similar to experimental result was adopted. The analytical result is shown in **Fig.9**. 1/15 analytical results of material property that longitudinally faced were similar to experimental value, as **Fig.9** showed. Therefore, 1/15 for the longitudinal material property at material property of E_2 and E_3 was adopted.

Table.2 Outline of Fiber bundle.

Material of fiber bundle	Acrylic fibers
Count of fiber bundle (Denier)	1850
Number of filaments in fiber bundle	1150
Number of twists of fiber bundle (T/m)	400

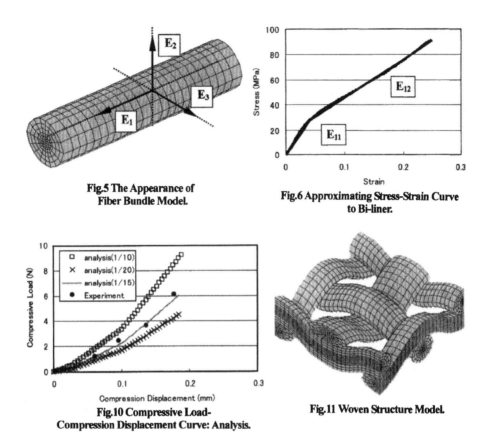

Fig.5 The Appearance of
Fiber Bundle Model.

Fig.6 Approximating Stress-Strain Curve
to Bi-liner.

Fig.10 Compressive Load-
Compression Displacement Curve: Analysis.

Fig.11 Woven Structure Model.

Model of woven structure

The fiber bundle model was combined in order to express the woven structure. The model of the woven structure is the smallest unit that can express the woven structure. The appearance of the model is shown in **Fig.10**. The contact condition was given between each fiber bundle model. This condition is for considering the effect of the contact of the fiber bundle fellow by the extension of textile. The shape of fiber bundle model in the textile was made to agree with actual thing. The internal structure of the textile was measured by observation of cross-section in actual textile. The model of the textile for the analysis was built in two types. The following were built: Model-1 considering the compressive property of fiber bundle and Model-2 that does not consider the compressive property of fiber bundle. Those two type models are for confirming the effectiveness of the compressive property of fiber bundle. The material property of cross sectional direction (E_2, E_3) of Model-1 set 1/15 values for longitudinal material property.

The material property of cross sectional direction (E_2, E_3) of Model-2 set the value equal to longitudinal material property. The material property set at each model is shown at the **Table.3**. The numerical analysis was carried out using the model of two types.

Table.3 Material Property.

	Model-1	Model-2
E_1:E_{11}, E_{12} (MPa)	655,296	655,296
E_2 (MPa)	$E_1/15$	E_1
E_3 (MPa)	$E_1/15$	E_1

ANALYTICAL RESULT AND DISCUSSIONCON

The load-displacement curve of analytical result and experimental result is shown in **Fig.11**. Two types of analytical result were compared with experimental result. Because of comparing, the gradient in the beginning of two types of curves agreed with the experimental result. This reason is that the effect of the contact of fiber bundle fellow in the deformation beginning is small. Afterwards, the error for the experimental result occurs in Model-2, when the deformation advances. The error increases with the advance of the deformation. In the meantime, Model-1 has expressed well the experimental value in deformation later stage. This was because the compressive property of fiber bundle was not considered in the Model-2. The effect of the contact of the fiber bundle fellow increases with the.advance of the deformation. Model-2 could be not correspondent to the effect of the contact of this fiber bundle fellow. Model-1 could be correspondent to these effects, as the compression property of fiber bundle was considered. In short, in the numerical analysis, it was important to consider compressive property of fiber bundle and was effective.

Moreover, Model-1 showed the tendency in two stages as well as the experiment. However, there was a little error on the tendency in the second stage. Then, the change of model-1 internal structure was focused. The change of model-1 internal construction is shown in **Fig.12**. The change of the crimp angle shows the tendency equal to the experimental result. However, the change of the aspect ratio of fiber bundle cross-section showed the tendency unlike the experimental result. The reason of this error, the material property of cross-sectional direction (E_2, E_3) did not consider the non-linearity. This study was considered like the following. The cause of the error in the second stage was the setting of the material property of fiber bundle cross-section. In short, it is necessary to reexamine the parameter introduced into the model.

By the result of the analysis, the following were confirmed: It is important to consider woven structure and compressive property of fiber bundle.

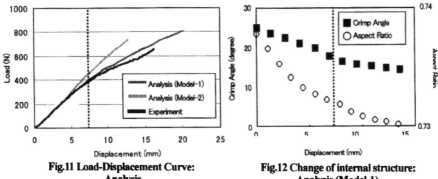

Fig.11 Load-Displacement Curve: Analysis.

Fig.12 Change of internal structure: Analysis (Model-1).

CONCLUSION

The fiber bundle in textile affects mechanical prosperities of the textile in property and change of shape that the self has. It is necessary to consider plane not only deformation behavior but also three-dimensional deformation behavior of textile inside, when the whole deformation behavior of textile was considered. In this study, the effect of the textile which compressive property of fiber bundle gave on mechanical properties was confirmed. In addition, in the model for the numerical analysis of the textile, it was important to consider the compressive property of fiber bundle, and it confirmed effect.

REFERENCE

[1] T. Hirai and T. Senba, "Analytical Study on the Tensile Properties of Plain Weaves Considering the Heterogeneous Compressive Properties of the Yarn. Part 1: Finite Element Analysis on the Biaxial Tensile Properties of Woven Fabrics.", Journal of the Textile Machinery Society of Japan Vol.33, 1980, T139-T149.

[2] T. Hirai, S. Fukui and T. Senba, "Yarn Tensile Property for Analysis. Finite Element Analysis on the Biaxial Tensile Properties of Woven Fabrics (Part 2).", Journal of the Textile Machinery Society of Japan, Vol.35, 1982, T79-T86.

[3] K. Sakakibara, A. Yokoyama and H. Hamada, "Deformation Mechanism of Texile under Uni and Bi-axial tensile Load.", Proceedings of 7th Japan International SAMPE Symposium, T. Ishikawa and S. Sugimoto, eds., Tokyo, Japan, 2001, 705-708.

Numerical simulation model for crashing behavior of FRP tube under Impact loading

Motoharu TATEISHI
MSC JAPAN LTD.
2-39Akasaka 5-choume
Minato-ku Tokyo 107-0052,Japan

Yoshihiro KITADA and Dr. Atsushi YOKOYAMA
Division of Advanced Fibro Science, Kyoto Institute of Technology
Matsugasaki, Sakyo-ku, Kyoto 606-8585,Japan
yokoyama@ipc.kit.ac.jp

Abstract

This paper reports the results from an experimental study on the crushing behavior of FRP tubes under quasi-static axial compression. FRP is known by the characteristic destruction called progressive crushing. And FRP exhibits good energy absorption performance. Because during this crush test, the axial load goes up to a maximum value and subsequently settled to a value. Results from a numerical analysis also are presented. This analysis model is considered the delaminations and friction. In this study, the characteristic crushing behavior called progressive crushing of FRP is investigated by FEM analysis.

Introduction

FRP have been used in the automotive industry primarily to reduce the overall weight of the vehicle, which results in fuel economy. So far they have been used mainly for secondary structures such as grill opening panels, door panels, bumpers, etc. The next step for FRP application in the automotive industry is using into primary structures. So FRP is said that is suitable material of parts for crashworthy structural automobiles recent years. Because FRP is considered that absorb collision energy larger than metals. On the other hand, in an actual design, FRP is hard to use it. Since the mechanical properties of FRP is very complicated. Then, the computer analyses, such as FEM, attract attention. If the mechanical characteristic can be expected in numerical analysis, application of FRP will expand. Mr. Hiroshi SAITO has issued the conclusion like the following in the proceedings (5). Although in both rectangular and circular pipes, fracture aspects were almost same in static and impact test. Thus in this study, the static analysis carried out to estimate crushing behavior of the impact.

Progressive crushing

Fig.1 (a) show load-displacement curve of FRP tubes under quasi-static axial compression. In this figure the load value increases as soon as starting compression, and after that, approximately constant load valued are shown stably. When it has such curve, an energy absorption performance is good, since the integration value becomes large. And FRP pipes show fracture modes as shown in Fig.1 (b) by making taper at the one end. This figure was quoted from the report of Mr. Hiroshi SAITO (1). Destruction of FRP is very complicated as seen. This destruction is combined fiber fracture, delaminations and so on. It is thought that the cause of such destruction occurring is existence of delaminations. This fracture mode is called progressive crushing.

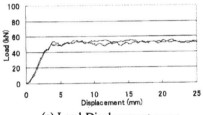
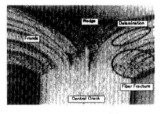

(a) Load-Displacement curve   (b) Fracture aspect

Fig.1 Load-Displacement curve and the cross sectional photograph of crushed specimen

Experiment Method

In this study, FRP tubes were examined as one of the energy absorption member. So it is necessary to lower the price to use FRP for motor vehicle. Then, there is the contrivance to decrease the process of production. The experiment was carried out using the initiator jig like Fig.2 (a). The purpose of using the jigs is initiation to occur progressive crushing without conducting the taper. The crashing behavior changes by the value of "R" in the Fig.2 (a). In this time, test jig with R=3mm was used for stable destruction. Because it was solved the best value of R is 3mm by our experiments. Compression load was applied to axial direction at constant speed quasi-statically by rate=5mm/min. The geometry of specimen is shown in Fig.2 (b). The FRP tube is made with epoxy and glass fiber by sheet winding method. And the experimental setup is shown in Fig.2 (c).

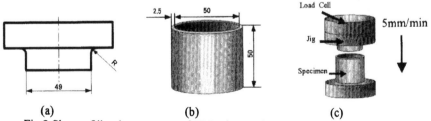

(a)   (b)   (c)

Fig.2 Shape of jig, the geometry of FRP tube specimen and experimental setup

Experimental Investigation

Photograph of after testing specimen by using jig and the load-displacement curves is shown in Fig.3. As the tube is compressed, the walls of the tube undergo axial compressive straining and large circumferential strains, leading to failure in the form of multiple matrix cracking, fiber fracture, delaminations and so on. The fronds were made and begin to buckle by jig compulsory. As a result, the Load-Displacement curve has a peak load and a large drop in initial stage. But in next stage it shows a stable load level. Therefore, by using the jigs, the FRP tubes show progressive crushing behavior under the compression loading, and it has the large energy absorption value.

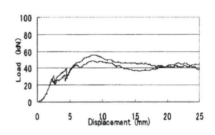

(a) Fracture aspect   (b) Load-Displacement curve

Fig.3 Photograph of after testing specimen using jig and Load-Displacement curves

Analytical Model

It is very difficult that the numerical model of FRP is made. Then, it was noticed that a primary reason of generating progressive crushing is the layered structure. The specimen and jig were expressed two-dimensional by the axisymmetric formulation analysis for purpose of the reduction in the calculation load in the analytical model. The finite element model of FRP tube is shown in Fig.4 (a). The condition of the large displacement analysis is used, because this analysis is containing large plastic deformation. And MARC is used this analysis in order to use the finite element method.

The each material property in this model was based on results at experimental value, and it is calculated by anisotropic elasticity theory in "The composite material engineering" (6) to considered to orthotropic material properties. The each material constant is shown in Table .1.

The feature of this analysis is that model have laminated structure. In order to consider the delaminations of FRP, the region where delaminations occur was assumed to be two or three places, and it calculated as three-layers or four-layers laminated structure like a Fig.4 (b). In this time, by existing the nodes double, and the inter-laminer shear stress was expressed by defining the separation force when the node separates from the element.

This conceptual figure is shown in Fig.4 (b). Here the separation force value was used the strength of matrix.

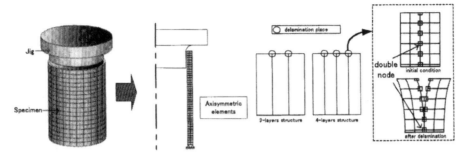

(a) Finite element model of FRP tube   (b) Conceptual figure

Fig.4 The definition of analysis model

Table.1 Material properties

E_{LL}	E_{ZZ}	E_{TT}	G_{LZ}	G_{ZT}	G_{TL}	Yield stress	ν_{LZ}	ν_{ZT}	ν_{TL}
17GPa	3.7GPa	17GPa	2.7GPa	2.7GPa	0.86Gpa	200MPa	0.30	0.65	0.30

Friction Definition

The surface area increases, when delaminations have occurred. Therefore, the consideration of friction is important, since the effect of the friction seems to increase. The calculation becomes difficult if the frictional force was changed with the discontinuous value. Then it is assumed the Coulomb friction model in following form (a) and applied.

$$\sigma_{fr} \leq -\mu\sigma_n \frac{2}{\pi} \tan^{-1}\left(\frac{v_r}{c}\right) \cdot t \qquad (a)$$

where σ_n : the normal direction stress
            σ_{fr} : the tangential stress or the frictional force
            μ : the friction coefficient
            t : the normal vector of the relative speed direction
            v_r : the relative slip speed
            c : a optional constant

The graph using this mathematical expression is shown in Fig.5 when ρ_n is 1, μ is 0.3 and various values of c. The vertical axis is frictional force and the horizontal axis is relative speed in this graph. Thus, this analysis is calculated by having assumed that it was continuous change like these. The adherence state is expressed by sudden change of the zero's neighborhood. And in the slide state load value is kept constantly. The most suitable value, as the actual friction, and calculable value is 0.01. It is adopted to c this time. From these results, fracture processes and mechanisms were discussed. It will be able to prove how influences the laminated structure to crushing behavior. And it will find how the friction of the between layer and layer or jig and specimen affects the energy absorption.

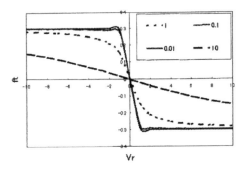

Fig.5 Approximation of an adherence-slide state of friction

Comparison of an experiment and analysis

At first, the analysis result is compared with an experiment results. Fig.6 is the load-displacement curve of simulation and experiment. Both of analysis results show the same tendency as an experiment result, by consideration laminate structure, friction between layers and specimen-jig, and orthotropic material properties. Especially the aspects that load reaches a peak at once and fall after that are similar to the load-displacement carve of this experiment. It is also in the same level in load value of experiment and analysis. Therefore, How influenced a laminated structure or friction to load value and the energy absorption was examined.

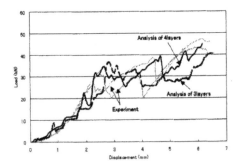

Fig.6 Load-Displacement curve of simulation

The effect of layer structure

Three analysis models are compared to see influence of laminate structure. One is the model with a single layer. The others are the models with three layers and four layers. These results are shown in a Fig.7. Those distributions are expressed by Von Mises stress when displacement quantity of jig is 2.5mm. And in those figures, the black area has reached the plastic region. Fig.7 (a) shows that only a tip deforms plastically. The destruction of a composite material cannot be expressed by this analysis. The model which is considered to laminated structure is shown Fig.7 (b) and (c). Thus, it turns out that delaminations were proceeding, and the stress is distributed with each layer, and reinforced regions reaches in a plastic stress. In this way destruction is advancing continuously. Therefore, it said that the laminated structure cause to happen progressive crushing. This cause is that the tubes were much stiffer in the axial direction and in circumferential direction than thickness direction.

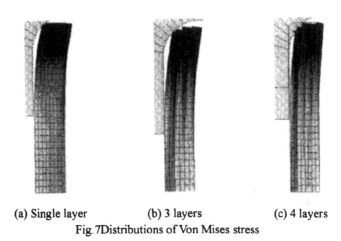

(a) Single layer   (b) 3 layers   (c) 4 layers

Fig. 7Distributions of Von Mises stress

The effect of the friction

The friction is important factor. It is anticipated large influence, since the laminated structure has many face area. Then the analysis carried out as follows 3 cases in order to examine the effects of the friction to search the influence degree of friction to energy absorption. Each case is following.

L: The friction between layers and jig-specimen was considered.

M: The friction between jig-specimen was considered.

N: The friction was not considered.

The friction coefficient in this analysis made to be 0.2 between layers and 0.3 between jig-specimen. This model of four layers structure is used for this analysis. Analysis result in Fig.6 corresponds with Case L.

The load-displacement curve acquired on these conditions is shown Fig.8 (a). Thus, it turns out that the load value is increase by considering frictions. Here the amount of energy absorption which integrates with the load of 0~6mm is compared each cases in Fig.8 (b). This figure said that the amount of energy absorption of Case L increases Case L from Case M by about 50%. Moreover, The way of the value in consideration of friction approaches an experiment value.

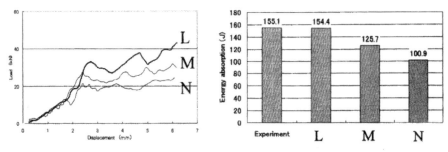

(a) The load-displacement curve each cases   (b) The amount of energy absorptions
Fig.8 The effect of the friction

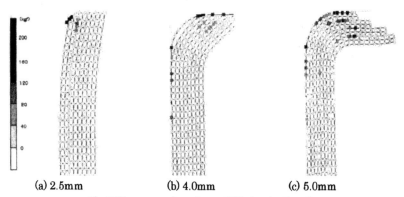

(a) 2.5mm   (b) 4.0mm   (c) 5.0mm
Fig.9 The generating place of friction in Case L

Moreover, the generating place of friction in Case L is shown in Fig. 9. This figures are the showed the friction force of each node by five levels from 0kgf to 200kgf. At the displacement 2.5mm in Fig.9 (a), it turns out that delamination was occurred inner side layers and friction was generated. At the displacement 4.0mm in Fig.9 (b), The place of delaminations was spread. Here, the values of friction force between layers were almost constant. And, the values of friction force between jig and specimen are higher than those. At the displacement 5.0mm in Fig.9 (c), buckling fracture was occurred and each layer came apart. The high friction force also into fronds is observed.

Conclusion

Complicated features of FRP, like a progressive crushing, can simulate by this analysis model. The analysis results could predict a reason why progressive crushing occurs is fragility of inter-laminer. And the reason why the continual destruction is generated efficiently is caused that too. Because of the stress is distributed at whole specimen by the crack preceding. The analysis results show that the overall energy absorption will be significantly affected by friction.

References

(1), Hiroshi SAITO, Ryuji INAI, Koji KAMEO, Hiroyuki HAMADA "Energy Absorption Properties of Composite", Proceedings of the Third Joint Canada-Japan Workshop on Composites, S. V. Hoa, H. Hamada, J. Lo, A. Yokoyama, eds., Kyoto, Japan, 2000, 198-202

(2), S J Bread and F-K Chang "Energy Absorption of Braided Composite tubes", Proceedings of Icrash2000-International Crashworthiness Conference, E C Chirwa and D Otte, eds., London, United Kingdom, 2000, 605-619

(3), D Graillet, J-P Ponthot, L Stainier "Augmented Lagrangian procedure for implicit computation of contact-impact between deformable bodies", Proceedings of Icrash2000-International Crashworthiness Conference, E C Chirwa and D Otte, eds., London, United Kingdom, 2000, 422-434

(4), Shu Ching Quek, Anthony M.Waas, Jennifer Hoffman, Venkatesh Agaram "The crushing response of braided and CSM glass reinforced composite tubes", Composite Structures 52, 2001, 103-112

(5). H. Saito, R. Inai, K.Kameo and H.Hamada "Energy Absorption Properties of Pultruded Rectangular Pipes", Proceedings of the 9th US-Japan Conference on Composite Materials, 383-389

(6). Tsuyoshi Hayashi "The composite material engineering", Hikagirenn press, 1971

Optimum Technique of Multiple Design Parameter for Die Design of Injection Molding using Discrete Optimization Technique

Kei TAMURA*, Akio TKAFUJI** and Atsushi YOKOYAMA***
Division of Advanced Fibro Science
Kyoto Institute of Technology
606-8585 Matsugasaki, Sakyo-ku, Kyoto, JAPAN
tamura01@ipc.kit.ac.jp*
takafuji.akio@sharp.co.jp**
yokoyama@ipc.kit.ac.jp***

ABSTRACT

The injection molds consume costs and time for manufacture. Therefore, it is important to predict the molding method. So, The CAE technology that injection molds design using the resin flow analysis.[1][2] This method is possible for checking the validity of a design. So, this technology cannot make the indication or the automatic correction of a point that improve design variables, such as the runner diameter, and the amount of resin, when fault arises in initial setting. Therefore, a design change is fixed by the trial and error of the engineer.

In this research, improvement of a genetic algorithm (GA) is advanced so that the multi-cavity molding of an injection molding can be automatically determined using the GA, which is one of the discrete optimization method.[3][4][5] First, the method of optimizing by dividing many design variables is proposed to the problem to which convergence speed falls with the increase in the number of design variables. Second, the optimization method that introduced the weight value is proposed on the calculation of objective function including the variable of which the property differs.

INTRODUCTION

The injection molding is the most widely used in the molding processing technology about injection-molds parts. The injection molding produces the various products such as the home electronics product and the electronic device that are around our life.

The injection molding has the following characteristic.

- The injection molding processing can be automated fully.
- The products of high geometric accuracy and complicated structure can be produced.
- The molded products with the same quality can be reduce production time by mass-production.
- The manufacturing of the injection molds takes a cost and time.

In the injection molding, it is difficult to measure the behavior of the resin inside the injection mold to be molding. By the troubles in molding process, such as generating with behavior, and remanufacturing of the injection mold, cost becomes high. Therefore, various CAE technologies are applied to the injection mold design.

However, though in this present technology, the confirmation of the validity of the design is possible, it is not possible to show how the injection molds may be improved, when the failure occurred for the initial setting. Therefore, trial and error methodology of the practical molding or the mold design change and know-how are required for engineer. Moreover, the multi-cavity mold can mass-produce the injection molding parts at once, and gives high productivity. However, if the number of cavity is increase, the difference has occurred to the filling up time of cavity. For the reason, it becomes difficult to make each cavity fill up simultaneously, and production efficiency falls.

The object of this study is to improve the productivity and reduce the production costs by achieving fill up simultaneously on the multi-cavity mold that has multiple variables. As an optimization method that automatically deduces optimum molding condition, the research has been advanced on optimization method that combined genetic algorithm (GA) with the resin flow analysis.

When the design variable increases, the following problem must be solved.

- The genetic operation of GA becomes late and the convergence speed with objective function declines.
- In order to treat the variable with difference dimension, a difference appears in the influence of calculation on the objective function, and the optimal solution is not obtained.

In this research, In order to solve this problem, the following two improved algorithms are proposed.

- The population of chromosome combination is reduced by dividing a design variable, and GA is performed separately.
- By introducing the weight function into the calculation of the objective function the variables balance of which the property differs is controlled.

By improving GA, it gives "usable solution " faster than previous algorithms.

THE OPTIMIZATION OF INJECTION MOLDING

In order to design the injection mold, it is necessary to consider many objective functions such as the condition in which the short shot does not arise, reduce cycle time, and cost reduction by the reducing the runner volume. In addition, it is difficult to analytically exactly optimize these design variables, because the flow of the resin in the cavity of mold is nonlinear physical phenomenon. GA is used in order to optimize the design variable in these trade-off relations using objective function of different property of the simultaneous filling and the runner volume in multi-cavity mold. The advantage using GA is the algorithm of the survival of the fittest by trial and error and selection, and is to be able to calculate "usable solution" in a short time.

In this research, GA is improved so that many design variables can be treated, and "usable solution" can be calculated in a short time using the objective function with which property differs.

GENETIC ALGORITHMS (GA)

The Genetic Algorithm (GA) is the optimization method to begin to find out a best molding condition automatically and is effective to the discrete design problem. The flow chart of GA is shown in Fig.1.

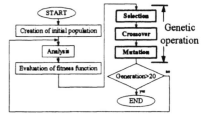

Fig.1 Flowchart of Genetic Algorithm

GA changes the design variable into the binary digit character line and operates of it as the chromosome that has design information. For this chromosome, it is left by choosing the chromosome with the high evaluation by carrying out the genetic operation (selection, crossover, mutation). The chromosome of a high evaluation value is made from repeating numbers of genetic operation, and it is possible to obtain an optimal condition.

THE PROBLEM OF GA

The model of the multi-cavity injection mold using for computer analysis is shown in Fig.2.

When the runner diameter is defined as a design variable, the range of the real number of a~b is taken. The range of the value that

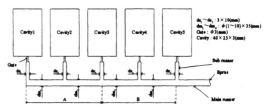

Fig.2 Schematic diagram of the 5 take cavity

the runner diameter can take is shown in the bit string of the n bit, when design variable shown by real number is handled in GA, as it is shown by Fig.3.-(a). It allocates the numerical value of 2^n that is shown at the bit string for the real number value of the design variable. It is necessary to take the bit string that shows one design variable long in order to decrease serration width of this real number. On the other hand, in treating two or more design variables, it uses the gene string that connected the design variable expressed in each bit string shown in Fig3-(b). When the number of design variable is m

which consists of string of the n bit, the gene string of $m \times n$ is used.

In order to determine the optimal design value of injection molding, when satisfying simultaneously the parameter in which property differs from simultaneous filling, comparison of the filling up time difference evaluated in the range for several seconds and other parameters is needed. In addition, in case of the search by GA, the design variable is randomly decided. Therefore, the technique that the size of term in the time and term of the parameter of which the property differs in the objective function dynamically suits is required.

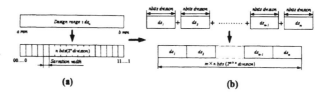

(a)   (b)

Fig.3 The discretization in the design range

THE APPLICATION OF GA TO THE DIVIDED DESIGN VARIABLE

In case of GA, when a bit string becomes long, convergence speed will fall and obtaining a solution needs more time. Then, by dividing two or more design variables used for GA, and performing genetic operation separately, it tried to solve this problem. As shown in Fig.4, a design variable is represented as bit series called a chromosome. When this bit series is l bit, the combination of a chromosome becomes 2^l way. The bit string length shortens by respectively dividing the bit string of l bit into bit string of m bit and n bit. The population of

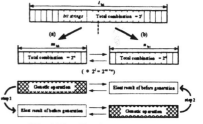

Fig.4 The division application of GA to the bit string

the chromosome combination becomes respectively 2^m and 2^n. The bit strings that perform genetic operation of GA can be shortened by dividing a design variable. Moreover, in order to operate GA by turns like Step1 and Step2 of Fig.4, while (a) is carrying out GA operation, (b) is being fixed only with the best value(elite solution) of a filling up time difference of previous generation. For the reason of above, population of the chromosome combination decreases and the improvement of the convergence speed becomes possible. This method is applied to the model in Fig.2

The objective function is represented as follows, when the main runner diameter is defined as design variable $dm_1 \sim dm_4$ and the filling up time of each cavity is defined as t.

$$f(dm_i, dm_j) = \frac{1}{n}\sum_{k=1}^{5}\sum_{l=1}^{5}\left|t_k - t_l\right| \qquad (n=\text{The combination of the Cavity number})$$

The objective function is the value that added the time difference between the filling up time of each cavity. The runner diameter values are optimized to make this objective function minimize. As shown in Fig.5, the main runner sets dm_1 and dm_2 to A, dm_3 and dm_4 to B. A,B individually carries out each operation of crossover and mutation within the genetic operation of GA as a design variable group in which each was separate, and it gets the elite solution. The process that has separately applied cross and mutation to the design variable group is shown in Fig.5. Then, crossover and mutation are performed only to B. At this time, A is only an elite solution of previous generation. Thus, when dividing and performing genetic operation to two or more design variables, another design variable is being fixed only with an elite solution of previous generation. For this reason, the population of the chromosome combination decreases. Therefore, it improves the convergence speed of a solution. The flow chart of this method is shown in Fig.6. As shown in Fig.6, by this method, a series of operations from Analysis(simulation module) to Mutation are performed in order to every design variable group.

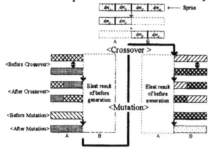

Fig.5 Crossover and Mutation of GA

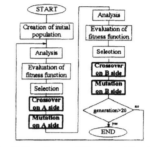

Fig.6 Flowchart of Genetic Algorithm

ANALYTICAL RESULT AND DISCUSSION I

The optimization of the runner diameter of the multi-cavity mold for injection molding, using the GA dividing design variable group technique, has been done(Case1). And, the optimization by the conventional method which does not divide a design variable was also computed(Case2), and it compared with the result of proposed method. The convergence situation of the objective function is shown in Fig.7.

Fig.7 shows Case1 continues rapid decrease from the beginning and it reaches the stable value at the early generation in comparison with Case2. This tendency is considered because of the population of chromosome combination individuals was decreased by dividing a design variable and performing GA separately. As the result, it was possible to obtain the convergence in the early generation. Therefore, when large numbers of design variables exist, it was to use this technique, and it became possible that "usable solution" was more quickly got. Nevertheless, before reaching 10 generations in Fig.7, the value of Case1 and Case2 is reversed. If Case1 and Case2 continue calculating, finally it will reach the same solution. However, in order to search for the best optimal result, simple Case1 is quicker than Case2 which steps on a complicated procedure proposed this time. For this reason, such an inversion phenomenon happens. The purpose of proposed technique is to obtain the "usable solution" for industry quickly. This result

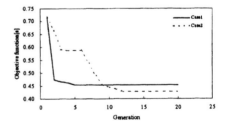

Fig.7 The comparison of the result

show the proposed method is satisfied this purpose.

THE METHOD OF CALCULATING THE OBJECTIVE FUNCTION WITE THE VARIABLE OF TWO OR MORE KINDS OF DIFFERENT CARACTER

The optimization method of multi-cavity mold for injection molding in which the multiple variables of which the property differs satisfy simultaneously optimum forming condition is considered. The schematic of analysis model is shown in Fig.8.

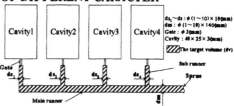

The design variable is defined as sub runner diameter $ds_1 \sim ds_4$, main runner

Fig.8 Schematic diagram of the 4 take cavity

diameter dm. In addition, the objective function is given by filling up time difference (t), and runner volume (dv) calculating from runner diameter. (Slant hatching area in Fig.8) The object for above is to control the runner volume for reducing the resin consumption and reduction of the production cost. For this model, it optimizes and the best result of a filling up time difference and runner volume is searched for as an elite solution. The value is small for a filling up time difference, when it is compared with the runner volume. Therefore, when it adds simply, the influence of runner volume commits the objective function strongly, and the influence of a filling up time difference is no longer taken hardly into consideration. Then, in this study, the weight function R was introduced into objective function in order to adjust the degree of the effect of filling up time difference and runner volume. The objective function of this method is set up as follows.

$$f(ds_1, ds_2, ds_3, ds_4, dm) = \sum_{i=1}^{4}\sum_{j=1}^{4}\left|t_i + t_j\right| + R\sum_{k=1}^{5}v_k \quad \left(\frac{\text{Elite filling up time}}{\text{Elite volume}}\right)$$

The right-hand side of objective function the first term is a filling up time difference of each cavity, and the second term is the runner volume. Weight function R would be able to adjust the effect of the runner volume on the objective function. GA obtains an optimal value by operating gene repeatedly even shat generation. Since a filling up time difference and runner volume change variously in the calculation process, it is impossible to set up a specific constant as a weight function Then, the ratio(t/V) of the elite solution of the filling up time difference and runner volume that are obtained by the calculation of GA with this research, were made the weight function in the calculation of the objective

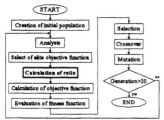

Relation between ratio and weight value	First term		Second term
$R < \dfrac{t}{V}$		<	O
$R = \dfrac{t}{V}$	△	≅	△
$R > \dfrac{t}{V}$	O	>	

Fig.10 Flowchart of Genetic Algorithm

Fig.9 Influence of the dignity value to the purpose function

function of the next generation. The effect of weight function R on the calculation of the objective function is shown in Fig.9. As shown in Fig.9, the effect of runner volume of the second term increases on the objective function, when R is smaller than t/V. It is possible that first term and second term of the objective function keep the relation in which a balance was removed, when R is equal to t/V. Then, the objective function becomes a result that the effect of a filling up time difference of the first term strengthens more, when R is bigger than t/V. Thus, the effect of the change of the weight function on the calculation of the objective function is big. In this research, weight function is. defined as $R=t/V$, to optimization while keeping the size of first term and second term. In addition, a weight function is suitably changed into the optimal value automatically by using the ratio of an elite solution of previous generation for a weight function The flow chart of the optimization method containing the objective function is shown in Fig.10.

ANALYTICAL RESULT AND DISCUSSION II

We have applied weight function for optimize method using two heterogeneous objective function such as filling up time difference and runner volume, while computing the balance of the influence to the objective function equally.(Case1) Moreover, calculation of the objective function by the conventional method was also performed.(Case2) The change of the filling up time difference and the runner volume in each method is shown in Fig.11,12. According to Fig.12 that it is as a result of Case2, runner volume is decreasing as the generation increases. However, the filling up time difference repeats fluctuation and tends to increase it from the middle. This is considered because the runner diameter became narrow in order to decrease runner volume, and the resin became difficult to flow. On the other hand, the results of Case1 show filling up time difference and runner volume are deceasing rapidly in Fig.11. The filling up time difference and runner volume are considered with since balance was maintained by the weight function which introduced this into the objective function, as a result of controlling the influence of calculation on the objective function. Thus, it became possible to optimize controlling simultaneously the variable from which the property differs of a filling up time difference and runner volume by introducing a weight function into the objective function. In addition, when comparing reaching speed to "usable solution" from the result with the objective function of Case1, Case2, it becomes like Fig14. Case1 which introduced the objective function shows reduction in a value rapidly

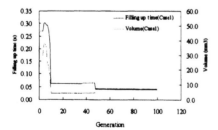

Fig.11 The convergence situation of
a design variable (Case1)

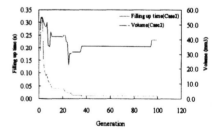

Fig.12 The convergence situation
of a design variable (Case2)

in the early generation, and has reached to "usable solution" about in the generation of the half of Case2 so that clearly from Fig.14. Therefore, this method made it possible to obtain "usable solution" to calculate early more, carrying out simultaneous control of each variable, when the variable from which parameter differs is treated with the objective function.

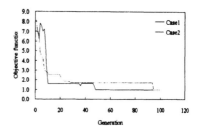

Fig.14 Comparison of convergence speed

CONCLUSIONS

In the traditional GA optimization procedure having a lot of design variables such as the multi-cavity injection mold, there are problems as follows.

- Since a bit string becomes long and the population of a chromosome combination increases by the increase in a design variable, convergence speed falls.
- Strength of influence in the calculation of each variable in the objective function constituted by the variable from which parameter differs.

In this research, we proposed the new optimizing method to solve each problems.

- When a design variable increased, it became possible by dividing a design variable and performing GA separately to obtain "usable solution" more early.
- By introducing the weight function into the objective function in order to satisfy simultaneously the variable of which multiple property differs, it is possible to get "usable solution" early more.

The results of this paper suggest that proposed method is the good procedure in the optimal design of the injection mold. .

REFERENCES

1. G. Titomanlio, V. Speranza and V. zbrucato, "On the Simulation of Thermoplastic Injection Moulding Prpcess", Intern. Polymer Processing XII(1997)1, 45-53.
2. R. Pantani and G. Titomanlio, "Analysis of Shrinkage Development of Injection Moulded PS Samples", Intern. Polymer Processing XIV(1999)2, 183-190.
3. DAVID E. GOLDBERG, GENETIC ALGORITHMS in Search, Optimization & Machine Learning, ADDISION-WESLEY PUBLISHING COMPANY, INC.
4. S. Yamane, A. Yokoyama, A. Takafuji and T. Kubo, "Optimization of Design for injection Molding Toward the Compound Problem Using GA", Proceedings of the eighth Japan Society of Polymer Processing, 2000, 327-328.
5. K. Tamura, A. Yokoyama, "Compound Method of Optimum Design Technology for Injection Molding", Proceedings of the ninth Japan Society of Polymer Processing, 2001, 59-60.

Application of the Discriminant Analysis using Mahalanobis Distance to Delamination Identification of CFRP

Dr. A.Iwasaki, Assistant Prof. A. Todoroki, Dr. Y. Shimamura, Prof. H. Kobayashi
Department of Mechanical Sciences and engineering
Tokyo Institute of Technology
2-12-1,Ohokayama, Meguro-ku
Tokyo, 152-8552, Japan
aiwasaki@ginza.mes.titech.ac.jp

ABSTRACT

The present paper proposes a new diagnostic tool for the structural health monitoring that employs a Mahalanobis Distance adopted in discriminant analysis. Structural health monitoring is a noticeable technology for aged civil structures. Most of the structural health monitoring systems adopts parametric method based on modeling or non-parametric method such as artificial neural networks or response surfaces. The conventional methods require FEM modeling of structure or a regression model. This modeling needs judgment of human, and it requires much costs. The present method does not require the process of modeling, in order to identify the damage level using the discriminant analysis. This suggest us, this technique is applicable to the health monitoring system, which identifies the damage of the structure, easily.

In the present paper, we developed the damage diagnostic methods using minimum Mahalanobis Distance as the method for identifying delamination from data. We applied this method to identifications of delamination crack of CFRP Plate. Delamination cracks are invisible and cause decrease of compression strength of laminated composites. Therefore, health-monitoring system is required for CFRP laminates. The present study adopts an electric potential method for health monitoring of graphite/epoxy laminated composites. The electric potential method does not cause strength reduction and can be applied existing structures by low cost. We compared this damage diagnostic method with Neural Network method. As a result, it was shown that this method is effective for identification of damages.

INTRODUCTION

CFRP have high specific tensile strength and stiffness, and are good for the major structure of airplanes and space instruments. The intensity between layers of CFRP is weak and a delamination crack between layers is produced by a comparatively slight impact. The delamination cracks are invisible and cause decrease of compression strength of laminated composites. Therefore, diagnostic method for the delamination crack of laminated composite is required. The present study adopts electric potential method for delamination diagnosis of graphite/epoxy laminated composites. Carbon fiber in a composite material has conductivity. In electric potential method, a damage of large area of the structure is diagnosed by conducting electricity to composite structure. Electric potential method does not cause strength reduction and can be applied to existing structures by low cost. The present study focuses on the damage identification method of the delamination crack using electric potential method.

Various researches about electric potential change of the composite structure have been proposed[1,2]. Effectiveness of electric potential method is proved in our former research [3]. In the research, delamination crack in a composite beam was identified from electric potential change between electrodes that attached to the surface of the composite beam. Relation between the delamination crack and electric potential change is identified from an inverse problem. The inverse problem in damage identification is one of optimization problem of estimation error minimization, and solved by common optimization tool such as neural network and genetic algorithm usually. But in general, these methods require try and error process to derive optimum solution, and demand much calculation and human cost.

Then, we propose a method that identifies the damage from statistical analysis for the purpose of constructing the simple method for damage identification. In the method, we use distinction analysis using Mahalanobis distance as statistical distance indicator. Using discriminant analysis, damage identification that requires only simple calculation is constructed.

In this research, this method is applied to identification of delamination crack of CFRP Plate using electric potential method. Result is compared with a result that derived from inverse problem using neural network, and effectiveness of this method is investigated from the result.

DISTINCTION ANALYSIS USING MAHALANOBIS DISTANCE

Discriminant analysis is one of statistical method to distinguish the group to which an individual piece belongs. Mahalanobis distance[4] is statistical indicator for the discriminant analysis that shows good result when distribution of each group is different. Mahalanobis distance shows distance from each data group calculated from distribution of the data of each group.

The discriminant analysis using Mahalanobis distance demands complicated calculation compared with the case where linear discriminant function is used. But in the case of a variance-covariance matrix of each group are different, discriminant analysis using Mahalanobis distance shows good accuracy than using linear discriminant function.

In the discriminant analysis using Mahalanobis distance, two or more discriminant groups are created from reference data. And the group, which an arbitrary data belonging to, is decided as the group that takes minimum Mahalanobis distance. Where the data groups which created from reference data is called as standard spaces. Generally, Mahalanobis distance from a standard space i is represented with following formula.

$$D_i = \frac{1}{T}\sum_{l=1}^{T}\sum_{k=1}^{T}\left(X_{il} - \overline{X_{il}}\right)S_{lk}^{-1}\left(X_{ik} - \overline{X_{ik}}\right) \tag{1}$$

Where D_i is Mahalanobis distance from the standard space i. X_{ij} is standardization of predictor valuable $x_j(i: 1\sim N, j: 1\sim T)$ derived as follow.

$$X_{ij} = \frac{x_j - \overline{x}_{ij}}{\sigma_{x_j}} \tag{2}$$

Where N is number of standard space and T is freedom of data. S is standardized variance-covariance matrix derived as follow.

$$S_{lk} = \frac{1}{T}\sum_{i=1}^{T}\left(X_{il} - \overline{X}_l\right)\left(X_{ik} - \overline{X}_k\right) \tag{3}$$

Where $\overline{X}_l$ is average of X_{il} ($i: 1\sim n$). Where n is number of reference data of standard space i.

Probability density of Mahalanobis distance follows χ^2-distribution and probability density function is defined as follow.

$$p_i = \frac{1}{2^{(T/2)}\Gamma(T/2)} x^{\left(\frac{T}{2}\right)-1} \cdot e^{-\frac{1}{2}D_i} \tag{4}$$

Where p_i is probability density.

Mahalanobis distance takes less value when the individual piece is close to the standard space, and takes large value when it is far from the standard space. Because of that attitude, the data is judged as belonging to the standard space that takes minimum Mahalanobis distance in this method.

DELAMINATION IDENTIFICATION OF CFRP PLATE
USING ELECTRIC POTENTIAL METHOD

Specimen for experiment

As mentioned before, the new damage diagnostic method is applied to delamination identification of a CFRP Plate, and the effectiveness of the method is experimentally investigated here. Specimen configuration of the present study is shown in Figure 1. The specimen is a CFRP Plate with a thickness of 1.4mm and stacking sequence of the specimen is $[0_2/90_2]s$.

In order to measure the electric potential change which was caused by delamination crack, eight electrodes are mounted on the single-sided surface of specimen. 5 electrodes are mounted to the direction of 0 degree, and 2 electrodes are mounted to the direction of 90 degree as shown in Figure 1. Copper foils which were co-cured with CFRP Plate are used as electrodes.

The delamination crack is created between layers using indentation fracture method from electrode side. Since stacking sequence of specimen is $[0_2/90_2]s$, delamination crack is made in interlaminar of 0/90 layer of electrode side. C-scan is used for measurement of the delamination crack.

In the present study, location of delamination crack is defined as center of delamination crack which was observed from C-scan image, and size is defined as max length of C-scan image.

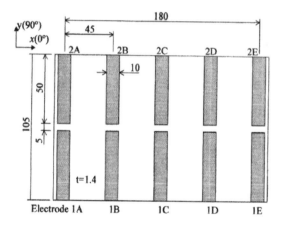

Figure.1 Specimen configuration

Procedure of identification of delamination crack

Procedure of identification of delamination crack is follow.
(1) Measure the electric potential change in the various cases of size and location of delamination crack. And named these data as reference data.
(2) Define the location and size levels of delamination crack, and create the standard space of each level from the reference data.
(3) Calculate the statistic values of each standard space such as variance-covariance matrix.
(4) Derive Mahalanobis distances of presuming data to the standard spaces.
(5) Discriminate the level from the Mahalanobis distance to which the presuming data belonging to.

In this research, 64 data measured from experiments are used as reference data and 12 data are used as new data. As shown in Table 1, each data consists of the position of x-direction, the position of y-direction, size of delamination crack, and eight electric potential changes between electrodes. Location and size level of delamination are defined as Table 2. Location of x direction is divided to four levels and y direction is divided to two levels. Size is divided to three levels.

The number of data which divided to each level is shown in Table 3. As shown in Table 3, delamination crack is created almost equally except size which is difficult to control.

Table 1 Data for estimation

Test No.	Location x[mm]	Location y[mm]	Size [mm]	1A-1B	1B-1C	1C-1D	1D-1E	2A-2B	2B-2C	2C-2D	2D-2E
1	68.69	24.88	9.23	9	19	12	3	1	2	1	3
2	158.5	77.5	11.61	1	8	16	14	4	11	154	280
3	159.91	27	23.25	-2	0	63	90	2	1	0	3
⋮	⋮	⋮	⋮						⋮		
64	112	83.73	22.3	-3	-1	-1	1	0	131	170	37
ND1	20.54	84.89	14.97	-4	-2	0	-1	74	27	0	1
ND2	114.6	24.03	13.35	-3	51	76	25	-3	-4	-3	0
⋮	⋮	⋮	⋮						⋮		
ND12	110.49	81.09	17.73	-6	-5	-4	-1	-5	14	55	48

Table 2 Location and size level of delamination

	Level [mm] 1	2	3	4
x	$0 \leq x < 45$	$45 \leq x < 90$	$90 \leq x < 135$	$135 \leq x < 180$
y	$0 \leq y < 52.5$	$52.5 \leq y < 105$		
Size	$13 \leq size < 17$	$17 \leq size < 21$	$21 \leq size < 25$	

Table 3 Level distributions of measured data

		Level 1	2	3	4
Reference Data	x	16	18	15	15
	y	28	38	---	---
	Size	31	14	19	---
New Data	x	3	3	3	3
	y	5	7	---	---
	Size	6	3	3	---

Identification results

Identification of location of delamination crack

Mahalanobis distance to standard space of each level has calculated at first. Table 4 shows average value of Mahalanobis distance to each standard space of reference data and new data. The average values of Mahalanobis distance to the actual levels are precisely smaller than that to another location level as shown in Table 4. This is approved to both reference data and new data. By this result, we confirm that Mahalanobis distance to actual level takes short distance at the location diagnosis of delamination crack.

Then Table 5 shows reliability of estimation on location identification of delamination cracks by this method. Damage identification is succeeded by the performance of 95 or more percent for reference data and succeeded by the performance of 83 or more percent for new data. On the basis of the performance result, the new diagnostic method provides high performance in location identification of delamination crack.

Table 4 Average of Mahalanobis Distance (Location)

		Average of Mahalanobis Distance(MD)	
		MD from actual location level	MD from another location level
Reference	x	7.5	7420
Data	y	7.75	890
New Data	x	22.8	1670
	y	41.3	212

Table 5 Reliability of estimation (Location)

		Reliability of Estimation
Reference	x	95.3%
Data	y	98.4%
New Data	x	83.3%
	y	83.3%

Identification of size of delamination crack

Table 6 shows average value of the Mahalanobis distance to each standard space of reference data and new data. Average value is not shows a clear difference compared with the case of location, and it means that identification of size level by this method has little difficulty. Table 7 shows reliability of estimation on size identification of delamination cracks by this method. Damage identification is succeeded by the performance of 63 percent for reference data and 42 percent for new data. But average value to actual level is sufficiently smaller than value to another level, and it is considered that it is improvable increasing number of the reference data.

Table 6 Average of Mahalanobis Distance (Size)

	Average of Mahalanobis Distance(MD)	
	MD from actual location level	MD from another location level
Reference Size	7.63	41.2
New Size	5.73	5.68

Table 7 Reliability of estimation (Size)

	Reliability of Estimation
Reference Size	62.5%
New Size	41.7%

COMPARISON WITH A NEURAL NETWORK

For investigation of effectiveness of this method, identification of delamination crack using neural network is performed in this chapter. Learning vector quantization neural network(LVQ)[5] are used for diagnostic method. LVQ is the supervised self-organization neural network which is mostly use for pattern recognition. OLVQ, which performs automatic optimization of a learning rate, is used for learning method. The number of neuron of each layer and training iterations for the delamination identification is shown in Table 8.

Then Table 9 shows reliability of estimation on location and size identification of delamination cracks by LVQ. As shown in the table, this method and LVQ show almost equivalent diagnostic accuracy. But in general, LVQ method requires trial and error process to setup an optimum neural network form and number of training iterations to derive optimum result. As the reason, the new diagnostic method using the discriminant analysis which does not require such a preprocessing is effective for identification of delamination crack of CFRP plate.

Table 8 Configuration of LVQ neural network

	x	y	Size
Number of neuron on each layer			
Input layer	8	8	8
Competitive layer	11	4	7
Otput layer	4	2	3
Number of training iterations	500	240	560

Table 9 Reliability of estimation (neural network)

		Reliability of Estimation
Reference Data	x	87.3%
	y	93.8%
	Size	68.8%
New Data	x	66.7%
	y	91.7%
	Size	50.0%

CONCLUSIONS

By conducting discriminant analysis using Mahalanobis distance at identification of delamination crack of CFRP Plate, the present paper describes the new damage diagnostic method with only simple calculation. Main topics of conclusion are following.

(1) Discriminant analysis using Mahalanobis distance is effective for damage identification of the structure.

(2) The new damage diagnostic method shows almost equivalent accuracy compared with the damage diagnostic method using neural network.

(3) The new damage diagnostic method requires only simple calculation compared with neural network and useful for damage diagnostic method for structural health monitoring.

REFERENCES

Edited Conference Proceedings

1. J.Abry, S.Bochard, A.Chateauminois, M.Salvia and G.Giraud, "In situ monitoring of flexural fatigue damage in CFRP laminates by electric resistance measurements", Proceedings of 4th ESSM and 2nd MIMRconference, Harrogate, 1998. p. 389–396

Journal References

2. P.Irving and C.Thiagarajan, "Fatigue damage characterization in graphite fibre composite materials using an electric potential technique", Smart Materials and Structures, No.7,(1998),456–66.
3. A.Todoroki and Y.Tanaka,"Delamination identification of cross-ply graphite/epoxy composite beams using electric resistance change method", Composite Science and Technology, Vol.62, No.5,(2002), 629-639
4. D.Maesschalck, D.Jouan-Rimbaud and D.L. Massart, "Tutorial The Mahalanobis distance", Chemometrics and Intelligent Laboratory Systems, Vol.50, (2000),1–18

Book References

5. T.Kohonen, Advanced Neural Networks, Elsevier, Amsterdam, 1990

CODE DEVELOPMENT

Canadian research on FRP in construction

A. Ghani Razaqpur
Department of Civil and Environmental Engineering
Carleton University, Ottawa, Ontario, Canada K1S 5B6
ghani_razaqpur@carleton.ca

ABSTRACT

Canadian research on FRP applications in construction is described. The research is carried out mainly in Canadian universities, or in some Canadian federal government laboratories, and is focused on the design, construction, and repair/strengthening of reinforced concrete structures. The research on new structures focuses primarily on bridges, particularly on bridge decks, while research related to repair/strengthening involves bridges and buildings. The FRP of choice in Canada is CFRP, with some research involving GFRP. The research results have led to the development of FRP design codes for both buildings and bridges and to a number of demonstration projects and field applications. To date, two highway bridges have been constructed in Canada in which some of the bridge girders are prestressed with CFRP tendons. In addition, a number of other bridges have been constructed where the deck slab and barrier walls have been partially or entirely reinforced with CFRP or GFRP bars. CFRP sheets and laminates have been used to repair/strengthen concrete structures for increased flexural, axial, and shear strength.

INTRODUCTION

The corrosion and deterioration of steel reinforced concrete structures in aggressive environments motivate interest in FRP applications in construction. The FRP in construction research in Canada started almost twelve years ago, and the initial efforts to gauge the state of the art in Europe, Japan and the United States were supported by the Canadian federal government. A wide range of research initiatives followed these fact-finding activities. The federal government approved the establishment of a centre of excellence related to FRP research in construction. The centre, called ISIS (Intelligent Sensing and Innovative Structures) proposed three principal areas of research: (a) FRP in new construction, (b) FRP in repair and rehabilitation of structures and (c) intelligent monitoring of structures, combining fibre optic sensor technologies and FRP. A fuller account of relatively recent ISIS research can be found in a special issue of the Canadian Journal of Civil Engineering (1). Since that publication, no significant shift in emphasis has occurred. In addition, researchers at the National Research Council of Canada and at other Canadian universities not affiliated with ISIS have been conducting wide-ranging research on FRP.

Canadian research has been focused primarily on bridges, but Public Works and Government Services Canada (PWGSC), whose real properties section is mainly interested in advanced and innovative building-related technologies, has taken the lead with respect to FRP research and applications in buildings. Furthermore, PWGSC, in collaboration with the Canadian Standards Association (CSA), took a major initiative by embarking on the development of a full-fledged standard for FRP applications in building structures. As others will present the details of this standard in this workshop, it will not be discussed further here.

In the following, a broad overview of the Canadian FRP research activities in construction will be provided. Due to time and space limitations, only the salient research activities will be described. Subsequently, a research project in which the author has been involved over the past four years will be discussed.

FRP IN CONSTRUCTION

The majority of FRP research in construction deals with application of FRP to concrete structures, either as internal reinforcement in lieu of steel bars and tendons, or as externally bonded sheets and laminates. External reinforcement is used to either increase the strength and energy absorption capacity of structures or to restore their original capacity which may have been reduced by environmental damage, such as corrosion of reinforcement, or by accidental damage caused by impact and collision.

Strength increase may be required due to change in a building occupancy or due to increase of live load on bridges. Generally, flexural and shear strength increase is required. Flexural strength increase is achieved by bonding one or more layers of FRP sheets or laminates to the tension face of a beam or slab while shear strength is effected by applying U-shaped stirrups to the web of beams and girders. Externally applied stirrups are made of flexible FRP sheets, generally CFRP, and bonded to the concrete surface by means of epoxy. Columns are strengthened by wrapping FRP sheets around them and bonding the sheet to the concrete surface. This technology has been extensively investigated, but it would appear that in the vast majority of cases circular columns have been tested. It is now accepted that for circular columns confinement with FRP is highly éffective in increasing column strength and energy absorption capacity. Both CFRP and GFRP have been found to be effective for this purpose.

As for the design of new structures, FRP reinforcement has been used to reinforce mainly bridge deck slabs. Although two bridges with some of their girders prestressed with CFRP tendons have been built in Canada (2), the philosophy of complete replacement of internal reinforcing and prestressing steel by FRP reinforcement and tendons has not been received enthusiastically by the construction community. There are several reasons for this lack of enthusiasm: (a) There is still inadequate empirical evidence with regard to the long term durability of FRP in aggressive environments and under sustained high stresses. (b) There are not sufficiently detailed design codes currently available to guide designers in selecting the proper materials and design methods. (c) FRP is still costly compared to conventional reinforcing/prestressing steel. (d) In major structural elements, the risks associated with failure of FRP reinforcement are perceived to be unacceptable as a trade off for long-term durability. The latter is a particularly important concern and can only be assuaged by the implementation of more well conceived and properly executed and monitored demonstration projects.

SOME RECENT CANADIAN RESEARCH ON FRP IN CONSTRUCTION

The research in Canada covers a wide range of topics related to repair (external reinforcement) and new construction (internal reinforcement). The research is carried out primarily in Canadian universities and nearly all major universities have active research programs on FRP. Here are some of the topics which have been or are being investigated:

1. Behaviour and strength of bridge decks reinforced with FRP (3,4)
2. GFRP poles for transmission lines (5)
3. FRP shear reinforcement for concrete members (6,7,8,9)
4. Deflection of FRP reinforced concrete members (10,11)
5. Effect of steel corrosion on structures externally repaired with FRP (12,13,14)
6. Effect of temperature on concrete structures reinforced with FRP bars (15)
7. Resistance to freezing and thawing of fiber-reinforced polymer concrete (16)
8. Shear and flexural strengthening of beams, girders and slabs (17,18,19)
9. Strengthening of concrete walls (20)
10. Use of FRP grids as ties in reinforced concrete columns (21)
11. Behaviour and strength of columns externally confined by FRP sheets (22,23)
12. Punching shear strength of FRP reinforced flat slabs (24)
13. Blast resistance of concrete structures strengthened with FRP (25)
14. Fire resistance of FRP repaired structures (26)

CANADIAN FIELD APPLICATIONS OF FRP

A number of jurisdictions in Canada have used FRP as partial or full reinforcement in bridge deck slabs. These include the Kent County Bridge, the Crowchild Bridge in Alberta, the Wotton Bridge and the Joffre Bridge in Quebec, and the Taylor Bridge in Manitoba (2, 27). The Taylor Bridge (2) comprises five simply supported equal spans of 33 m each, with 8 prestressed I-girders per span. Four of the forty I-girders in this bridge were prestressed with carbon fibre tendons and reinforced with CFRP stirrups. Glass FRP bars were used to reinforce the barrier walls.

The deck slabs of the bridges in Quebec were reinforced with a combination of CFRP grids and GFRP reinforcing bars while in the Crowchild Bridge in Alberta GFRP bars were used to reinforce the overhanging cantilever portions of the deck slab. On the other hand, The Ramsay Creek Bridge in Ottawa is a box bridge and is reinforced with CFRP grids, known as NEFMAC. Half of the entire box was reinforced with carbon fibre grids while the other half was reinforced with epoxy coated steel bars. Figure 1 shows the reinforcement of the top slab of the latter bridge.

External reinforcement in the form of sheets and laminates is particularly popular because installation procedures are simple, economically it is competitive, it can be applied in physically constrictive spaces, the installation time is short and the risks associated with its failure are generally acceptable. Accordingly, in a number of projects, particularly bridges, CFRP sheets have been used to strengthen flexural or shear deficient members or to confine concrete columns and piers. Examples of this kind of application include the Clearwater Creek Bridge in Alberta , the Champlain Bridge in Montreal, the Saint-Emile-de-l'Energie Bridge in Quebec, and the Leslie Street Bridge in Toronto. For examples of other repair projects, see references (27,28).

Figure 1 – CFRP Grid and Epoxy Coated Steel Reinforcement in Ramsay Creek Bridge

BLAST RESISTANCE OF FRP REINFORCED PANELS

Due to recent developments, researchers at Carleton University and The Canadian Explosives Research Laboratory tested concrete panels externally reinforced with FRP sheets and laminates to study their resistance to blast loads. Eighteen $1000 \times 1000 \times 70$ mm reinforced concrete panels were made of 40 MPa concrete. Each panel was reinforced with two orthogonal layers of top and bottom steel mesh reinforcement, with an average reinforcement ratio of 0.003 and yield strength of 400 MPa. Five of the panels were used as control and were not retrofitted with any FRP. Four panels were identically retrofitted with four 500 mm wide unidirectional glass FRP (GFRP) sheets, with two sheets applied in a cross shape to the top surface and the other two sheets similarly applied to the bottom surface, Figure 3. The sheets were bonded to the concrete by means of epoxy. Five other

Figure 2 – Typical CFRP and GFRP Retrofitted Panels Subjected to Blast Load

panels were similarly retrofitted with uni-directional carbon (CFRP) sheets. The remaining four panels were retrofitted with CFRP laminate strips. The strips were 80 mm wide and were applied diagonally, akin to X-brace, to the bottom and top surfaces of each panel. Four of the control panels and twelve of the FRP retrofitted panels were subjected to various blast pressures, emanating from 13.4, 22.4 or 33.4-kg of ANFO explosive at a standoff distance of 3.0 m. The blast wave characteristics, including incident and reflected pressures and impulses were measured and recorded. The central deflection and strains in the reinforcing steel and the concrete/FRP surfaces were also measured and recorded. The post-blast damage and mode of failure of each panel were observed, and panels that were not completely damaged were subsequently statically tested to find their residual strength.

The test data, as shown in Figure 3, revealed that the panels retrofitted with either the GFRP or the CFRP sheets had higher blast resistance and they generally performed better than the control or non-retrofitted panels, but the panels with the CFRP laminates did not.

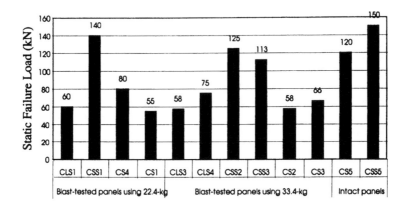

Figure 3 – Comparison of Post-blast Static Residual Strength of CFRP Retrofitted CLS
and CSS Panels with those of Non-retrofitted CS Panels

CONCLUSIONS

The scope of Canadian FRP research related to construction is quite wide and the subject is investigated in a large number of universities. The consequence has been the implementation of this new technology in a number of demonstration projects throughout Canada. The field applications have included both new construction and repair/strengthening of existing structures. The vast majority of these structures have been reinforced concrete and it is expected that applications in the near future will continue to focus on this type of structures.

REFERENCES

1. Canadian Journal for Civil Engineering, Volume. 27, No. 5, 2000, 839-1045.
2. Sami Rizkalla and Pierre Labossiere, "Structural Engineering with FRP – in Canada", Concrete International, October 1999 Issue, American Concrete Institute, .25-28.
3. T. Hassan, A. Abdelrahman, G. Tadros and S. Rizkalla, "Fibre Reinforced Polymer Reinforcing Bars for Bridge Decks", Canadian Journal of Civil Engineering, Volume 27, No. 5, 2000, 839-849.
4. B. Benmokrane, H. Rahman, P. Mukhopadhyaya, R. Masmoudi, B. Zhang, I. Lord and G. Tadros, "Fiber-Optic Sensors Monitor FRP-Reinforced Bridge", Concrete International, June 2001 Issue, American Concrete Institute, 33-38.
5. Sherif Ibrahim, Dimos Polyzois and Sherif K. Hassan, "Development of Glass Fibre Reinforced Plastic Poles for Transmission and Distribution Lines, Canadian Journal of Civil Engineering, Volume 27, No. 5, 2000, 850-858.

6.   Emile Shehata, Ryan Murphy and Sami Rizkalla, "Fibre Reinforced Polymer Shear Reinforcement for Concrete Members: Behaviour and Design Guidelines", Canadian Journal of Civil Engineering, Volume 27, No. 5, 2000, 859-872.

7.   S. Greenaway and A. Selley, "Contribution of Concrete to the Shear Reistance of Carbon Fibre Reinforced Beams without Stirrups", Research Report, Department of Civil and Environmental Engineering, Carleton University, Ottawa, 2001.

8.   A.G. Razaqpur, O.B. Isgor, M.S. Cheung and A. Wiseman, "Background to the Shear Design Provisions of the Proposed Canadian Standard for FRP Reinforced Concrete Structures", Proceddings of Composites in Construction, J. Figueiras, L. L. Juvandes and R. Faria, eds. Porto, Portugal, 2001, 403-408.

9.   A.G. Razaqpur and D. Mostofinejad, "Experimental Study of Shear Behavior of Continuous Beams Reinforced with Carbon Fiber Reinforced Polymer", Proceedings of Fourth International Symposium on Fiber Reinforced Polymer Reinforcement for Reinforced Concrete Structures, SP-118, C. W. Dolan, S.H. Rizkalla and A. Nanni, eds., American Concrete Institute, 1999, 169-178.

10.   T. Hall and A. Ghali, "Long-term Deflection Prediction of Concrete Members Reinforced with Glass Fibre Reinforced Polymer Bars", Canadian Journal of Civil Engineering, Volume 27, No. 5, 2000, 890-898.

11.   A.Ghani Razaqpur and O. Burkan Isgor, "Methods for Calculating Deflection of FRP Reinforced Concrete Structures", Procedings of the Third International Conference on Advanced Composite Materials in Bridges and Structures, J.L. Humar and A.G. Razaqpur, eds., Ottawa, Canada, 2000, 371-378.

12.   J.F. Bonacci and M. Maaleji, "Externally Bonded Fiber-Reinforced Polymer for Rehabilitation of Corrosion Damaged Concrete Beams", ACI Structural Journal, Volume 97, No. 5, 2000, 703-711.

13.   C. Lee, J.F. Bonacci, M.D.A. Thomas, M. Maalaji, S. Khajehpou, N. Hearn, S. Pantazapoulou and S. Sheikh, " Accelerated Corrosion and Repair of Reinforced Concrete Columns using Carbon Fibre Reinforced Polymer Sheets", Canadian Journal of Civil Engineering, Volume 27, No. 5, 2000, 941-948.

14.   K.A. Soudki and T.G. Sherwood, "Behaviour of Reinforced Concrete Beams Strengthened with Carbon Fibre Reinforced Polymer Laminates Subjected to Corrosion Damage", Canadian Journal of Civil Engineering, Volume 27, No. 5, 2000, 1005-1010.

15.   M.M. Elbadry, H. Abdalla and Amin Ghali, "Effects of Temperature on the Behaviour of Fibre Reinforced Polymer Reinforced Concrete Members: Experimental Studies", Canadian Journal of Civil Engineering, Volume 27, No. 5, 2000, 993-1004.

16.   L.A.Bisby and M.F. Green," Resistance to Freezing and Thawing of Fiber-Reinforced Polymer-Concrete Bond", ACI Structural Journal, Volume 99, No. 2, 2002, 215-223.

17.   C. Deniaud and J.R.Cheng, "Shear Behavior of Reinforced Concrete T-Beams with Externally Bonded Fiber-Reinforced Polymer Sheets," ACI Structural Journal, Volume 98, No. 3, 2001, 386-394.

18.   Wight, R.G., Erki, M.A. and El-Hacha, R. (2001), "CFRP Sheet Strengthening Damaged Continuous Reinforced Concrete Beams," Proceedings of FRP Composites in Civil Engineering, CICE 2001, Hong Kong, 433-440.

19.   P. Labossiere, K.W. Neale, P. Rochette, M. Demers, P. Lamothe, P. Lapierre and G. Desagne', "Fibre Reinforced Polymer Strengthening of the Saint-Emelie-de-L'Energie Bridge: Design, Instrumentation, and Field Testing, Canadian Journal of Civil Engineering, Volume 27, No. 5, 2000, 916-927.

20.   J.C. Lombard, D.T. Lau, D.T. and J.L. Humar, "Seismic Strengthening and Repair of Reinforced Concrete Shear Walls," CD-ROM Proceedings of the 12th World Conference on Earthquake Engineering, Auckland, New Zealand, 2000.

21.   M. Saatcioglu and K. Sharbatdar, "Use of FRP Reinforcement in Concrete Columns", Procedings of the Third International Conference on Advanced Composite Materials in Bridges and Structures, J.L. Humar and A.G. Razaqpur, eds., Ottawa, Canada, 2000, 363-370.

22.   M. Theriault and K. Neale, "Design Equations for Axially Loaded Reinforced Concrete Columns Strengthened with Fibre Reinforced Polymer Wraps", Canadian Journal of Civil Engineering, Volume 27, No. 5, 2000, 1011-1020.

23.   S.A. Sheikh and G Yau, "Seismic Behavior of Concrete Columns Confined with Steel and Fiber Reinforced Polymer", ACI Structural Journal, Volume 99, No. 1, 2002, 72-80.

24.   A.R. Zaghloul, "Behaviour and Strength of CFRP Reinforced Flat Plate Interior Slab-Column Connections Subjected to Shear and Unbalanced Moments",Master of Engineering Science Thesis, Department of Civil and Environmental Engineering, Carleton University, Ottawa, Ontario, 2002.

25.   A.F. Tolba, "Response of FRP-Retrofitted Reinforced Concrete Panels to Blast Loading," Doctoral Dissertation, Department of Civil and Environmental Engineering, Carleton University, Ottawa, Ontario, 2001.

26.   Bisby, L.A., Green, M.F., and Kodur, V.K. (2001), "Fire Behaviour of FRP Wrapped Reinforced Concrete Columns," CD-ROM Proceedings of Structural Faults & Repair 01 Conference, London, England, 2001.

27.   A.A. Mufti, P. Labossiere and K. Neale, "Recent Bridge Applications of FRP in Canada," Structural Engineering International, IABSE, Volume 12, No. 2, 2002, 96-98.

28.   G.W. Wight, M.A. Erki and P.J. Heffernan, " Canadian Federal Interest in FRP," Structural Engineering International, Journal of IABSE, Volume 12, No. 2, 2002, 99-101.

CSA standard S806 - Design and construction of building components with fibre reinforced polymers - A current overview

Dr. M. S. Cheung
Technology Directorate
Public Works & Government Services Canada
11 Laurier Street
Hull, Quebec, Canada K1A 0S5
moe.cheung@pwgsc.gc.ca

ABSTRACT

This paper provides an up-date on the Canadian Standards Association document CSA-S806, "Design and Construction of Building Components with Fibre Reinforced Polymers". This new CSA standard was originally planned to be published in late 2001. However, due to some unforseen administrative problems, publication of this standard has been delayed until later this year. The standard has been balloted and approved by the Technical Committee S806 as well as the CSA Steering Committee on Structures. Currently, it is undergoing final editorial review by the publication unit of the Canadian Standard Association.

The provisions in this standard address not only the design of new buildings, but also retrofit and strengthening of existing building components. The paper highlights the contents of the document as well as its technical background. This paper outlines the approach taken in the development of the standard, summarizes the information and decisions pertaining to design/retrofit provisions and principles, and touches on the material testing methodologies. Some relevant information relating to the Canadian building practices are also included in annexes of this document. The contents of this standard are divided into 6 subject areas; basic limit states design principles, material testing procedures, design provisions for reinforced and prestressed components, repair and retrofit of existing components, strengthening earthquake deficient elements, and site preparation and construction procedures. The above 6 subject areas are presented in 11 sections with the format and notations similar to other Canadian codes of practice.

INTRODUCTION

The use of fibre reinforced composites in the aerospace, military and the automotive industries is now well established. For example, fibre reinforced composites are used routinely and extensively in structural components of both military and commercial aircraft. The use of these materials in Civil Engineering structures is increasingly coming to the fore; it has reached the stage where it is now economically feasible to use them in some components of buildings, if the life cycle costing of the structure is considered. Although, a number of research institutes and some construction firms around world are interested in FRP research, field demonstrations and applications, no code/standards are available for guiding the design and construction of such structures at this moment. Japan and U.S. are currently developing some technical guidelines in this area. However, they are not legal documents to be used for building permit purpose. Public Works and Government Services Canada(PWGSC), concerned with the high cost of repairs and maintenance caused by premature deterioration, was one of the first of Canada's major building owners to recognize the need to address this issue and to support the investigation of other alternative construction materials. As part of its material research program, PWGSC initiated, supported and provided technical input into the development of CSA-S806, "Design and Construction of Building Components with Fibre Reinforced Polymers".

The Canadian Standards Association(CSA) is one of several organizations in Canada that develops standards pertaining to design and construction of physical infrastructure. CSA is a member of the Standards Council of Canada. The membership of CSA-S806 technical committee assigned to write the standard was selected carefully to include representative experts from different sectors within Canada and from other countries.

Early work on the CSA standard drew heavily from the PWGSC's research results along with Architect Institute of Japan's (AIJ) guidelines. The CSA standard evolved, however, to be significantly different in scope, focus, breadth and depth. The format and approach of the CSA standard on "Design and Construction of Building Components with Fibre Reinforced Polymers" follows very closely with the Canadian design standard for concrete, CSA-A23.1-3 documents (1,2,3) .

The purpose of this paper are two fold: 1) to provide an update on the development of the standard and 2) to introduce it to the construction industry before official publication. The paper will touch upon some background and discussions on the specific provisions and approaches taken in the development of the standard, with a view to receiving comments and feedbacks from especially those engineers who are familiar with the materials under discussion.

CONTENTS, STRUCTURE AND SCOPE

Contents and Structure

The contents of the CSA standard are divided into 6 subject areas; basic limit states design principles, material testing procedures, design provisions for reinforced and prestressed components, repair and retrofit of existing components, strengthening earthquake deficient elements, and site preparation and construction procedures. The above 6 subject areas are presented in 11 chapters with the format and notations similar to the CSA A23.3 and National Building Code of Canada(NBCC). While the design provisions and methodologies are provided in the body of the standard, detailed analysis, testing methods, excerpts from other codes and standards and reference materials are included in the appendices and annexes. This CSA standard is designed as a standalone document which covers all the necessary information for design of building components with FRP.

It is also intended to develop a companion commentary volume of the standard which will provide crucial information on the explanation of requirements in the standard.

Scope

The CSA standard largely reflects the current state of the knowledge in this field and focuses on areas where use of FRP materials have proven to be advantageous over conventional materials in reducing premature deteriorations, maintenance cost and construction/renovation time.

The building components covered by this standard include columns, slabs, reinforced and prestressed beams(with bars, mats, grids, sheets and tendons), parking garage slabs, balconies and exterior wall cladding, etc.

LIMIT STATES DESIGN

Design Philosophy

The use of FRP materials as a reinforcement for concrete beams and slabs requires the development of design procedures that ensure adequate safety from catastrophic failure and prevent excessive deformations affecting the serviceability of the structure. With steel reinforcing, a confident level of safety can be provided by specifying that a section's flexural strength be less than its balanced flexural strength. This ensures the steel will yield before the concrete crushes, therein, guaranteeing a ductile failure. The result is the ability of the failed flexural members to absorb large amounts of energy through plastic straining in the reinforcing steel. FRP materials respond linearly and elastically to failure at which point a brittle rupture occurs. As a result, failure, whether the result of shear or flexural, is sudden and brittle. The standard recognizes the unfavourable property of this material, it incorporates this parameter in analytical and flexural design procedures by lowering flexural capacity reduction factors to be more compatible with the specific performance limitations of FRP materials.

Ductility

The traditional definition of ductility is the ratio of the total deformation or strain at failure to the deformation at yielding. FRP reinforcements have a linear stress versus strain relationship to failure. Therefore, by the above definition, the behaviour of FRP reinforced members can not be considered ductile. In order to avoid this dilemma, the standard uses another term "deformability - c/d" to define the equivalency of the minimum ductility requirements. As one of the design criteria, all FRP reinforced concrete sections must be designed such that failure of the section is initiated by failure by crushing of the concrete in the compression zone. To ensure this, the standard specifies that the moment-curvature relation of FRP reinforced concrete members must be assumed to be tri-linear with the slope of the three segments being $E_c I_g$, zero and E_c Icr. Also for flexural members at ultimate limit states, the extreme compressive strain in concrete must be assumed to be 0.0035 and

$$(c/d) \geq 7(7 + 2000 \, \epsilon_{Fu}) \tag{1}$$

Where c is the distance from the extreme compression fibre to neutral axis; d is the distance from the extreme compression fibre to the centroid of longitudinal tensile force; and ϵ_{Fu} is ultimate strain of FRP reinforcement.

Load Combinations

The effect of factored loads acting on a member, its cross section, and its connections to other members in terms of moment, shear, and torsion must be computed from factored loads and forces and combining with appropriate factors. Since the seismic design of FRP members requires special considerations, two types of load combinations are required to be considered in the design. 1) Combinations not including earthquake and 2) Combinations including earthquakes. A section of the standard with specific details is devoted to the load and load combinations.

MATERIAL PROPERTIES AND TESTING METHODS

Material Properties of RFP

FRP composites are a material system. It consists of any combination of two or more separate materials having an identifiable interface between them. The most commonly used FRP material are composed with a matrix of polymeric material reinforced by fibres or other reinforcement with a discernible aspect ratio of length to thickness. Therefore, the material properties of a FRP composite is quite different from the traditional construction materials, such as steel and concrete, in that, it depends on the characters of polymer, fibre, and their interface properties (4,5).

When considering material properties of FRP bars or tendons, the following points are worthy of note: First, a FRP bar and/or FRP sheet is anisotropical material, with one axis being the strong axis. Secondly, unlike steel, the mechanical properties of FRPs vary significantly from one product to another and thirdly, FRPs, like other composites, are effected by such factors as the loading history and duration, temperature and moisture (6,7).

Testing Methods

In view of the variable properties and characteristics of FRP material, specific testing methods for such materials are specified in the standard. This includes the following eight areas of testing requirement specified in the Annexes of the standard.

FRP internal reinforcement

1)   Determination of cross-sectional Area of FRP Reinforcement,
2)   Anchor for Testing FRP Specimens Under Monotonic and Sustained Tension,
3)   Test Method for Tensile Properties of FRP Reinforcements,
4)   Test Method for Development Length of FRP Reinforcements,
5)   Test Method for FRP Bent Bars and Stirrups.

Surface bonded FRP reinforcement

6)   Test Method for Direct Tension Pull off Test,
7)   Test Method for Tension Test of Flat Specimen,
8)   Test Method for Bond Strength of FRP Rods by Pull-out Testing.

FLEXURAL AND SHEAR DESIGN

Flexural Consideration

Unlike steel reinforcement, no constant tensile force may be assumed after yield point. This is because the FRP reinforcement continues to gain strength with increasing strain until its final rupture. The only condition of known forces in a FRP reinforced flexural member is the balance condition where the concrete fails in compression at the same time that the FRP ruptures. In order to ensure that sudden ruptures do not occur, the design must meet the requirement in Eqn. 1 of controlling the compression centroid of the flexural member.

If the above condition is met, compression failure of the concrete will occur first and it ensures that a less catastrophic failure will occur.

Cracking control

Excessive cracking is undesirable because it reduces stiffness and possibility of deterioration, and adversely affects the appearance. Since FRP reinforcements are not subject to the same corrosion mechanisms as steel, the crack width limitations established by CSA A23.3 are not applicable. For FRP reinforced concrete members, the standard specifies the crack limitations of 0.5 mm for exterior exposure, and 0.7 for interior exposure, except that for structures subjected to abnormally aggressive exposure conditions or designed to be water-tight, smaller allowable crack widths shall be adopted. In the standard, a Z factor as shown in Eqn. 2, was introduced for cracking controls.

$$Z = k_b (E_s / E_F) f_F \sqrt[3]{d_c} A \qquad (2)$$

Z must not exceed 45000 N/mm for interior exposure and 38000 N/mm for exterior exposures. Where K_b is a coefficient dependent on the reinforcing bar bond characteristics; E_s and E_F are modulus of elasticity of reinforcement and modulus of elasticity of longitudinal reinforcement respectively; d_c is distance from extreme tension fibre to the centre of the longitudinal bar; and A is effective area of concrete surrounding the flexural tension reinforcement and extending from the extreme tension fibre to the centroid of the flexural tension reinforcement and an equal distance past the centroid, divided by the number of bars.

Deflection Control

The deflection of FRP reinforced members will be greater than comparable steel reinforced members because of the lower modulus of elasticity of the FRP. This leads to greater strains to achieve comparable stress levels and to lower transformed moment of inertia. Deflection limitations are independent of the concrete strength and reinforcement. Therefore, similar deflection limitations as proposed by CSA A23.3 are adopted by this standard.

When calculating the defection for FRP reinforced members, the moment-curvature relation must be assumed to be tri-linear with the slope of the three segments being $E_c I_g$, zero and $E_c I_{cr}$. This effectively softens the section and results in a reasonable deflection prediction for FRP reinforced members.

Shear Consideration

The vast majority of the research data is for members that are not shear critical. There are a very small number of tests with FRP shear reinforcement. Experimental results of shear anchorage indicate that the stirrups will fail in the corners due to premature fracture at the bend. The few tests that have been completed with FRP stirrups suggest that the shear resistance is less than predicted. This may be due to the large cracks that result from the lower modulus of elasticity of the stirrups. In view of the above reason, the standard provides a very conservative approach in dealing with shear. The standard provides specific guidance for members not subjected to significant axial tension including members with FRP flexural and shear reinforcement as well as members with FRP flexural reinforcement and steel shear reinforcement. In the standard, a minimum shear reinforcement clause was introduced for flexural members where the factored shear force, V_f exceeds $0.5 V_c + \varphi_F V_p$ or the factored torsion, T_f, exceeds $0.25 T_{cr}$.

PROVISIONS OF SEISMIC DESIGN

FRP materials can be effectively used for seismic design or strengthening seismically deficient structures or structural components. However, when using FRP materials, structural systems and structural components must be designed and detailed with particular recognition of the effects of the differences in mechanical characteristics between the fibre reinforced plastic materials and reinforcing steel on their behaviour and performance during earthquake. These differences include the lack of ductile behaviour from the essentially linear elastic stress-strain relationship of the materials until rupture, lower modulus of elasticity and higher ultimate strength than steel, resulting in significantly different stiffness, damping and energy dissipation characteristics for structures reinforced with FRP materials.

The seismic design provisions have been focussed on the shear strength enhancement, concrete confinement as well as lap splice clamping, major advantages of using FRP. The standard provides specific design criteria and design equations for 3 different types of FRP applications. They include 1) Design requirements for column retrofit and rehabilitation (8), 2) Design requirements for shear wall retrofit and rehabilitation(9) and 3) FRP reinforcement for concrete confinement in new construction (10).

OTHER DESIGN CONSIDERATIONS

Thermal Stress

Reinforced concrete itself is a composite material, where the reinforcement acts as the strengthening fibre and the concrete as the matrix. It is therefore imperative that the behaviour under thermal stresses for the two materials be similar so that the differential deformations of concrete and the reinforcement are minimized. However, unlike normal steel reinforcement, the coefficient of linear thermal expansion of FRP reinforcements is widely different from that of concrete. Thermal stress analysis using theory of elasticity indicates that concrete could crack around some FRP reinforcing bars due to the environmental temperature rise expected in the Canadian climate. Therefore, the standard recognizes that the thermal stress must be considered in the design of FRP reinforced concrete members.

Fire Resistance

The fire resistance of FRP reinforced concrete slabs can be determined similarly to that for steel-reinforced concrete slabs as specified in the National Building Code of Canada. From a parametric study, it was found that the fire resistance of FRP reinforced concrete slabs depends on the critical temperature of FRP reinforcement, the thickness of the concrete cover and the type of aggregate in the concrete mix. The critical temperature is defined as the temperature at which the reinforcement loses enough of its strength, typically 50%, that the applied load can no longer be supported. For reinforcing steel, the critical temperature is 593 degree C. For FRP reinforcement, however, it depends on the type and composition of FRP, and hence the critical temperature must be obtained from the manufacturer's data.

In lieu of actual test data, the fire resistance of FRP-reinforced concrete slabs can be established for a given critical temperature specified by the FRP manufactures. Alternatively, the standard provides a series of design charts which can be used to obtain the relevant concrete cover thickness to FRP reinforcement for a required fire resistance rating.

Ultraviolet Rays

FRP Composites can be damaged by ultraviolet rays present in sunlight. These rays will cause chemical reactions in the polymeric matrices which can lead to a degradation of their properties. Therefore, when using FRP composites for building cladding, railing and other applications which are directly exposed to sunlight, protective coating or appropriate additives to the resin must be considered in order to reduce the damage caused by ultraviolet rays. There is no problem when FRP elements are used as internal reinforcement for concrete structures.

CONSTRUCTION REQUIREMENTS

The characteristics and mechanical properties of FRP materials are significantly different from steel reinforcements. Specific procedures for handling, storage, transport and installation of FRP materials must be developed to protect them against damage or deterioration during the construction period. In addition, due to the lower modulus of elasticity of FRP bars, bar supports and their spacing must be properly designed in order to prevent displacement before and during concreting. To prevent flotation of FRP bars during placement of concrete, sufficient tie-downs must also be provided. In this case, supports and tie-downs must also be plastic or other non-corroding materials in order to avoid corrosion concentration problems.

The standard specifies unique construction procedures and requirements for the use of FRP materials as reinforcement including FRP bars and sheets.

CONCLUSIONS

CSA S806 is the first standard developed in Canada for building applications using fibre reinforced polymers. This standard reflects the current state of knowledge in this area and it's format follows closely to CSA A23.3, the Canadian concrete design practices. The use of FRP in civil engineering applications is relatively new. A number of areas requires further research, especially in the area of long term durability and mode of failure at ultimate limit states. However, the standard did not compromise the safety and long term performance. For those areas lacking information, knowledge and/or practical experience, the design criteria specified in the standard are intended to be conservative. It is hoped that those areas will be improved in the future edition of the standard from research results and findings. After the first edition of this standard is published, the CSA S806 committee will be actively working with construction industry and research community to identify future research needs and prepare for the development of the next edition of the standard.

REFERENCES

1. CSA A23.1, Concrete Materials and Methods of Concrete Construction, Canadian Standards Association, 2000

2. CSA A23.2, Concrete Materials and Methods of Concrete Construction Methods of Test for Concrete, Canadian Standards Association, 2000

3. CSA A23.3, Design of Concrete Structures, Canadian Standards Association, 1994

4. S. Faza and H. V. S. GangaRao, Bending Response of Beams Reinforced with FRP Rebars for Varying Concrete Strengths. Advanced Composite Materials in Civil Engineering Structures, ASCE, 262-270, 1991

5. H. Mutsuyoshi, K. Vehara and A. Machida, Mechanical Properties and Design Methods of Concrete Beams Reinforced with Carbon Fiber Reinforced Plastics, Transactions of the Japan Concrete Institute, 12, 231-238, 1990

6. S. Akihama, T. Suenaga and H Nakagawa, Mechanical Properties of Three-Dimensional Fabric Reinforced Members, Proceedings of JCI, 10:.2, 677-682, 1988

7. M. S. Cheung, Georges Akhras and Wenchang Li, Progressive Failure Analysis of Composite Plates by the Finite Strip Method, Journal of Computer Methods in

Applied Mechanics and Engineering **124**, 49-61, 1995

8.   A. G. Razaqpur, D. Svecova and M. S. Cheung, Analysis of Concrete Structures
        Strengthened with Fibre-Reinforced-Plastic Reinforcement, Proceedings CSCE
        Annual Conference, 563-572., 1994

9.   J.D.Berset, Strengthening of Reinforced Concrete Structures for Shear using
        Composite Materials, M.Sc. thesis, Massachusetts Institute of Technology, 1992

10.   D. A. Taylor, A. H. Rahman and M. S. Cheung, Durability of Fibre-Reinforced
        Plastic Rebars in Concrete Structures, Fiber Reinforced Concrete-Modern
        Developments, UBC, 379-393, 1995

FATIGUE, FRACTURE, IMPACT

Progressive failure analysis of composite laminates subjected to combined in-plane loadings

Rajamohan Ganesan and Daying Zhang
Concordia Center for Composites
Department of Mechanical and Industrial Engineering
Concordia University
Montreal, Quebec. H3G 1M8

ABSTRACT

The objective of this work is to study the progressive failure of symmetric and un-symmetric laminates that are subjected to bi-axial compressive loads and combined shear loads and compressive loads. The finite element formulation is developed based on the first-order shear deformation theory and incorporates the effects of the geometric non-linearity based on von Karman's theory. The tensor polynomial form of Tsai-Hill criterion is used to predict the failure of the lamina and the maximum stress criterion is used to predict the onset of delamination. The failure progression in symmetric laminates $(+ -45/0/90)_{2s}$, $(+ -45)_{4s}$, and $(0/90)_{4s}$ and un-symmetric laminates $(+ -45/0/90)_4$, $(+ -45)_8$, and $(0/90)_8$ is investigated in detail. A parametric study is conducted to characterize the effects of stacking sequence, aspect ratio of the plate, and fiber orientation on the maximum deflection, the first-ply failure load, and the ultimate load.

387

INTRODUCTION

Composite laminates can sustain considerable loading after the first occurrence of localized damage such as matrix cracking, fiber breakage and delamination. The early investigations related to the failure of laminated plates include the work by Reddy and Pandey (1). They presented the finite element procedure for the prediction of linear first-ply failure loads of composite laminates subjected to transverse and in-plane tensile loading. The study by Reddy and Reddy (2) used the first order shear deformation theory in the finite element modeling to conduct the linear and nonlinear failure analyses. Kam and Sher (3) studied the nonlinear behavior and the first-ply failure strength of centrally loaded laminated composite plates with semi-clamped edges using a method developed from the von Karman-Mindlin plate theory in conjunction with the Ritz method. There are a few investigations available in the literature on the progressive failure analysis of laminates subjected to uni-axial compression (4) and in-plane positive and negative shear loads (5).

METHODOLOGY

In the present study, the finite element formulation is developed based on the first order shear deformation theory. The nine-node Lagrange element that has five degrees of freedom per node is employed in the analysis. The effects of geometric nonlinearity are incorporated based on von Karman's theory. The nonlinear algebraic equations are solved using the Newton-Raphson technique. The stresses are calculated at the nodal points. All the six stress components are calculated at each nodal point. However, in order to predict the failure of a lamina only five stress components that are three in-plane stresses and two transverse shear stresses are used in the failure criterion. To predict the onset of delamination, two transverse shear stress components and one transverse normal stress component are used in the Tsai-Hill criterion. Delamination at any interface is said to have occurred when any of the transverse stress components in any of the two layers adjacent to the interface becomes equal to or greater than its corresponding strength. The ply failure is said to have occurred when the state of stress at any point within the lamina satisfies the selected failure criterion. The first–ply failure refers to the first instant at which one or more than one plies fail at the same load. After the first–ply failure, the progressive failure analysis is carried out using progressive failure development procedure that is appropriate to the tensor polynomial form of the Tsai-Hill criterion. The tangent stiffness coefficients are obtained using numerical integration. The order of the Gauss quadrature used is based on the selective integration rules, i.e. the 3×3 rule is used for terms associated with membrane and bending actions while the 2×2 rule is used for terms associated with

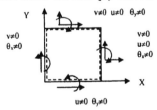

Fig. 1 The boundary conditions for the plate

transverse shear response. Shear correction factors of value $K_1 = K_2 = 5/6$ are used. The dimensions of the laminate used in the parametric study are 279 mm × 279 mm× 2.16 mm. The boundary conditions used are shown in Fig. 1.

FAILURE CRITERIA AND PROGRESSIVE FAILURE DEVELOPMENT

The most general polynomial failure criterion as proposed by Tsai (6) is expressed as

$$F_1\sigma_1 + F_2\sigma_2 + F_3\sigma_3 + 2F_{12}\sigma_1\sigma_2 + 2F_{13}\sigma_1\sigma_3 + 2F_{23}\sigma_2\sigma_3 +$$
$$F_{11}\sigma_1^2 + F_{22}\sigma_2^2 + F_{33}\sigma_3^2 + F_{44}\sigma_4^2 + F_{55}\sigma_5^2 + F_{66}\sigma_6^2 + \cdots \geq 1 \tag{1}$$

Particular cases of the above criterion differ from one another by their strength tensors F_{ij}. Hence, various degenerate cases of the tensor polynomial criterion can be obtained by substituting the appropriate tensor strength factors F_{ij} in Equation (1). In the above expression $\sigma_1, \sigma_2, \sigma_3$ are the normal stress components; $\sigma_4, \sigma_5,$ and σ_6 are the shear stress components $\tau_{23}, \tau_{31},$ and τ_{12} respectively. In the Tsai-Hill criterion

$$F_1 = F_2 = F_3 = 0; \quad F_{11} = \frac{1}{X^2}; \quad F_{22} = \frac{1}{Y^2}; \quad F_{33} = \frac{1}{Z^2};$$
$$F_{44} = \frac{1}{R^2}; \quad F_{55} = \frac{1}{S^2}; \quad F_{66} = \frac{1}{T^2}; \quad F_{12} = -\frac{1}{2}\left(\frac{1}{X^2} + \frac{1}{Y^2} - \frac{1}{Z^2}\right); \tag{2}$$
$$F_{13} = -\frac{1}{2}\left(\frac{1}{Z^2} + \frac{1}{X^2} - \frac{1}{Y^2}\right); \quad F_{23} = -\frac{1}{2}\left(\frac{1}{Y^2} + \frac{1}{Z^2} - \frac{1}{X^2}\right)$$

In the above expressions X_t and Y_t are the tensile strengths of the lamina in the fiber direction and in the direction transverse to it, respectively; X_c and Y_c are the corresponding compressive strengths. Z_t and Z_c are the tensile strength and the compressive strength, respectively, in the principal direction 3 of the lamina. R and T are the shear strengths of lamina in planes 2-3 and 1-2 respectively. The shear strength in plane 1-3 is designated by S. The failure of the lamina is said to have occurred when the state of stress at any nodal point within the lamina (at mid-thickness) satisfies the tensor polynomial form of the Tsai-Hill criterion in which the terms associated with the normal stress component in the third principal material direction are omitted. Delamination at any interface between two adjacent layers is said to have occurred when any of the three transverse stress components (adjacent to the interface) in any of the two layers becomes equal to or greater than its corresponding allowable strength. In this regard, the interlaminar shear strengths are taken to be the same as the transverse interlaminar normal strengths of the lamina, while the transverse interlaminar normal strength is taken to be the tensile strength of the lamina in the principal material direction-3.

The procedure used for studying the progressive failure is given in the following. At each load step, nodal point stresses are used in the Tsai-Hill failure criterion. If failure occurs at a nodal point in a layer under combined in-plane loads (bi-axial compressions and shear loads), a reduction in the appropriate lamina moduli is introduced everywhere in the lamina as per the mode of failure which causes the change in the overall laminate stiffness. The terms given below are used to determine the failure modes.

$$H_1 = F_1\sigma_1 + F_{11}\sigma_1^2; H_2 = F_2\sigma_2 + F_{22}\sigma_2^2;$$

$$H_4 = F_{44}\sigma_4^2; H_5 = F_{55}\sigma_5^2; H_6 = F_{66}\sigma_6^2 \tag{3}$$

The largest H_i term is selected to represent the dominant failure mode and the corresponding modulus is reduced to zero. H_1 Corresponds to the modulus E_1; H_2 to E_2; H_4 to G_{23}; H_5 to G_{13} and H_6 to G_{12}. After the convergence in the iteration in the nonlinear analysis is achieved, the stresses at the nodal points on the mid-plane of each layer and at its interfaces with the adjacent layers are calculated. The stresses are transformed to the stresses in the principal directions. The failure indices H1, H2, ... are calculated. If failure occurs the appropriate lamina moduli are reduced and the corresponding laminate stiffness is calculated. The nonlinear analysis is re-started at the same load step. If no failure occurs, the analysis proceeds to the next load step. Final failure is said to have occurred when delamination occurs or when the plate is no longer able to carry any further load because of very large deflection.

PARAMETRIC STUDY

The properties of the composite material used in the parametric study are presented in Table 1.

Table 1   Material properties of T300/5208 graphite-epoxy

Mechanical properties	Values	Strength properties	Values
E_1	132.58 GPa	X_t	1.515 GPa
E_2	10.8 GPa	X_c	1.697 GPa
E_3	10.8GPa	$Y_t = Z_t$	43.8 MPa
$G_{12} = G_{13}$	5.7 GPa	$Y_c = Z_c$	43.8 MPa
$v_{12} = v_{13}$	0.24	R	67.6 MPa
v_{23}	0.49	S=T	86.9 MPa

The symmetric laminate with configuration $(+-45/0/90)_{2s}$ that is subjected to (a) uni-axial compressive loading and (b) in-plane positive shear loading is analyzed first. The first-ply and ultimate failure loads have been calculated for the cases (a) and (b), and these values are compared with that given in the works of Singh and Kumar (4, 5). The complete load-deflection curves until ultimate failure for both the cases (a) and (b) are given in Figures 2 and 3. In Figures 2 and 3, a and b denote the side length of the laminate, h denotes the thickness of the laminate, w_c denotes the central deflection, w_{max} denotes the maximum deflection of the laminate, and N_x is the applied uni-axial compressive load per unit length. The values of the non-dimensionalized quantity corresponding to the failure, $N_x b^2/E_2 h^3$, is given in Tables 2 and 3.

	First-ply failure load	Ultimate failure load
Ref. (4)	58.09	77.45
Present	55.78	81.53

	First-ply failure load	Ultimate failure load
Ref. (5)	59.38	116.18
Present	61.68	115.32

Table 2 First-ply and ultimate failure loads for the (+-45/0/90)$_{2s}$ laminate subjected to uni-axial compressive loading

Table 3 First-ply and ultimate failure loads for the (+-45/0/90)$_{2s}$ laminate subjected to in-plane positive shear loading

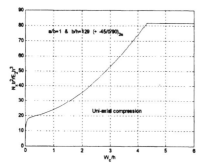

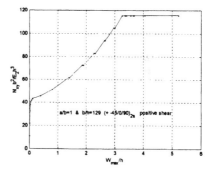

Fig. 2 Load versus maximum deflection curve for (+-45/0/90)$_{2s}$ laminate under uni-axial compression

Fig. 3 Load versus the maximum deflection curve for (+-45/0/90)$_{2s}$ laminate under positive shear loading

Figure 4 shows the load versus the maximum deflection response of various laminates under the action of bi-axial compression combined with in-plane positive shear loads. It is observed that the largest strength is exhibited by the (+-45)$_{4s}$ laminate within the deflection range $w_{max}/h \leq 0.6$ and by the (+ -45/0/90)$_{2s}$ laminate within the deflection range $W_{max}/h > 0.6$. It is also observed that the (0/90)$_{4s}$ laminate shows the least strength for a fixed value of the maximum deflection within the range $W_{max}/h < 1.9$. However, there is a drastic increase in the strength of this laminate in the deflection range $W_{max}/h > 1.9$, and this is attributed to a substantial increase in the axial stiffness. Figure 5 shows the load versus the maximum deflection response of the symmetric laminates (+ -45/0/90)$_{2s}$, (+ -45)$_{4s}$, and (0/90)$_{4s}$, and the un-symmetric laminates (+ -45/0/90)$_4$, (+ -45)$_8$, and (0/90)$_8$ under the action of bi-axial compression combined with in-plane negative shear loads. It is observed that the variation of the maximum deflection with the load for the symmetric laminate (0/90)$_{4s}$ almost coincides with the variation of the maximum deflection with the load for the un-symmetric laminate (0/90)$_8$. That is, the stacking sequence does not have a strong influence on the response of the laminate.

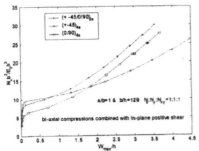

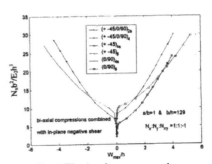

Fig.4 The load versus maximum deflection plot for laminates under the action of bi-axial compression combined with in-plane positive shear loads

Fig.5 The load versus maximum deflection plot for symmetric and un-symmetric laminates under the action of bi-axial compression combined with in-plane negative shear loads

Figures 6 and 7 show the load versus maximum deflection curve for a square laminate with configuration $(+ -45/0/90)_{2s}$ for different load ratios $N_x : N_y : N_{xy}$. The value of N_{xy} is fixed and the values of N_x and N_y are changed. It is seen that for a fixed value of the maximum deflection, the first-ply failure loads and ultimate failure loads decrease with increasing load ratios.

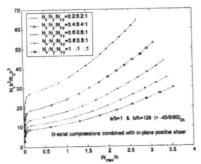

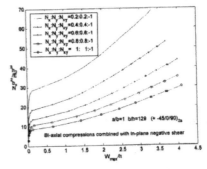

Fig. 6 The load versus maximum deflection curve for the $(+ -45/0/90)_{2s}$ laminate for different load ratios in the case of bi-axial compression combined with in-plane positive shear loads

Fig. 7 The load versus maximum deflection curve for the $(+ -45/0/90)_{2s}$ laminate for different load ratios in the case of bi-axial compression combined with in-plane negative shear loads

Figures 8 and 9 show the variation of the first-ply failure load and the ultimate failure load for $(+ -\theta)_{4s}$ laminates with different fiber orientation angles and subjected to bi-axial compression combined with positive and negative shear loadings. The results show that the variations in response are symmetric with respect to 45^0 fiber orientation. Peak values of the first-ply failure load and ultimate failure load are predicted to occur for 45^0 fiber orientation. For $(0/0)_{4s}$ and $(15/-15)_{4s}$ laminates, the values of the first-ply failure loads are equal to the ultimate failure loads. Figure 10 and Figure 11 show the variation in these loads with aspect ratio for the $(+ -45/0/90)_{2s}$ laminate. It is seen that the first-ply failure load and the ultimate failure load decrease with increasing values of aspect ratio.

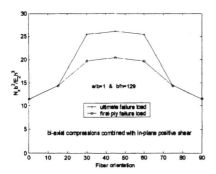

Fig.8 Variation of the first-ply failure load and the ultimate load of (+ - θ)₄ₛ laminates under the action of bi-axial compression combined with in-plane positive shear loading.

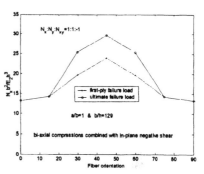

Fig.9 Variation of the first-ply failure load and the ultimate load of (+ - θ)₄ₛ laminates under the action of bi-axial compression combined with in-plane negative shear loading.

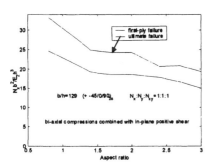

Fig.10 Variation of the first-ply and ultimate loads of (+ -45/0/90)₂ₛ laminate with the aspect ratio under the action of bi-axial compression combined with positive shear loading

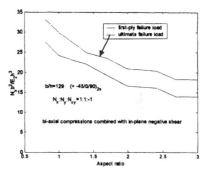

Fig.11 Variation of the first-ply and ultimate failure loads of (+ -45/0/90)₂ₛ laminate with the aspect ratio under the action of bi-axial compression combined with negative shear loading

CONCLUSIONS

The response of laminates beyond first-ply failure under the action of bi-axial compression and combined bi-axial compression and shear are investigated in the present work. The effects of transverse shear deformation and geometric non-linearity are incorporated in the finite element formulation. Both the symmetric and un-symmetric laminates are considered in the parametric study. The parametric study reveals the following.

- The maximum difference between the first-ply failure loads and the ultimate loads is strongly dependent on the laminate configuration and the aspect ratio.
- The failure mode corresponding to the first-ply failure is associated with the localized matrix cracking irrespective of the laminate configuration and the aspect ratio. The failure occurs primarily due to the in-plane normal stresses acting in the direction transverse to the fiber direction.
- The most critical points corresponding to the first-ply failure lie near the loaded edges of the plate.
- In all the $(+-45/0/90)_{2s}$ laminates with different aspect ratios under the action of bi-axial compression combined with in-plane shear loading, the ultimate failure mode is delamination.

REFERENCES

1. J. N. Reddy and A. K. Pandey, "A first-ply failure analysis of composite laminates", Computers and Structures, 1987, Volume 25, pp. 371-393.
2. Y. S. N. Reddy and J. N. Reddy, "Linear and non-linear failure analysis of composite laminates with transverse shear", Composites Science and Technology, 1992, Volume 44, pp. 227-255.
3. T. Y. Kam and H. F. Sher, "Nonlinear and first-ply failure analysis of laminated composite cross-ply plates", J. of Composite Materials, 1995, V. 29, pp. 463-485.
4. S. B. Singh, A. Kumar and N. G. R. Iyengar, "Progressive failure of symmetric laminates under uni-axial compression", Structural Engineering and Mechanics, 1997, Volume 5, pp. 433-450.
5. S. B. Singh and A. Kumar, "Postbuckling response and failure of symmetric laminates under in-plane shear", Composites Science and Technology, 1998, Volume 58, pp. 1949-1960.
6. S. Tsai, *Composites Design*, SAMPE Publication, 1986.

Buckling Analysis of Composite Laminated Plate under Ribs

Qing-Qing NI and Jia XIE

Division of Advanced Fibro-Science
Kyoto Institute of Technology
Matsugasaki, Sakyo-ku, Kyoto
606-8585, Japan
nqq@ipc.kit.ac.jp

ABSTRACT

In this paper, the buckling analysis of symmetrically laminated composite plates with ribs was proposed. The higher-order shear deformation theory was taken into account, and the pb-2 Ritz displacement functions was introduced, which consisted of the product of a basic function and a two dimensional polynomial function and could be employed to represent an arbitrary edge condition and rib structures. Based on the proposed method, the buckling behavior of symmetrically laminated composite plates under line, circle or ellipse ribs was investigated under the biaxial compression loading. Numerical results were presented to show the effects of lamination angles, edge conditions, aspect ratio, and rib shapes on critical buckling load. It was confirmed that the present method was an effective one and could be widely used to analyze the buckling load and to predict the buckling characteristics of laminated composite plates under rib structures.

INTRODUCTION

Fiber reinforced laminated composites are being increasingly used in modern engineering applications, due to their high specific strength and modulus. However, in such applications, buckling phenomenon may often be observed. Buckling phenomenon is critically dangerous to structural components because the buckling of composite plates usually occurs at a lower applied stress and generates large deformation. This led to a focus on the study of buckling behavior in composite materials (1, 2, 3).

In the buckling analysis of laminated composite plates, several shear deformation theories have been proposed due to the lower out-of-plane shear modulus and/or the significant thickness of the plate. The first-order shear deformation theory accounted for the transverse shear deformation, but only the constant distribution of transverse shear strain throughout the thickness was predicted (4,5,6), whereas the higher-order shear deformation theory accounted for: (1) the transverse shear deformation; (2) the parabolic variation of the transverse shear throughout the thickness; (3) vanishing shear stresses at the plate surfaces, and it eliminates the need for shear correction factors. (2,7)

In the present paper, both of the higher-order shear deformation plate theory and the two-dimensional Ritz displacement functions to represent an arbitrary edge condition were utilized to analyze the buckling behavior of symmetrically laminated composite plates with rib structures under biaxial compression loading, and the effect of the parameters such as the aspect ratio, angle of fiber orientation, boundary conditions, rib shapes and so on were investigated.

THEORETICAL ANALYSIS

Consider a laminated plate of length a, width b, and thickness t as shown in Fig. 1 in a typical coordinate system. The higher-order shear deformation theory developed by Reddy and others yields the following displacement fields:

$$u_z(x,y,z) = u_0(x,y) + z\theta_x(x,y) - \frac{4}{3t^2}z^3\left[\theta_x(x,y) + \frac{\partial w}{\partial x}\right]$$

$$v_z(x,y,z) = v_0(x,y) + z\theta_y(x,y) - \frac{4}{3t^2}z^3\left[\theta_y(x,y) + \frac{\partial w}{\partial y}\right] \quad (1)$$

$$w_z(x,y,z) = w_0(x,y)$$

where $u(x, y, z)$ and $v(x, y, z)$ are the in-plane displacements at any point (x, y, z), $u_0(x, y)$ and $v_0(x, y)$ denote the in-plane displacements of the point $(x, y, 0)$ on the mid-plane, and $w_0(x, y)$ is the deflection assumed to be constant in the thickness direction z. $\theta_x(x, y)$ and $\theta_y(x, y)$ are the rotations of the normal to the mid-plane about the y-axis and x-axis, respectively. For symmetrically laminates, the in-plane displacements $u_0(x, y)$ and $v_0(x, y)$ could be neglected because the coupling rigidity was zero.

The total potential energy for a composite plate was given by

$$\Pi = \frac{1}{2}\int_V \Delta\varepsilon_L^T\left[\overline{Q}\right]\Delta\varepsilon_L dV + \int_V \tau^T \Delta\varepsilon_N dV \quad (2)$$

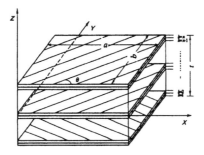

Figure 1. A schematic representation of laminated composite plates.

where ε_L, ε_N were the strains of the linear and non-linear items, respectively. τ^T was the stress tensor. For the in-plane biaxial compression loading, $\tau^T = \{-N_x/t \ -N_y/t \ 0 \ 0 \ 0\}$. N_x, N_y are the applied compression load in x and y direction, respectively.

The pb-2 Ritz function was used in the present analysis to obviate the tedious task of choosing the form of the infinite series or trigonometric functions to suit the edge support conditions. $w_0(x, y)$, $\theta_x(x, y)$ and $\theta_y(x, y)$ may be parameterized by equation (3)

$$w_0(x,y) = \sum_{i=1}^{m} c_i \phi_i(x,y)$$

$$\theta_x(x,y) = \sum_{i=1}^{n} d_i \psi_{xi}(x,y)$$ (3)

$$\theta_y(x,y) = \sum_{i=1}^{n} e_i \psi_{yi}(x,y)$$

in which,

$$\phi_i(x,y) = f_i(x,y)\phi_1(x,y)$$
$$\psi_{xi}(x,y) = g_{xi}(x,y)\psi_{x1}(x,y)$$ (4)
$$\psi_{yi}(x,y) = g_{yi}(x,y)\psi_{y1}(x,y)$$

where, c_i, d_i, e_i were the unknown coefficients and the Ritz function, $w_0(x, y)$, $\theta_x(x, y)$ and $\theta_y(x, y)$ were taken as the product of the boundary function, $\phi_1, \psi_{x1}, \psi_{y1}$, and polynomial function, f_i, g_{xi}, g_{yi}

Several configurations of ribs (see Fig. 2) and supporting edge conditions for the laminated plates were taken into account for the present study. To consider the rib structures, the boundary functions were defined as follows:

$$\phi_1 = \left\{ \prod_{i=1}^{4} [\Gamma_i(x,y)]^{p_i} \right\} \left\{ \prod_{j=1}^{n_1} [\Lambda_j(x,y)]^{p_j} \right\}$$

$$\psi_{x1} = \prod_{i=1}^{4} [\Gamma_i(x,y)]^{p_i}$$

$$\psi_{y1} = \prod_{i=1}^{4} [\Gamma_i(x,y)]^{p_i}$$ (5)

where $\Gamma_i(x,y)$ is the boundary equation of the ith supporting edge, and Ω_i depends on the support edge condition as listed in Table I.

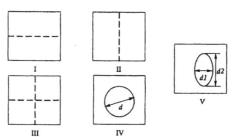

Figure 2. Rectangular plates under representative rib structures.

Table I. The value of Ω_i for various supported conditions

	$\Omega_i = 1$	$\Omega_i = 0$
ϕ_1	Clamped or Simply	Free
ψ_{x1}	Clamped or x- direction simply	Free or y-direction simply
ψ_{y1}	Clamped or y direction simply	Free or x-direction simply

$\Lambda_j(x,y)$ was the equation of the jth rib, n_i was the number of straight line/circular rib. If the jth rib existed, $\Phi_j=1$; if the jth rib was removed, $\Phi_j=0$. The equations, $\Lambda_j(x,y)$, of Fig. 2 are shown in Table II.

Table II. The equations of $\Lambda_j(x,y)$ for five types of rib structure

type	$\Lambda_j(x,y)$
I	$\Lambda_1(x,y) = y - b/2$
II	$\Lambda_1(x,y) = y - a/2$
III	$\Lambda_1(x,y) = y - b/2 \quad \Lambda_2(x,y) = y - a/2$
IV	$\Lambda_1(x,y) = (x - a/2)^2 + (y - b/2)^2 - (d/2)^2$
V	$\Lambda_1(x,y) = \dfrac{(x - a/2)^2}{(d_1/2)^2} + \dfrac{(y - b/2)^2}{(d_2/2)^2} - 1$

Substitute the above displacement function into equation (2), and minimize the total energy by differentiation, and then the coefficients, c_i, d_i, e_i, could be determined. Here, the normalized coordinates $\xi = x/a$, $\eta = y/b$ were used.

$$
\begin{bmatrix}
[K_{cc}] & [K_{cd}] & [K_{ce}] \\
 & [K_{dd}] & [K_{de}] \\
\text{Symmetric} & & [K_{ee}]
\end{bmatrix}
\begin{Bmatrix} \{c\} \\ \{d\} \\ \{e\} \end{Bmatrix}
=
\begin{Bmatrix} 0 \\ 0 \\ 0 \end{Bmatrix}
\tag{6}
$$

By solving equation (6) as an eigenvalue problem, the critical buckling load could be obtained.

NUMERICAL RESULTS AND DISCUSSION

Here, some numerical results were presented for symmetrically laminated plates under line and circular ribs. In the computations, the following mechanical properties of a lamina were used.

$a/b = 1, b/t = 10, \quad G_{12} = 0.5E_2, \quad G_{13} = 0.5E_2, \quad G_{23} = 0.3E_2, \quad E_1/E_2 = 25, \quad \nu_{12} = 0.25.$

ν_{21} was determined by the reciprocal theorem. For the convenience, the non-dimensional buckling load was defined by:

$$K = Nb^2 / \pi^2 D \tag{7}$$

where $D = E_2 t^3 / [12(1 - \nu_{12}^2)]$.

Accuracy and Convergence Study

To demonstrate the accuracy and convergence of the proposed method, a comparison study was carried out for isotropic plates under ribs with simply supported 4 edges condition due to no exact solution existed for anisotropic plates under ribs. Table III shows the computational results for in-plane biaxial compression loading, which were comparable with the values available in the reference (3). A very close agreement was observed. For the convenience of description, three capital letters were used to represent the edge situation of the plate. The symbols S, C, and F represented simply supported, clamped, and free edges, respectively. The convergence of the non-dimensional buckling load, K, of symmetrically cross-ply laminated plates under rib type I and III with two different edge conditions was listed in the Table IV with

Table III. Non-dimensional buckling load, K, of isotropic plates under rib II and rib IV structures

	II (SSSS)			IV (SSSS)	
a/b	Present	Reference (3)	d	Present	Reference (3)
0.5	16.9996	17.000	0.2	8.31917	8.2873
1.0	4.99997	5.0000	0.4	9.88775	9.7900
1.5	2.77776	2.7778	0.6	9.56354	9.5684
2.0	1.99997	2.0000	0.8	7.39190	7.3947
2.5	1.6400	1.64004	0.9	6.42107	6.4226

Table IV. The convergence of non-dimensional buckling load, K, of symmetrically cross-ply
(0/90)₂ laminated square plates under rib type I and III

Polynomial terms m	I		III	
	SSSS	SSFF	SSSS	SSFF
49	11.6947	1.9407	18.5464	5.1224
64	11.6947	1.9389	18.5459	5.1087
81	11.6945	1.9372	18.5471	5.0985
100	11.6946	1.9370	18.5464	5.0896

the increment of the polynomial terms, m. It was also found that the convergence of the solution
for the buckling load was quite well.

Symmetrically Laminated Plates under Line Ribs

At first, symmetrically angle-ply laminated plates under line ribs were analyzed by the
proposed method. Figure 3 shows the non-dimensional buckling loads of plates under lib type I,
II, III and no rib structure, respectively. The boundary condition was simply supported 4 edges. It
could be seen that the buckling loads were non-symmetrical about the lamination angle from
$\theta=0°$ to $\theta=90°$ because of the effect of the lib type I and II, while it was symmetrical and has a
maximum buckling load at the lamination angle of 45 degrees under type III rib due to its
symmetry. The buckling load at any lamination angle was much larger than that without rib
supports (×) except at lamination angle $\theta=0°$ (type I) and $\theta=90°$ (type II). The effect of various
aspect ratios, a/b, on the buckling load under the rib type I with simply supported 4 edges was
presented in Fig. 4. It shows that for $a/b=0.5$ the buckling load had a maximum value at
lamination angle $\theta=45°$ and symmetrical, while for the other aspect ratios, the buckling load
became unsymmetrical and the lamination angle θ played a more and more role on the bucking

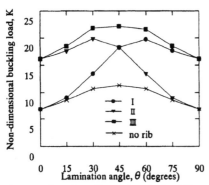

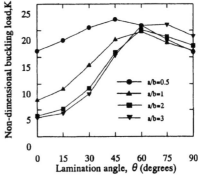

Figure. 3. Non-dimensional buckling
loads of symmetrically angle-ply
laminated plates $(\theta/-\theta)_2$ under rib I, II,
III and no rib.

Figure. 4. Effect of aspect ratios, a/b, on
non-dimensional buckling load for
symmetrically angle-ply laminated plates
$(\theta/-\theta)_2$ under rib I.

Table V. Non-dimensional buckling loads, K, of symmetrically cross-ply laminated plates (0/90)₂
under rib I with different edge support

a/b	SSSS	CCCC	SSFF	CCCF
0.5	18.5465	20.3849	2.0237	13.6460
1.0	11.6947	15.6386	1.9389	13.5865
1.5	10.5801	14.8348	1.8994	13.2665
2.0	10.3710	14.6729	1.8819	12.9710
2.5	10.3318	14.5430	1.8729	12.8525

curves with the increment of the aspect ratio, a/b. Table V shows the buckling loads of symmetrically cross-ply laminated plates under the rib type I with several different edge supports.

Symmetrically Laminated Plates under Circular Ribs

In this section, the target was to analyze the buckling behavior of the symmetrically angle-ply laminated plates under circular ribs. Figure 5 shows the buckling loads of laminated plates under circle rib with simply supported 4 edges. It was obvious that the buckling loads to be symmetrical with a maximum value at the lamination angle of 45°, and when the normalized diameter of the circle rib increased from 0.2 to 0.6, the buckling load became larger. The reason may be considered as the enhanced area of the plates relatively expanded with the increment of diameter, and this led to the increment of the buckling load. The results in the symmetrically angle-ply laminated plates under ellipse rib as shown in Fig. 2 were illustrated in Fig. 6 by changing one of the diameters of ellipse rib (d_2) under the four edges simply support. It could be seen that the buckling loads fell into two groups: one was $d_1 > d_2 = 0.4$, and the other one was $d_1 < d_2 = 0.4$. In the case of $d_1 > d_2 = 0.4$, the buckling loads for the lamination angle $\theta < 45°$ were larger than those at $\theta > 45°$. However, it turned to be opposite in the case of $d_1 < d_2 = 0.4$. When $d_1 = 0.4$, and $d_2 = 0.8$, the buckling load was K=19.9 ($\theta = 0°$), and K=10.1 ($\theta = 90°$), respectively,

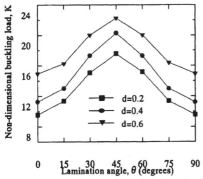

Figure. 5. Non-dimensional buckling loads of symmetrically angle-ply laminated plates ($\theta/-\theta$)₂ under different diameters of circle rib structures.

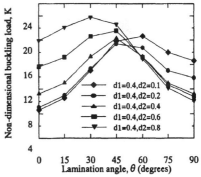

Figure. 6. Non-dimensional buckling loads of symmetrically angle-ply laminated plates ($\theta/-\theta$)₂ under different long/short diameters of ellipse rib structures.

nearly 50% reduced. It indicated that the importance of the lamination angle and the size of the ellipse rib on the design of symmetrically angle-ply laminated plates under ellipse rib.

CONCLUSIONS

In this paper, a widely useful method was developed for the buckling analysis of symmetrically laminated composite plates under line and circular ribs subjected to in-plane biaxial compression loading. It should be pointed out that in the present method, the higher-order shear deformation theory and the pb-2 Ritz displacement functions, especially the treatment of rib structure were taken into account, so that more accurate results and simply computation could be expected for buckling analysis. As an example, the optimal lamination angle at which the non-dimensional buckling load took the maximum may be obtained for symmetrically laminated plates under different rib structures.

REFERENCES

Book References

1.   Timoshenko, S. P. & Gere, J. M., Theory of Elastic Stability (2nd edn). McGraw-Hill, New York, 1961.

Edited Conference Proceedings

2.   Q. Q. Ni, J. Xie and Z. Maekawa, "Buckling Analysis of Laminated Composite Plates using Higher-order Shear Deformation Theory", The 13th International Conference of Composite Material, 2001.

Journal References

3.   K. M. Liew and C. M. Wang, "Elastic buckling of Rectangular Plates with Curved Internal Supports", J, Struct. Engng ASCE 118(6), 1480-1493, (1992).

4.   G. B. Chai and K. H. Hoon, "Buckling of generally laminated composite plates", Composite Sci. and Tech., 45 (1992), 125-133.

5.   G. Akhras, M. S. Cheung and W. Li, "Static and vibration analysis of anisotropic laminated plates by finite strip method", Int. J. Solids Struct. 30(22), 3129-3137 (1993).

6.   Q. Q. Ni, "Buckling Analysis of Laminated Composite Plates Compressed in Two Perpendicular Directions with Out-of-Plane Shear Deformation", J. Soc. Mat. Sci., Japan, Vol.46, 1362-1368 (1997).

7.   J. N. Reddy, "A simply higher-order theory for laminated composite plates", J. Appl. Mech., Vol.51, 745 (1984).

Interfacial fracturing and debonding failure modes in FRP-strengthened concrete structures

Zhishen Wu
Department of Urban and Civil Engineering, Ibaraki University, 4-12-1 Nakanarusawa-cho, Hitachi, 316-8511, Japan

Jun Yin
Research Organization for Information Science and Technology (RIST), 2-2-54 Naka-Meguro, Meguro-ku, Tokyo 153-0061, Japan

Takashi Ishikawa
Advanced Composite Evaluation Technology Center, National Aerospace Laboratory of Japan, 6-13-1 Ohsawa, Mitaka, Tokyo 181-0015, Japan

Mikio Iizuka
Research Organization for Information Science and Technology (RIST), 2-2-54 Naka-Meguro, Meguro-ku, Tokyo 153-0061, Japan

ABSTRACT

The strengthening of fiber reinforced polymers (FRP) to concrete structures is significantly limited due to the interfacial debonding along the FRP-concrete bond interface. Experimental work shows that debonding fracture, in most cases, occurs through the interfacial concrete adjacent to bond interface. Although this interfacial debonding is a shear-dominated fracture that resembles mode II fracture, the interfacial concrete crack should be considered to initiate in mode I fracture followed by the aggregate interlock along the crack surface. To model such a fracture, smeared crack models are used to describe the fracturing process. But they cannot predict the load-carrying capacity correctly. So, a displacement discontinuity model is adopted. In this paper, the emphasis is laid on the study of interfacial fracture that occurs along the interfacial concrete layer. A finite element analysis is performed, in which the debonding propagation in concrete is modeled by both smeared crack models and the displacement discontinuity model. The results are compared to mode II fracture assumption. The effect of concrete fracture energy on load-carrying capacity and shear transfer ability of bond interface is investigated.

INTRODUCTION

The external bonding technique by using fiber reinforced polymer (FRP) composites has been demonstrated remarkably efficient in strengthening and upgrading structurally inadequate or damaged concrete structures. However, an important fracturing behavior, interfacial debonding, that occurs along FRP-concrete bond significantly limits and reduces FRP strengthening performance. So, it is very necessary to understand the FRP-concrete interfacial debonding process.

In recent years, many efforts on this research have been made. Täljsten (1) studied on the interfacial debonding of carbon fiber sheets bonded concrete prism by nonlinear fracture mechanics. It was assumed that the interfacial debonding resembled mode II fracture in pure shear manner. Given a linear local shear stress vs. relative displacement τ–δ relationship, the load-carrying capacity of bond interface can be predict. Yuan et al.(2) developed Täljsten's work by using several τ–δ relationships including softening. The corresponding shear stress distributions along the bond interface were quantitatively obtained. It addresses that load-carrying capacity is uniquely dependent on the interfacial fracture energy G_f^{II}, and FRP properties, but not on the shapes of τ–δ relationships. In experimental work, Yoshizawa et al.(3) and Wu et al.(4) carried single/double-lap shear tests of FRP bonded concrete prism to identify the interfacial fracture energy G_f^{II}. A finite element analysis was also carried by Yin et al.(5), in which the effects of interfacial bond strength and τ–δ relationship on shear stress distribution along the bond interface were presented.

All the previous works assume the interfacial debonding as a mode II fracture. Such an assumption has provided a quantitative expression between load-carrying capacity of FRP-concrete bond and interfacial fracture energy. However, many experiments record that the interfacial debonding usually occurs and propagates in concrete adjacent to bond interface. And concrete crack usually initiates as mode I fracture. Therefore, the debonding behavior, to a certain extent, should relate to the concrete property. In this paper, we focus on the debonding behavior through interfacial concrete adjacent to FRP-concrete bond through finite element simulation. The conventional smeared crack models and a displacement discontinuity model are adopted. Comparing results by mode II fracture assumption, it is found that the smeared crack models can not correctly express the fracture energy release. And the displacement discontinuity model is more reasonable, which considers the fracture energy release in the directions both normal and parallel to the crack surface. In addition, it also gives a quantitative relation of fracture energy with the mode II fracture assumption.

INTERFACIAL DEBONDING BEHAVIORS

The force transfer between FRP and concrete is achieved by adhesive resin, through which the concrete matrix is strengthened. For simplicity, the debonding failure along FRP-concrete bond interface is generally modeled as mode II fracture. However, looking at the debonded FRP sheet, it is found that a layer of concrete, with thickness from 2mm to 10mm, is usually stuck on it, as presented in Figure 1. In addition, the finite element simulation of FRP-strengthened concrete beams by Wu and Yin(6) also reported the interfacial debonding failure through a layer of cracked concrete elements, in form of mode I fracture.

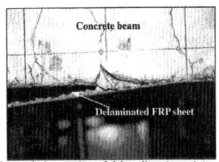

Figure 1 Experimental observation of debonding through interfacial concrete

Although FRP strengthening can prevent flexural concrete cracks from further propagation, the stress transfer by adhesive layer heightens the stress level of uncracked concrete adjacent to the bond interface. Once the concrete stress of reaches its tensile strength, the interfacial concrete will start to crack. In this case, the crack orientation is usually 45^0 inclined to the bond surface. With the further stress transfer, more interfacial concrete cracks, so that a macro interfacial debonding occurs eventually. Different from the mode II fracture assumption, crack in concrete is considered to initiate as mode I fracture with subsequent aggregate interlock along the crack surface, as shown in Figure 2.

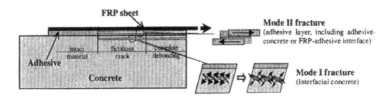

Figure 2 Fracture mechanisms of interfacial debonding

Mode II Fracture Assumption-Macro Representation of Debonding

The interfacial debonding behavior can be expressed by a $\tau-\delta_t$ function. A simple linear ascending-descending $\tau-\delta_t$ relationship is usually adopted, in which interfacial fracture energy G_f^{II} is introduced. According to the works of Täljsten[1], Yuan et al.[2], The load-carrying capacity of FRP-concrete bond interface can be approximately expressed as a function of the stiffness of FRP sheets ($E_{frp} \times t_{frp}$) and interfacial fracture energy G_f^{II}.

$$P_{max} = b_{FRP} \sqrt{2 E_{FRP}\, t_{FRP}\, G_f^{II}} \tag{1}$$

Mode I Fracture Assumption-Micro Representation of Debonding in Concrete

Smeared Crack Models

There are two kinds of conventional smeared crack models (7). One is the rotating crack model, which assumes that the crack normal continuously rotates with the changing axis of principal stress. The other is the fixed crack model, in which a constant shear retention factor β is introduced to account for the shear transfer effect like aggregate interlock. In both crack models, a linear softening curve is used to describe the normal stress degradation on crack surface.

Displacement Discontinuity Model

This crack model with embedded displacement discontinuity within a finite element was originally developed for simulating the localized crack propagation in concrete structure. The detailed finite element formulation can be referred to (8). In crack normal, the linear softening relation similar to smeared crack models is used. Along the crack surface, a softening model is also assumed by introducing the shear fracture energy G_f^{IIs}.

FINITE ELEMENT ANALYSIS

The interfacial debonding behavior of a concrete prism strengthened with FRP sheets is simulated. The schematic sketch and finite element mesh are presented in Figure 3, where L is FRP sheets bond length, $t_c=$ 80mm, and $t_{FRP}=$ 0.1mm are the thickness of concrete prism and FRP sheets, respectively. A unit out-of-plane bond width $b_c=b_{FRP}=1$mm is assumed.

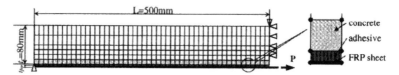

Figure 3 Finite element discretization of FRP bonded concrete prism

In the FE simulation, the concrete prism, FRP sheets and FRP-concrete bond interface are discretized by 4-node plane stress elements, truss elements and line-to-line interface elements respectively. If cracks occur in concrete, the smeared crack models or the displacement discontinuity model is triggered. FRP sheet is kept linear elastic until rupture and debonding behavior of interface element follows the τ–δ_t relationship in previous section. Properties of each material are summarized as follows. For concrete, Young's modulus $E_c=2.5\times10^4$ MPa, Poisson ratio $\nu=0.15$, tensile strength is $f_t=2.0$MPa. For FRP sheets, Young's modulus $E_{FRP}=2.3\times10^5$MPa, rupture stress of FRP sheet is 3.35×10^3MPa. The bond strength of FRP-concrete interface is $f_b=3.0$MPa, initial stiffness is $K_b=160$N/mm^3, and interfacial mode II fracture energy $G_f^{II}=1.2$N/mm, referred to the experiment by Yoshizawa et al.(3).

Mode II Fracture

Firstly, the interfacial debonding is assumed to propagate only within interface elements as a pure mode II fracture. To avoid the interaction of possible concrete crack, the concrete element is enforced to be elastic. Varying interfacial fracture energy $G_f^{II}=0.25$, 0.62 and 1.2N/mm, the load-displacement relations are obtained, as shown in Figure 4. The load-carrying capacity of FRP-concrete bond are 109N, 170N and 237N, respectively. Compared to the theoretical solutions calculated by Eq.(1), it can be seen that the ultimate load-carrying capacity of simulation results match well. It is seen that mode II fracture assumption can basically describe the debonding behaviors along FRP-concrete interface in a macro level.

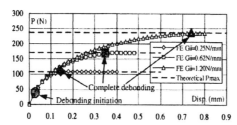

Figure 4 load-displacement relations with varying G_f^{II}

Mode I Fracture

For the debonding propagation through interfacial concrete, concrete crack is initiated as mode I fracture. It is assumed that the interfacial concrete crack is the only source resulting in the debonding failure. Hence, the interface elements are enforced to be linear elastic. Also, because the concrete crack occurs adjacently to the bond interface, only one layer of concrete elements are allowed to crack, as shown in Figure 3, while the rest are kept elastic.

Rotating Smeared Crack Model

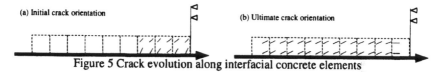

Figure 5 Crack evolution along interfacial concrete elements

By using the rotating smeared crack model, the evolution of crack orientations at each cracked element is presented in Figure 5. Concrete cracks initiate in the direction of inclined 45^0 to the loading axis. With continuing load, the crack orientations gradually rotate and ultimately approach to the horizontal direction. It means that fracture starts in tension and subsequently proceeds in tension-shear. This means the rotating crack model is applicable to explain the evolution of interfacial concrete crack that finally results in the debonding failure.

With G_f^I=0.2, and 1.0N/mm, Figure 6 gives the simulation results of load-displacement relationship by rotating crack model and comparison to the mode II fracture assumption with G_f^{II}=1.2N/mm. According to the mode II fracture assumption, G_f^{II}=1.2N/mm is regarded as an average value predicted from the experimental data (3). On the other hand, for common concrete G_f^I is around 0.1~0.2N/mm, based on recent experimental research of uniaxial concrete tension test (9). Expectantly, the same bond load-carrying capacity should have been obtained no matter whichever interfacial fracture assumption is applied. However, result shows big difference. For rotating crack model, G_f^I

must be increased to 1.0N/mm, an unpractical value of concrete fracture energy, to obtain the same load-carrying capacity of bond. It is because that rotating crack model neglects the aggregate interlock on crack surface so as to underestimate load-carrying capacity of bond.

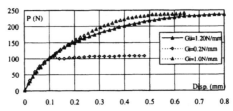

Figure 6 Fracture energy release of mode I and mode II fracture assumption

Fixed Smeaed Crack Model

Can fixed smeared crack model that treats the aggregate interlock by a constant shear retention factor β obtain better simulation result? With fixed $G_f^I = 0.2$N/mm, Figure 7 shows β's influence with β=0, 0.001, 0.005, 0.01. It is seen that when β is zero or very small, the load-carrying capacity is low and similar to that of rotating crack model. When β increases to 0.005 and 0.01, the aggregate interlock behavior is apparently seen. However, the load intends to increase continuously but not approaches to a constant. This is not the true fracturing behavior. The aggregate interlock is overestimated due to the constant accumulation of residuel shear stress along the crack surface by a constant shear modulus. It is suggested that the shear retention factor does have its physical meaning to describe the aggregate interlock behavior. But a non-zero constant shear retention factor β easily leads to the shear stress accumulation on the crack surface, thus resulting in incorrect output.

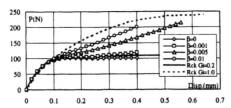

Figure 7 Load-displacement curves with different shear retention factor β

Displacement Discontinuity Model

Considering the disadvantages of the both smeared crack models, the displacement discontinuity model is used, in which the shear behavior along crack surface follows a softening curve by introducing shear fracture energy G_f^{IIs}. By fixing the concrete fracture energy at $G_f^I=0.2$N/mm as used in previous simulations and varying the shear fracture energy G_f^{IIs}, the load-carrying capacity of FRP-bonded concrete prism increases, and the ultimate value approaches a constant, as shown in Figure 8.

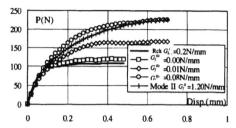

Figure 8 Load-displacement curves with varying G_f^{IIs}

By comparing to the simulation results by rotating smeared crack model and mode II fracture assumption, it can be seen that when G_f^{IIs} is zero the load-carrying capacity lies on its lower limit similar to that of rotating smeared crack model, while the increasing G_f^{IIs} enhances the load-carrying capacity. When G_f^{IIs} =0.08N/mm, the same level of load-carrying capacity obtained by mode II fracture assumption with G_f^{II}=1.2N/mm can be reached with unchanged G_f^{I}=0.2N/mm. It implies that the shear fracture energy G_f^{IIs} reflects the shear transfer ability of concrete when the shear fracture is dominated. Through a numerical fitting by displacement discontinuity model, a relationship between mode II fracture and mode I fracture in concrete followed by shear transfer is provided in Figure 9.

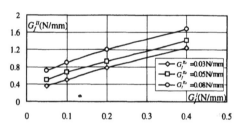

Figure 9 G_f^{I}-G_f^{II} relations with varying G_f^{IIs}

In this sense, if the debonding fracture occurs along interfacial concrete, both fracture assumptions, mode II fracture and mode I fracture in concrete, should lead to the same load-carrying capacity of the bond. The introduction of shearing fracture energy on crack surface after the mode I crack initiation in concrete bridges these two assumption. For different concrete, the quantitative relation may be different.

CONCLUSIONS

The debonding through the interfacial concrete in FRP-strengthened concrete structures is studied. This study is necessary now that it provides a new view of the debonding mechanism in FRP-strengthened concrete structure more physically. Through a finite element analysis using different crack models, the following conclusions are drawn:

1) Rotating smeared crack model can physically describe the concrete crack evolution during debonding propagation process. But it underestimates the real aggregate interlock effect on the crack surface.

2) Fixed crack model considers the shear transfer by a constant shear retention factor β. But a big value of β may lead to accumulated residue shear stress on the crack surface that never vanishes, thus overestimating the load-carrying capacity of FRP-concrete bond.

3) In the case of shear-dominated fracture, the fracture energy release along the crack surface plays an important role to describe the ability of shear transfer. Displacement discontinuity model that introduces shear fracture energy for shear transfer on crack surface provides a solution. By comparing the mode II fracture assumption and mode I fracture assumption, the shear fracture energy is identified, and the relation between these two debonding fracture assumption is obtained.

REFERENCES

1   Täljsten, B.: Strengthening of concrete prisms using the plate-bonding technique. International Journal of Fracture, 82:3, pp.253-266, 1996.

2   Yuan, H., Wu, Z.S. and Yoshizawa, H.: Theoretical solution on interfacial stress transfer of externally bonded steel/composite laminates, J. of Struc. Mech. And Earthquake Eng. JSCE, pp.27-39. 2001.4.

3   Yoshizawa, H., Wu, Z.S., Yuan, H. and Kanakubo, T.: Study on FRP-concrete interface bond performance, Journal of Material, Concrete Structures and Pavements, JSCE, 49:662, pp.105-119, 2000.

4   Wu, Z.S., Yuan, H., Yoshizawa, H. and Kanakubo, T.: Experimental/analytical study on interfacial fracture energy and fracture propagation along FRP-concrete interface, ACI International, Special Publication 201, pp.133-152, 2001.

5   Yin, J. and Wu, Z.S.: Interface crack propagation in FRP-strengthened concrete structures using nonlinear fracture mechanics, FRPRCS-4, ACI International SP 188, edited by Dolan et al., pp.1035-1047, 1999.

6   Wu, Z.S. and Yin, J.: Fracturing behaviors of FRP-strengthened concrete structures, Engineering Fracture Mechanics(submitted)

7   Borst, R. de: Some recent developments in computational modelling of concrete fracture, Int. J. Fracture Mechanics, 86, pp.5-36, 1997.

8   Wu, Z. & Yin, J.: Analysis of mixed-mode brittle fractures by mixed finite element with internal displacement discontinuities", Proceedings of Conference on Computational Engineering and Science, JSCES, 3, pp.841-844, 1998.

9   Lenke, L. R. and Gerstle, W. H.: Tension test of stress versus crack opening displacement using cylindrical concrete specimens, ACI International SP 201, edited by Vipulanandan and Gerstle, pp.189-199, 2001.

Damage monitoring of glass/epoxy laminates with luminance of backlight of EL device

Associate Professor Akira Todoroki, and Dr. Yoshinobu Shimamura
Department of Mechanical Sciences and Engineering
Tokyo Institute of Technology
2-12-1, Ohokayama, Meguro-ku, Tokyo 152-8552, Japan
E-mail: atodorok@ginza.mes.titech.ac.jp (Akira Todoroki)

Yasuyuki Tanaka
Graduate student of Tokyo Institute of Technology

ABSTRACT

Authors have proposed a new damage monitoring system that employs luminance of EL backlight for transparent composites such as glass/epoxy composites. For the transparent composites, damage is easily found by visual inspections. In the cases that the structures are sealed or huge structures, however, the visual inspections are difficult to perform. The present paper adopts a system using change of luminance of transmitted light. When the composite structures are damaged, the damage reduces luminance of the transmitted light from the backside. The present study adopts an Electro Luminescent device (EL) as the backlight. In the present study, the method is applied to the glass/epoxy and Al_2O_3/epoxy composites fabricated from the unidirectional glass/epoxy prepreg. For these non-fabric composites, the silhouette of the matrix cracks is quite different from the damage silhouette of the fabric composites. In the present study, therefore, the method is applied to the unidirectional composites: glass/epoxy and Al_2O_3/epoxy composites. Specimens were made from the cross-ply laminates. Tensile tests are conducted using rectangular specimens with open holes fabricated from cross-ply glass/epoxy and Al_2O_3/epoxy composites. The luminance of the transmitted light of the EL device increases with the increase of the elastic deformation of the transparent composites by tensile loading. The damage is monitored without loading with measurements of the luminance of transmitted light of the EL device.

INTRODUCTION

Damage of transparent or semi-transparent composites such as glass/epoxy composites or aluminum-oxide-ceramic-fiber/epoxy (Al_2O_3/epoxy) composites reduces its visual transparency owing to the scattering of transmitted light by the generated damage. The damage, therefore, can usually easily be detected visually. For super conductive coil support structures of MAGLEV, however, the visual inspections are sometimes quite difficult because the structure of the coil is completely sealed to obtain high thermal insulation. For large glass/epoxy composite fans of wind plants, the fan blade itself is too large, and the visual inspections of damages after dismantling the fan blades are not cost effective. To detect damages of composite structures, recently, health-monitoring systems using embedded fiber optic sensors have been attempted [1-3]. The health monitoring systems using fiber optic sensors can measure strain data at multiple points. Damages, however, cannot be detected without loading. This means that the damages of the coil support structures of MAGLEV can not be detected before running and the damage of the fan blades for wind plants can not detected when the wind plants are stopped. Although these problems are not fateful for fiber optic sensors, the problems surely require additional cumbersome approaches. Furthermore, some researches have reported that embedding fiber optic sensors may cause strength reduction [4,5].

For transparent or semi-transparent plastic materials and composites, a luminance pattern of transmitted light through the thickness is a very helpful approach to detect damages [6-8]. Since the micro cracks cause scattering of the transmitting light, the damaged zones can be easily identified visually as a dark area by the composites. Aoyama et al. have revealed that the brightness pattern of the transmitted light are applicable for a new non-destructive inspection tool using a CCD camera and proposed a pattern recognition system for coil support components of MAGLEV [8]. The method is very simple and easy to handle, but CCD cameras are not always available for health monitoring systems. Moreover, uniform light sources are very difficult for curved complicated structural components.

Electro luminescent (EL) devices can emit uniform-plane-light, and the stiffness of the device is very low and flexible. The EL devices are usually applied to backlight sources of liquid crystal displays such as pocket bells and cellular phones. The thickness is less than 0.1 mm, and can emit uniform-plane-light by charging AC current to the devices. The size is not limited and the configuration is changeable just by cutting a sheet of the EL device. The thinnest EL device is very flexible, and can be mounted on curved surfaces.

In our previous study [9], a new damage monitoring system employs change of luminance of transmitted light from an EL backlight has been proposed. The system adopts the EL devices as backlight source for static and fatigue damage monitoring of fabric glass/epoxy composites. An EL device is mounted on a surface of a rectangular specimen with an open hole notch, and the luminance of the transmitted light is measured from the other side of the specimen. The luminance of the transmitted light is measured with optical sensors, and the damage at the notch root is monitored by the change of measured luminance of the transmitted light.

In the present study, this system is applied to cross-ply glass/epoxy and Al_2O_3/epoxy composite laminates. The cross-ply laminates are produced from stacking unidirectional prepreg sheet, and the rectangular specimens are fabricated from the laminates. An EL device is mounted on a rectangular plate specimen, and the luminance of the transmitted light is measured with optical sensors during a static tension test. With the unidirectional laminates, luminance change during elastic loading is measured.

HEALTH MONITORING SYSTEM

Electro Luminescent devices

Two kinds of electro luminescent materials are available; Inorganic and organic. A widely used commercially available EL device is an inorganic EL device. The typical component of the inorganic EL is ZnS. ZnS with some impurities such as Al and Cu is a kind of semi-conductor materials. ZnS with the impurities has two electron levels of a donor level and an acceptor level. By charging alternating current, electron moves between the two energy levels, and the energy equals to the difference between the energy levels is emitted as light. Since the EL devices emit light owing to the energy difference between the two electron levels, the luminescence of EL devices does not generate heat and emission gas. Inside structure of the EL device is illustrated in Fig.1. In EL devices, ZnS is powder and the particles are bonded by adhesive. Thickness of the luminescent layer is approximately from 20 mm to 100mm. The surface of luminescent layer is electrically insulated and the layer is sandwiched by two electrodes. One of the electrodes is ITO (Indium Tin Oxide) film layer that is electro conductive and transparent film. By charging AC current between the two electrodes, the luminescent layer emits cold light through the transparent electrode layer ITO. In the case of the soft back electrode layer, the total stiffness of the EL device is soft just like a sheet of paper. EL

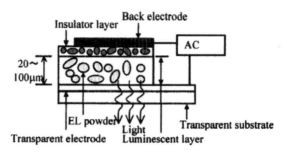

Figure.1 Inside structure of EL device

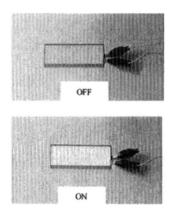

Figure.2 EL devices

devices can emit almost uniform plane light, but the luminance of the EL device is n ot constant. The luminance decreases with the increase of luminescent time. EL devices used for pocket bells are shown in Fig.2.

Monitoring system with EL backlight

A new damage monitoring system is proposed in the previous study [9], and the schematic image is shown in Fig.3. In this figure, a notched plate specimen is shown as an example for structural health monitoring. An EL device is mounted on one of the specimen surface. On the other surface, transmitted light at the point where damage is monitored is transferred using plastic large core optical fiber to a luminance meter (optical sensor), and the luminance of the transmitted light is measured with the luminance meter. Since the luminance of the EL device is not constant owing to aging of

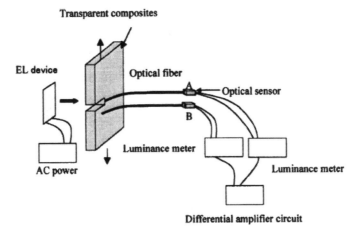

Figure.3 Damage monitoring system with EL backlight

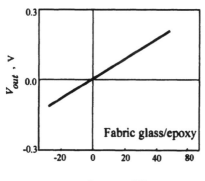

Stress σ , MPa

Figure.4 Luminance change of fabric glass/epoxy laminates due to loading in static tension and compression tests taken from the reference [9].

the EL device, reference light is also measured as shown by optical fiber B in Fig. 3. The reference point must be selected from the points where damage is not created. If there is no such a place, we can adopt dummy specimen that is made from the same composite plate and is not loaded. By comparison of the difference of luminance between these two points, we can detect change of luminance of the transmitted light. Since the reference light is adopted in the system, we can recognize the change owing to the damage even if aging reduces the transparency of semi-transparent composites.

As described before, the EL device is very flexible. This system, therefore, can be applied to curved surfaces like shell structures. Since the EL devices do not generate

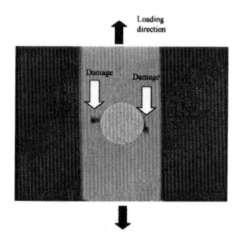

Figure.5 Typical fatigue damage observed in the test of σ=100MPa at N/N$_f$=99% fabric glass/epoxy taken from the reference [9].

heat, this system can be applied to cryogenic structures such as super conductive coil support structures for MAGLEV. All instruments required are not expensive. This system is very attractive for semi-transparent composites for damage monitoring.

In our previous study [9], the system was applied to monitor fatigue damage around an open hole notch of fabric glass/epoxy composites. The results has shown that the luminance of the transmitted light is decreased with the increase of the applied tensile stress in elastic deformation region, and that is increased with the increase of applied compressive stress even in the elastic deformation region as shown in Fig.4. This variation of the luminance of the transmitted light owing to the applied stress made us monitor the damage at unloaded condition. Figure 5 shows the shadow produced by the fatigue damage with the EL backlight. Fatigue damage of the fabric glass/epoxy composites is observed as dark shadows on both ends of the open hole notch.

EXPERIMENTAL METHOD

Specimens

Material used is unidirectional glass/epoxy prepreg and unidirectional Al$_2$O$_3$/epoxy prepreg. Stacking sequence of specimens is [0$_2$/90$_2$]s. Curing condition is 130°×2hr for glass/epoxy composites and 120°×2hr for Al$_2$O$_3$/epoxy composites. After the curing, rectangular specimens are fabricated: 200mm length, 25mm and width 0.5

mm. Thickness of the laminates of 8 plies is approximately 0.5mm. The material properties of this glass/epoxy composites are as follows: E_L=33.6GPa, E_T=14.2GPa, G_{LT}=3.14Gpa, ν_{LT}=0.32, σ_L=924MPa, σ_T=49.6MPa. . The material properties of this Al$_2$O$_3$/epoxy composites are as follows: E_L=94.1GPa, E_T=10.0GPa, ν_{LT}=0.30, σ_L=992MPa, σ_T=60.7MPa.

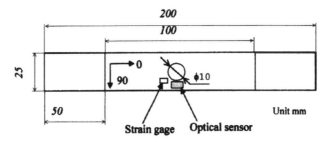

Figure.6 Specimen configuration of open hole notch

Specimen configurations are shown in Fig6. The open-hole-notched specimens are used for measurements of the luminance change during tensile tests. This test is performed to confirm the applicability of the method for the non-uniform matrix cracking around the stress concentrated region. An EL device is mounted near the open hole, and an optical sensor to measure the luminance of the transmitted light is mounted on the edge of the open hole. Strain gage is attached near the open hole as shown in Fig.6 to detect the damage initiation with high sensitivity.

Test procedure

In the present study, static tension tests are performed in room temperature. To measure the luminance change of transmitted light, an EL device is mounted on the specimen surface with adhesive, and an optical sensor is directly attached on the other surface. Tension tests are conducted with measurements of the luminance of the transmitted light with an optical sensor described later. These tests are conducted using a closed-loop material-testing machine produced by MTS under displacement control of 1.5mm/min.

Photo-diodes of BS500B by Sharp Co. are used for the optical sensors here. To measure the luminance of transmitted light using the photo-diodes, a luminance meter circuit shown in Fig. 7 is employed. Using this circuit, change of luminance is converted

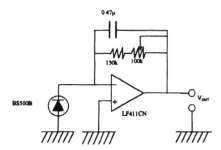

Figure.7   Optical power meter circuit

to electric voltage change. The actual output signal is input into an amplifier, and the magnitude of amplifier is 500. The increase of the output, therefore, means the increase of the transmitted light power in the present study.

RSULTS AND DISCUSSION

Typical images of the both types of the specimens are shown in Fig.8 (a) and (b). The luminance change of the transmitted light and the stress-strain relationship is shown in Fig. 9 (a) and (b). The abscissa is applied strain and the ordinate is measured output of the optical sensor Vout and the nominal stress.

For the glass/epoxy composites, under the applied strain of 5000μ, the stress-strain relation is linear, and no damage is observed as shown in Fig.9 (a). This region is a complete elastic deformation region, and the luminance of the transmitted light is also linearly increasing as the same as the previous test results in the elastic deformation region. Over the applied strain of 5000μ, the stress-strain curve deviates from the linear relation. This deviation is caused owing to the matrix cracking generated from the ends of the open hole notch. This matrix cracking significantly reduces the luminance of the transmitted light Vout as shown in Fig.9 (a).

(a) Glass/epoxy (220MPa)   (b) Al$_2$O$_3$/epoxy (250MPa)
Figure 8 Typical images of damages taken with EL backlight

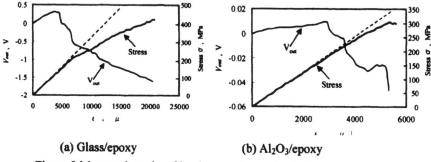

(a) Glass/epoxy   (b) Al₂O₃/epoxy

Figure 9 Measured results of luminance change due to the damage

Almost at the applied strain of 13000μ, longitudinal split cracks of surface 0° plies are observed at the both ends of the notch. The luminance of the transmitted light Vout, however, does not show significant change at this point. This is because the optical sensor is attached at the point slightly apart from the exact end of the notch. This cause that the optical sensor attached at the end of the notch cannot detect the longitudinal split cracks. The longitudinal split cracks could be detected when other optical sensors are attached at the points over or under the notch ends.

For the Al₂O₃/epoxy, however, smaller change of the measured luminance is observed under the linear deformation region: the linear deformation region is under 2500μ. Over the damage initiation, the luminance is reduced. The reduction of the luminance, however, is significantly smaller than that of the glass/epoxy composites. This small luminance reduction is caused by the small damage around the open hole as shown in Fig.8 (b). Although the luminance change is small, the luminance change is available for damage detections.

These results clearly show that the matrix cracks around the stress concentration points can be detected with the luminance change of the transmitted light using an EL backlight. Since the matrix crack does not vanish even when the applied load is completely unloaded, this method can detect the damage without loading.

CONCLUSIONS

(1)   Even for the cross-ply laminates of unidirectional glass/epoxy and Al₂O₃/epoxy composites, the luminance of transmitted light of an EL backlight increases with the increase of applied tensile load in the elastic deformation region.

(2)   This method is applicable to detect matrix cracking at the stress concentration points of transparent composites.

References

1. Rodney Alan Badcock and Gerard Franklyn Fernando, An intensity-based optical fibre sensor for fatigue damage detection in advanced fibre-reinforced composites, Smart Materials and Structures, 4 ,1995, 223-230.
2. Chia- Chen Chang and Jim Sirkis, Impact-Induced Damage of Laminated Graphite/Epoxy Composites Monitoring Using Embedded In-Line Fiber Etalon Optic Sensors, Journal of Intelligent Material Systems and Structures, 8 ,1997, 829-841
3. Chia-Chen Chang and James S Sirkis, Design of fiber optic sensor systems for low velocity impact detection, Smart Materials and Structures, 7 ,1998,166-177.
4. D.C. Seo and J.J. Lee, Effect of embedded optical fiber sensors on transverse crack spacing of smart composite structures, Composite Structures, 32, 1995,51-58.
5. D.C.Lee, J.J.Lee and S.J.Yun, The mechanical characteristics of smart composite structures with embedded optical fiber sensors, Composite Structures, 32 ,1995,39-50.
6. A.G.Efimov, V.M.Parfeev, Kh.Kurzemnieks and S.Vkharitonov, study of Damage in Plastics by Optical Method, Mechanics of Composite Materials, 27, 6, 1992, 711-719.
7. Hiizu Hyakutake and Toshihiro Yamamoto, Damage near the notch root of notched FRP plates in static load- evaluation of damage by a luminance-measuring system, 4th International conference on Localized Damage, Computational Mechanics, 1996, 417-424
8. Aoyama H, Tanaka K, Watanabe H, and Takeda N, Health-monitoring technologies for alumina-fiber-reinforced plastics, Composite Structures,52,2001,523-531.
9. Todoroki A, Shimamura Y., Damage Monitoring for Semi-Transparent Composites using Luminance of EL Backlight,   JSME Int. J., Series A,43,1, 2000,76-82.

Mechanism of Impact Damage Accumulation in Composite Laminates

Hiroshi Suemasu
Department of Mechanical Engineering, Sophia University
7-1, Kioicho, Chiyoda-ku, Tokyo 102-8554, Japan
suemasu@sophia.ac.jp

Yuichiro Aoki
Graduate Student, Department of Mechanical Engineering, Sophia University
7-1, Kioicho, Chiyoda-ku, Tokyo 102-8554, Japan
aoki-y@sophia.ac.jp

Takashi Ishikawa
Advanced Composite Evaluation Technology Center, National Aerospace Laboratory
6-13-1, Ohsawa, Mitaka, Tokyo 181-0015, Japan

ABSTRACT

In order to study an impact damage accumulation problem in composite laminates, a special three-dimensional interface element is proposed to simulate the delamination propagation in laminates. The traction at the interface, which is a function of relative displacements of interface, can reduce to zero when the energy stored in element per unit area reaches critical total energy release rate. This element is incorporated in a commercially available finite element code and applied to the DCB and ENF fracture toughness test problems to demonstrate its validity. The solution converges smoothly and numerical results agree well with the theoretical ones when the element is sufficiently small compared to the significantly deformed area. Also, the present method is applied to damage accumulation problem of CFRP laminates subjected to transverse load and the results obtained are reasonable and interesting.

INTRODUCTION

Composite laminates are used in many engineering applications such as aerospace structures because of the need for reducing weight. But structures made of composite laminates are susceptible to impact damage [1-2]. The impact damages, being difficult to detect from outside of structures, may cause severe reduction of compressive strength (Compression After Impact: CAI)[3]. Design loads of the structures are often limited by the degraded compressive performance. Therefore, the mechanism of damage accumulation due to impact must be well understood in order to utilize composite laminated structures to their full advantage.

Many researchers have studied the failure process of composite laminates based on the damage mechanics [e.g.4]. To know the delamination propagation problem, however, the damage mechanics based analyses may not be sufficient and fracture mechanical theory must be incorporated. Suemasu et al. have explained a damage propagation problem in axi-symmetric plate subjected to quasi-static loading analytically and numerically [5,6]. Aoki et al. have numerically studied the stability of the delamination in composite laminates by three-dimensional finite element analysis [7,8].

In this paper, a special interface element is proposed to simulate the delamination propagation. This element is applied to the DCB and ENF problems to demonstrate its validity. Also, the element is applied to damage accumulation problem in CFRP laminates subjected to static transverse load to disclose the impact damage problem.

INTERFACE ELEMENT

An interface element is proposed to simulate the damage propagation. A 16-node element is developed to incorporate between the 20-node brick element (Figure 1). The elements, which have no thickness, are placed where delaminations are expected to propagate. In this figure, x, y, z and ξ, η are global and local coordinate system, respectively.

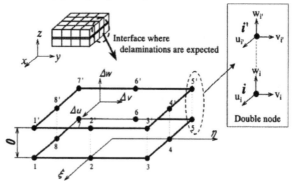

Figure 1. Concept of interface element

The relative displacement distributions in in-plane (Δu, Δv) and out-of plane (Δw) directions are defined as

$$\Delta u = \sum_{i=1}^{8} \phi_i(\xi,\eta)\Delta u_i, \quad \Delta v = \sum_{i=1}^{8} \phi_i(\xi,\eta)\Delta v_i, \quad \Delta w = \sum_{i=1}^{8} \phi_i(\xi,\eta)\Delta w_i$$

$$\Delta u_i = u_{i'} - u_i, \quad \Delta v_i = v_{i'} - v_i, \quad \Delta w_i = w_{i'} - w_i \tag{1}$$

where $\phi(\xi,\eta)$ is a shape function, and u_i, v_i, w_i and $u_{i'}$, $v_{i'}$, $w_{i'}$ indicate nodal displacement components at nodes i and i', respectively. Using these relations, the energy stored in the element per unit area is defined as

$$\Phi = G_t \left\{1 - \exp(-\varphi)\right\} + f\left(\left|\Delta w\right| - \Delta w\right)$$

$$\varphi = \alpha \Delta w^2 + \beta\left(\Delta u^2 + \Delta v^2\right) \tag{2}$$

where G_t is a critical total energy release rate, and α and β are the coefficients which is determined to express the relationship between relative displacement and traction. The second term in Eq. (2) is introduced to consider the contact problem. The term should be smoothly increasing function to make the finite element code converge well. The tractions in each direction at node i are given in the following forms.

$$T_{ui} = \frac{\partial \Phi}{\partial u_i}, \quad T_{vi} = \frac{\partial \Phi}{\partial v_i}, \quad T_{wi} = \frac{\partial \Phi}{\partial w_i} \quad (i = 1 \sim 8) \tag{3}$$

The stored energy and traction are plotted against a relative displacement Δw in Figure 2. The traction reaches an extreme value and starts to reduce to zero. The properties are determined by the coefficients α and β. This interface element is incorporated into the finite element code ABAQUS 5.8 by using the USER SUBROUTINE command.

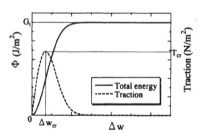

Figure 2. Relationship between potential energy and traction

NUMERICAL RESULTS

Crack propagations in a double cantilever beam and an end notched flexure specimens are simulated. Then, the element is applied to a damage accumulation problem of CFRP square laminates subjected to static transverse load is analyzed.

Numerical Simulation of Double Cantilever Beam Problem

The finite element mesh and loading condition is illustrated in Figure3. Dimensions of specimen are 150 mm × 25 mm, and thickness is 4 mm. The length of an initial crack is 30 mm. Elastic properties of the beam are E_L= 142 GPa, E_T = 10.8 GPa, G_{LT} = 5.49 GPa, G_{TT}= 3.72 GPa and v_{LT}=0.3, v_{TT}=0.45. The fiber direction is same as the beam axis. The value of critical total energy release rate G_t is 400 (J/m²). The present analysis is performed by a displacement control.

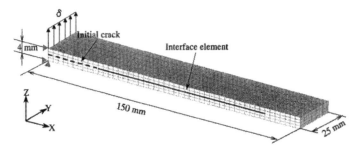

Figure 3. Specimen and finite element modeling

The applied load is plotted against crack opening displacement in Figure 4. When the critical relative displacements Δu_{cr}, Δv_{cr} and Δw_{cr} are too large, the stiffness of the beam reduces by some amount because the deformation of the interface element becomes relatively large compared to that of the beam. The area of process zone depends on the critical relative displacement. In the propagation stage, the load-displacement curves agree well with theoretical ones based on a beam theory (broken lines). The influence of mesh refinement is shown in Figure 5. When the mesh is not fine enough, the propagation process oscillates relating to element size. The distributions of the normal stress σ_z are shown in Figure 6. The high stress area appears at the crack tip area and moves rightward with the load increase. Stress in the newly delaminated surface is nearly zero. The process zone, that is, high stress area, shows an edge effect, crack front advances a little faster at the center portion.

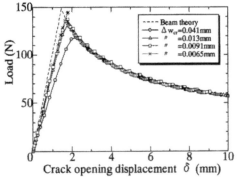

Figure4. Relationship between load and crack opening displacement for DCB specimen

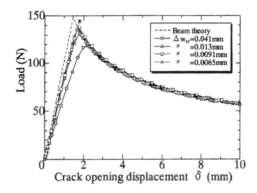

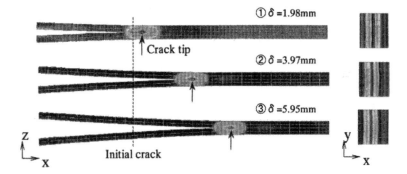

Figure5. Influence of mesh refinement

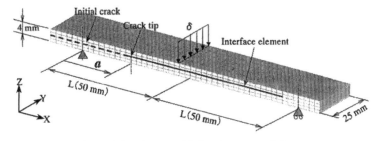

Figure 6. Distribution of the normal stress σ_z

Numerical Simulation of End Notched Flexure Problem

Let us consider an end notched flexure problem as a pure mode II situation (Figure 7). Material properties of beam are same as DCB specimens.

Figure 7. Specimen and finite element modeling

Figure 8 shows load-deflection relationship for ENF specimen. A broken line is the closed form solutions based on beam theory. The first stage OA is the load-deflection relationship of the beam with a crack of initial length a_0. Along the curve AB, the crack propagates from initial length to the center of the beam (a < L), where the energy release rate satisfies the condition $G_{II} = G_c$. Along the curve BC, crack advances from center point to the other support point (a > L), where the energy release rate also satisfies the condition $G_{II} = G_c$. The line OD is the load displacement curve of the completely split beam. The present numerical results agree well with the theoretical curves. When the load reaches a certain value, the load decreases abruptly accompanying unstable crack propagation.

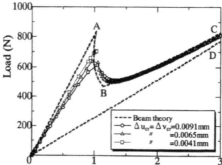

Figure 8. Relationships between center deflection and applied load for ENF specimen

Damage Accumulation in CFRP Laminates

The delamination propagation behaviors in composite laminates subjected to concentrated load at its center are analyzed by using the interface element. Dimensions of laminates is 100mm × 100mm, thickness is 4 mm. Two types of laminates, an isotropic plate and a cross-ply laminate $(0/90)_2$, are analyzed. These laminates have circular initial delaminations, whose radius is 2.5 mm, at its center. Because of the symmetry of the structure and loading conditions, only a quarter of the laminate is analyzed. Figure 9 shows a finite element mesh used in the analysis.

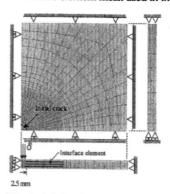

Figure 9 Typical finite element mesh for laminate

Figure 10 shows the relationship between the applied load and center deflection. The propagation of delamination occurs at the center of interlaminar plane with accompanying some reduction of the stiffness. With the load increase, delamination near the loading surface propagates next and finally delamination near bottom surface propagates. Delamination located at center of the plate is large. All the delaminations elongate in the fiber direction of layer below the interface. The processes of delamination propagations are shown in Figure 11.

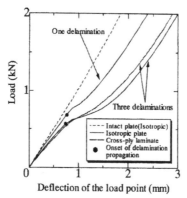

Figure 10. Relationships between center deflection and applied load

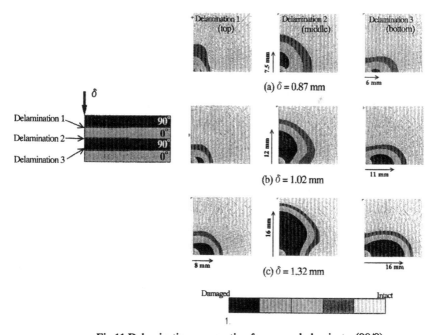

Fig.11 Delamination propagation for cross-ply laminate: $(90/0)_2$

CONCLUSION

We could propose an efficient interface element for crack-like damage propagation in three-dimensional problem. The convergence of the solution is smooth when the mesh size is small compared to the process zone and good agreement with the theoretical results is obtained for crack propagation problem for both DCB and ENF specimens. The present method was successfully applied to the multiple-delamination propagation problem in CFRP laminates. In order to define the various damage accumulation problem, the unloading process must be incorporated.

References
1. S. Liu and F. K Chang, Matrix Cracking Effect on Delamination Growth in Composite Laminates Induced by a Spherical Indenter, *J. Compos. Mater.* 28, 10, 940-977 (1994).
2. Shang-Lin Gao, Jang-Kyo Kim and Xing-Juan Xian, Characterization of Impact Damage in CFRPS Using a Scanning Acoustic Microscope, *Proc.ICCM-11*, Gold Coast, Australia, 14th-18th July (1997).
3. T. Ishikawa, S. Sugimoto, M. Matsushima and Y. Hayashi, Some Experimental Findings in CAI Tests of CF/PEEK and Conventional CF/EPOXY Flat Plates, *Compos. Sci. Technol.* 55, 349-362 (1995).
4. O. Allix and P. Ladevèze, Interlaminar modeling for the prediction of delamination, Composite Structures, 22, 235-242 (1992).
5. H. Suemasu and O. Majima, Multiple Delaminations and Their Severity in Circular Axisymmetric Plates Subjected to Transverse Loadings, *J. Compos. Mater.* 30, 4, 441-453 (1996).
6. H. Suemasu and O. Majima, Multiple Delaminations and Their Sevirity in Nonlinear Circular Plates Subjected to Transverse Loadings, *J. Compos. Mater.* 32, 2, 123-140 (1998).
7. Y.Aoki, H.Suemasu and O.Majima, Damage Accumulation in Composite Laminates during Quasi-Static Transverse Loading, *Adv. Compos. Mater.* 10, 2, 3, 219-228 (2001)
8. Y. Aoki and H.Suemasu, Fracture Mechanical Study on Mechanism of Damage Accumulation in Composite Laminates, Tran. JSME (A) 67, 667, 1563-1568 (2001).

Environmental effects on E-glass and C-glass fiber reinforced vinylester composites

Dr. T. Morii
Department of Materials Science and Engineering
Shonan Institute of Technology
1-1-25 Tsujido-Nishikaigan, Fujisawa 251-8511, Japan

Prof. H. Hamada and Ms. M. Mizoguchi
Advanced Fibro Science
Kyoto Institute of Technology
Matsugasaki, Sakyo-ku, Kyoto 606-8585, Japan

Dr. Y. Fujii
Seikow Chemical Engineering and Machinery, Ltd.
3-1-16 Shioe, Amagasaki 661-0976, Japan

ABSTRACT

This study dealt with the fracture behaviors of E-glass fiber and C-glass fiber plain woven fabric reinforced vinylester composites under static tensile stress after the exposure to hydrothermal and acid environments. The laminates were immersed in water at 95°C or acid solution at room temperature. After the immersion, the step-stress was applied to the specimen to failure by loading-unloading process and the fracture behavior was monitored simultaneously through the AE analyses. The effect of acid solution at room temperature on the laminate strength was poor in this paper (in immersion time less than 100 hours), however it induces the micro-damage during immersion and such damage derived the higher accumulation of AE event count. On the contrary, the immersion in hot water induced remarkable strength reduction in both laminates. However, AE results suggested that the degradation mechanism might be different depending on the fiber type.

INTRODUCTION

Glass fiber reinforced plastics (GFRP) have been widely used as the structural materials (chemical tanks/vessels, pipes, etc.) under severe environments such as hydrothermal, acid and alkaline environments. In general, E-glass fiber has been used as a reinforcement of GFRP for these applications. However, E-glass fiber is not stable under acid environment, and corrosion of the fiber occurs significantly. Fiber surfaces are attacked by acid and cracks appear on the surfaces (1). Such degradation induces catastrophic failure of GFRP structures under acid environment, and it is unfavorable phenomenon for actual applications. From this reason, application of acid resistant glass fiber (e.g. C-glass and ECR-glass fiber) is strongly demanded as the reinforcement of FRP structure under acid environment (2). In these fibers the content of Al_2O_3 is reduced, while the content of SiO_2 is increased (3). SiO_2 has strong structure against the attack by acid. Therefore it is considered that the GFRP reinforced by such fibers has higher acid corrosion resistance than the GFRP reinforced by E-glass fiber. For the present, however, the data for C-glass fiber reinforced plastics are very poor in order to apply this material as reliable structural material. In addition, it is important to know the strength degradation mechanism of such composites under hydrothermal and acid environments in order to improve the corrosion resistance of GFRP composites.

In our previous work, the creep behavior and material stability of E-glass and C-glass fiber woven fabric reinforced plastics under acid environment were discussed by using acoustic emission analyses (4). In this paper we have reported that the C-glass fiber had excellent creep resistance even under acid environment, while the E-glass fiber reinforced plastics showed undesirable residual strength reduction after creep stressing under acid environment. Such strength reduction may be related to the damage accumulation during creep stressing. Therefore it is important to evaluate damage accumulation behavior of these composites under stressing. Acoustic emission (AE) is one of the powerful tools in order to clarify the fracture behavior during stressing and real-time monitoring is possible by this tool.

From this background, this study discusses the fracture behaviors of E-glass fiber and C-glass fiber plain woven fabric reinforced vinylester composites under static tensile stress after the exposure to hydrothermal and acid environments. The step-stress was applied to the specimen to failure and the fracture behavior was monitored simultaneously through the AE analyses.

EXPERMENTAL PROCEDURE

Materials used were plain woven glass fabric as reinforcement and vinylester resin (R-806, Showa High Polymer Co., Ltd., Japan) as matrix. In order to discuss the influence of fiber type on fracture behavior after environmental exposure, E-glass fiber and C-glass fiber fabrics (YEM 2103-N7, Mie Textile Co., Ltd., Japan and WF230C100BS6, Nitto Boseki Co., Ltd., Japan, respectively) were adopted as the reinforcement. By stacking 12 plies of the same fiber fabrics, two kinds of laminates were fabricated by hand lay-up method.

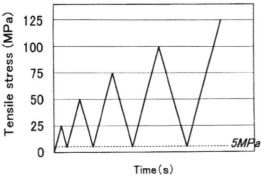

Figure 1 Static tensile test with step stress.

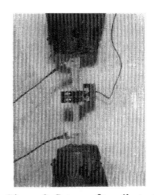

Figure 2 Set-up of tensile test.

In order to discuss the effects of environmental exposure on fracture behavior of both laminates, specimens were immersed in distilled water at 95°C or in acid solution (5% HNO_3) under room temperature. Immersion times were 0, 5, 10, 50 and 100 hours for both environmental immersions. After the fixed periods of immersion the surface of the specimen was wiped out by paper towel to remove drops of water before the mechanical testing.

After environmental immersion, static tensile test was performed. In order to discuss the reliable stress level for actual application, the maximum applied stress was increased step by step to final failure as schematically shown in Figure 1. The applied maximum stress was increased 25MPa step and minimum stress level was fixed at 5MPa. Static tensile test was performed by Instron type universal testing machine (AGS-1000B, Shimadzu Corporation, Japan) at a constant crosshead speed of 1mm/min in air at room temperature. During tensile test, acoustic emission (AE) from the specimen was monitored simultaneously in order to discuss the effects of environmental immersion and fiber type on fracture behavior. AE signal was monitored by 7600 series AE instrument (NF Corporation, Japan). Two AE transducers with the resonant frequency of 140 kHz were attached onto the specimen as shown in Figure 2, and AE was monitored with the total gain of 50dB and 70dB (different in each transducer) and the threshold of 100mV.

EXPERMENTAL RESULTS AND DISCUSSION

Table I summarized the static tensile strengths of E-glass and C-glass fiber

Table I Comparison of static tensile strengths for virgin specimens.

Laminate type	Tensile strength
E-glass fiber laminate	270 MPa
C-glass fiber laminate	208 MPa

of C-glass fiber laminate. In general, fiber strength of C-glass is lower than that of E-glass. Therefore the results of static strength of laminates reflected the fiber strengths.

Figure 3 shows the changes of the tensile strength after environmental immersion. The tensile strengths after the immersion in hot water showed significant reduction for both E-glass fiber and C-glass fiber laminates. The tendency of strength reduction was almost the same independent of the fiber type. The strength of E-glass fiber laminate showed about 25% reduction to the virgin specimen, while the strength of C-glass fiber laminate showed about 30% reduction. On the other hand, significant strength reductions did not appear after the immersion in acid solution under room temperature. The strength of C-glass fiber laminate kept almost constant even after immersion in acid solution. In E-glass fiber laminate the strength after immersion in acid solution decreased a few percent to the strength of virgin specimen. In our previous work the creep strength was reduced significantly in E-glass fiber laminate as shown in Figure 4 and the

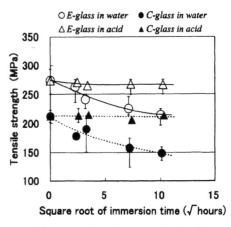

Figure 3   Changes of tensile strength after immersion in hot water or in acid solution.

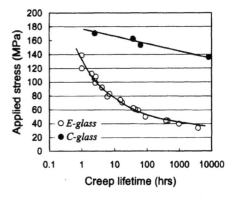

Figure 4   Creep lifetime in acid solution.

residual strength after creep stressing was also reduced even at low creep stress level (4). From these results, it is clear that the strength reduction under creep testing in acid environment is caused not only by acid but also by stress. The effect of only acid is considered not to be serious for the strength of glass fiber laminates.

Figures 5 and 6 show the stress history and change of AE event count of the virgin E-glass fiber and C-glass fiber laminates. In these figures the change of AE event count indicated the accumulation of AE event count in each stress steps. In E-glass fiber laminate AE count increased even at low stress level at early stage. In addition, the maximum cumulative count of each step was almost the same to maximum applied stress of 200MPa. In the following steps the maximum count increased with increasing the applied stress level. Generally, the Kaiser effect is often used in AE analysis in order to judge the healthiness of the material. In Kaiser effect AE count never increases if

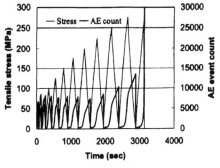

Figure 5 Stress history and change of AE event count of virgin E-glass fiber laminate.

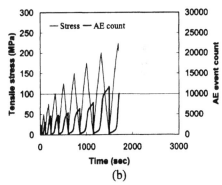

Figure 6 Stress history and change of AE event count of virgin C-glass fiber laminate.

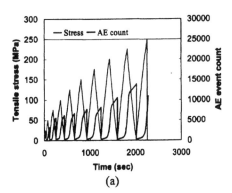

(a)

(b)

Figure 7 Stress histories and change of AE event counts of water immersed E-glass fiber laminates for (a) 10 hours and (b) 100 hours.

applied stress is lower than the previous stress history when material still has the healthiness of the material. The Kaiser effect still existed just before the final step in virgin E-glass fiber laminate. In virgin C-glass fiber laminate the accumulation of AE count was lower than the E-glass fiber laminate at low stress level. The maximum AE count in each step increased with increasing the maximum applied stress. Also in virgin C-glass fiber laminate the Kaiser effect still existed just before the final step.

Figure 7 shows the stress history and change of AE event count of the E-glass fiber laminates immersed in hot water for 10 and 100 hours. In these laminates the tensile strength gradually decreased with increasing the immersion time. In these laminates the maximum AE counts in one stress step showed higher values than the virgin laminate when the maximum applied stress reached 200MPa. In these laminates the Kaiser effect did not exist in this stress step. Therefore the serious damage already occurred in the previous stress step, namely in the step in the maximum applied stress of 150 MPa. Therefore it is considered that the reliable stress level after water immersion is

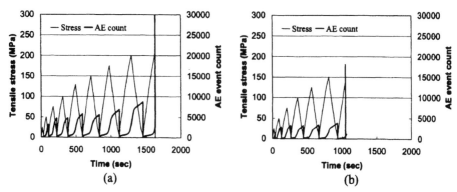

Figure 8 Stress histories and change of AE event counts of water immersed C-glass fiber laminates for (a) 10 hours and (b) 100 hours.

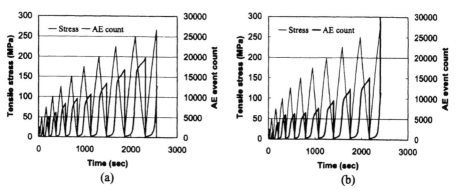

Figure 9 Stress histories and change of AE event counts of acid immersed E-glass fiber laminates for (a) 10 hours and (b) 100 hours.

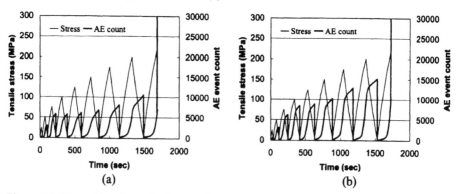

Figure 10 Stress histories and change of AE event counts of acid immersed C-glass fiber laminates for (a) 10 hours and (b) 100 hours.

about 150 MPa in E-glass fiber laminate. Figure 8 shows the stress history and change of AE event count of the C-glass fiber laminates immersed in hot water for 10 and 100 hours. In C-glass fiber laminates the tendency of change of AE count was different from E-glass fiber laminates. In C-glass fiber laminates the maximum cumulative AE counts in each stress steps gradually decreased with increasing the immersion time. This phenomenon might suggest that the damage already occurred due to water immersion (without stress) and such damage did not induce the accumulation of AE event count. Therefore it is considered that the degradation mechanism due to water immersion is different depending on fiber type.

Figure 9 and 10 show the stress history and change of AE event count of the E-glass fiber and C-glass fiber laminates immersed in acid solution for 10 and 100 hours. In E-glass fiber laminates the maximum cumulative AE counts in each stress step were higher than the virgin specimen. However, the tensile strength of these laminate did not seriously decreased after the immersion in acid solution. Therefore it is considered that the most of AE events were induced by the micro-damage derived from the acid attack to the laminate, and such micro-damage is not influenced to the laminate strength. The same tendency could be observed in C-glass fiber laminates. In the laminates immersed in water the maximum cumulative AE counts decreased with increasing the immersion time, while they increased with increasing the immersion time in the laminates immersed in acid solution. This means the different type of degradation inducing the micro-damage occurs in these laminates.

CONCLUSION

This study dealt with the fracture behaviors of E-glass fiber and C-glass fiber plain woven fabric reinforced vinylester composites under static tensile stress after the exposure to hydrothermal and acid environments. The step-stress was applied to the specimen to failure and the fracture behavior was monitored simultaneously through the AE analyses. The effect of acid solution at room temperature on the laminate strength was poor in this paper (in immersion time less than 100 hours), however it induces the micro-damage during immersion and such damage derived the higher accumulation of AE event count. On the contrary, the immersion in hot water induced remarkable strength reduction in both laminates. However, AE results suggested that the degradation mechanism might be different depending on the fiber type.

REFERENCES

1. Q. Qiu and M. Kumosa, "Corrosion of e-glass fibers in acidic environments", Composites Science & Technology, Volume 57, 1997, 497-507.
2. D. Santrach and R. Matzeg, "FRP Corrosion Resistance: The Role of the Glass Fibre Type", Polymers & Polymer Composites, Volume 1, 1993, 451A-465A.
3. D. Hull, An Introduction to Composite Materials, Cambridge University Press, Cambridge, 1981.

4. M. Mizoguchi, H. Hamada, T. Morii and Y. Fujii, "Acoustic Emission Analysis of Corrosion Resistance of Glass Fiber/Vinylester Composites under Acid Environment", Proceedings of the Seventh Japan International SAMPE Symposium and Exhibition, T. Ishikawa and S. Sugimoto, eds., Tokyo, Japan, 2001, 895-898.

Damage identification of composite laminates using anti-resonance frequency

Y. Shimamura,
Department of Mechanical Sciences and Engineering
Tokyo Institute of Technology
2-12-1 O-okayama, Meguro, Tokyo, 152-8552, Japan
yshimamu@ginza.mes.titech.ac.jp

T. Inada
Graduate student of Tokyo Institute of Technology

A. Todoroki and H. Kobayashi
Department of Mechanical Sciences and Engineering
Tokyo Institute of Technology

ABSTRACT

Detection and identification of damage in composite structures are very important because damage causes strength reduction of the structures. Vibration characteristics are often used to detect and identify damage because it is easy to measure. Many researchers use natural frequency to detect and identify damage. In their studies, it is shown that it is easy to detect damage using natural frequency, but it is difficult to identify damage location and size. In this study, a new method for damage identification using anti-resonance frequency is proposed. Though natural frequency is global parameter, anti-resonance frequency is local parameter. In the other words, anti-resonance frequency includes local information. Basic investigations of damage identification in composite beams are conducted. As a result, it is shown that anti-resonance frequency is useful to identify damage location and size.

INTRODUCTION

Recently, composite laminated structures have been applied to many structures of vehicles. Since interlaminar strength of composite laminated structure is relatively low, internal damage, such as delamination, can be easily induced in service. In order to assess integrity of the damaged structures, it is necessary to identify the size and location of the damage nondestructively.

The present study proposes a delamination identification method using resonance and anti-resonance frequency changes. For simplicity, delamination is modeled as stiffness degradation of the damaged part. In order to examine effectiveness of the present method, damage identification of a graphite/epoxy clamped-clamped beam was conducted in analysis. As a result, it is shown that there are good agreements between actual and predicted damage locations and sizes.

DAMAGE IDENTIFICATION METHOD

Many researchers have investigated damage identification methods using the resonance frequency changes (1-4). However, the resonance frequency change method alone may not be sufficient for unique identification of damage location because of its symmetric properties (4). For example, in symmetric structures like a clamped-clamped beam, the appearance of the resonance frequency change as a function of damage location is symmetric. In this case, two candidates for damage location are obtained from the resonance frequency changes, but it is impossible to select the correct one.

In the present study, a two-stage damage identification method using resonance and anti-resonance frequency changes is proposed. The damaged domain is first identified from the anti-resonance frequency changes, and damage location and size are identified from the resonance frequency changes to the next. In order to examine effectiveness of the present method, damage identification of a clamped-clamped beam is conducted in analysis.

Damaged Domain Identification Using Anti-Resonance Frequency Change

Damaged domain identification means identifying the domain in which damage is supposed to exist. In order to identify the damaged domain in a structure, anti-resonance frequency changes from an intact state is used.

Figure 1 shows a clamped-clamped composite beam with damage. The beam is 1.45mm thick, 19mm wide and 300mm long. ANSYS, a general-purpose finite element code, was used to analyze frequency response changes of the beam with damage. The stacking sequence of the beam is $[0_2/90_2]_S$, and damage is modeled as delamination at a $0°/90°$ interlayer. Table I shows material properties of uni-directional graphite/epoxy ply used in the analyses, where subscript 1 means the direction parallel to the fiber and subscript 2 and 3 means the direction perpendicular to the fiber.

The beam is divided into two spanwise domains as shown in Figure 1. Domain A is defined by $0<x/L<0.5$, and Domain B is defined by $0.5\leq x/L<1$. Existence of damage between actuating point and measuring point is detectable by anti-resonance frequency changes from an intact state. In order to use this characteristics of anti-resonance frequency changes for damaged domain identification, we set the actuating point as $x/L=0.05$ and the measuring point as $x/L=0.45$. Since the boundary of the domains corresponds to nodal point of the second flexural mode, its peak and the corresponding dip are undetectable on the frequency response measured at $x/L=0.5$. For this reason the measuring point is displaced from $x/L=0.5$ to $x/L=0.45$.

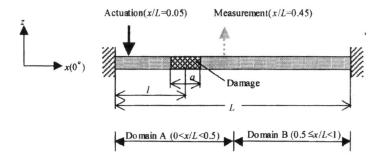

Figure 1 - Clamped-clamped composite beam with damage

Table I - Material properties of unidirectional graphite/epoxy ply

Longitudinal modulus E_1	91.5GPa
Transverse modulus $E_2(=E_3)$	8.5GPa
Longitudinal shear modulus $G_{12}(=G_{13})$	4.3GPa
Transverse shear modulus G_{23}	3.15GPa
Major Poisson's ratio $\nu_{12}(=\nu_{13})$	0.31
Transverse Poisson's ratio ν_{23}	0.35
Density ρ	1466 kg/m^3

In order to use the anti-resonance frequency changes for damaged domain identification, we use a non-dimensional parameter D as follows.

$$D_{ij} = \frac{\Delta f_{A,ij}}{\left|\Delta f_{R,i}\right| + \left|\Delta f_{R,j}\right|} \times 100 \quad (i \geq 1, \ j = i+1) \qquad (1)$$

where $\Delta f_{R,i}$ and $\Delta f_{R,j}$ mean the resonance freqnency changes of the ith and the jth mode respectively. $\Delta f_{A,ij}$ means the anti-resonance frequnecy change observed between the ith and the jth resonance peaks on the frequency response diagram. Figure 2 shows damaged domain identification example based on parameter D. In Figure 2, case I and case II show the results when damage is located in domain A, and case III shows the result when damage is located in domain B. Since the anti-resonance frequencies may decrease on case I, we set a threshold line of decrement as shown in Figure 2 with a dashed line. When at least one of the parameter D_{ij} is larger than the threshold , the damaged domain is identified as Domain A. When at least one of them is smaller than the line, the damaged domain is identified as Domain B.

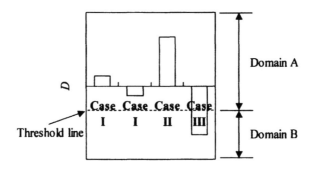

Figure 2 - Damaged domain identification example based on parameter D

Damaged Identification Using Resonance Frequency Changes

Figure 3 shows relations between damage location and the first three resonance frequencies of the flexural modes at a constant damage size (a/L=0.1) of the clamped-clamped beam (shown in Figure 1). We can see from Figure 3 that the resonance frequency changes as a function of damage location are symmetric to the center of the beam. For this reason, two candidates for damage location can be obtained from the resonance frequency changes, but it is impossible to select the correct one. In this study, we can obtain a unique damage location, since damage location and size are identified after the damaged domain identification.

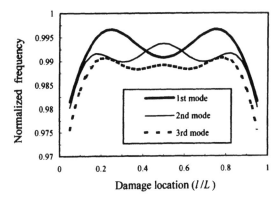

Figure 3 - Resonance frequency changes (a/L=0.1)

We use response surfaces for damage identification. Namely, response surfaces are used to obtain the approximate expressions of relation between damage parameters and the first three resonance frequencies of the flexural modes. Since the resonance frequency changes are symmetric to the domain boundary, we make only the response surfaces for damage identification for Domain A. For simplicity, we use polynomial response surfaces. For example, the quadratic polynomial response surface of three variables is expressed as follows.

$$y = \beta_0 + \beta_1 x_1 + \beta_2 x_2 + \beta_3 x_3 + \beta_4 x_1^2 + \beta_5 x_2^2 \\ + \beta_6 x_3^2 + \beta_7 x_1 x_2 + \beta_8 x_2 x_3 + \beta_9 x_3 x_1 \tag{2}$$

where y is damage location or size, and x_i (i=1,2,3) is the resonance frequency of each flexural mode. In order to obtain good approximate results, cubic polynomial expressions are adopted in practical. We make response surfaces of the damage location and size respectively from 35 data sets. Adjusted coefficients of decision of the response surfaces are about 0.98.

RESULTS AND DISCUSSION

In order to examine effectiveness of the present method, damage identification of a clamped-clamped beam is conducted based on analytical data. The damaged domain is identified first from the anti-resonance frequency changes, and the damage location and size are identified from the resonance frequency changes to the next.

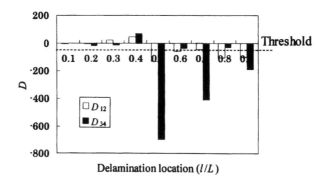

Figure 4 - Damaged domain identification results (a/L=0.05)

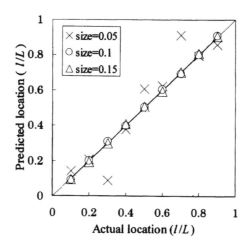

Figure 5 - Comparison between actual and predicted damaged locations (a/L=0.05)

Figure 4 shows damaged domain identification results of the clamped-clamped beam based on parameter D at normalized damage size is a/L=0.1. In the figure, the vertical axis is the value of parameter D, and the horizontal axis is damage location normalized by the beam length. We used two anti-resonance frequency changes for damaged domain identification. One is D_{12} and the other is D_{34}. By setting the appropriate threshold as shown in Figure 4, we can identify the damaged domain correctly. Based on the results of FEM analyses, we set the threshold as -50.

Figure 5 shows comparison between actual and predicted damage locations. For simplicity, we show only the case that the normalized damage size is a/L=0.1. Though a few data sets were not considered in the response surface of damage location, the identified values are well agreed to the actual ones. Though it is not shown in the paper, we can also identify the damage sizes precisely. The mean identification error of the damage sizes is 0.6%. The damage identification results based on analytical data clearly show effectiveness of the present damage identification method using resonance and anti-resonance frequency changes.

CONCLUSIONS

This study describes the damages identification method using resonance and anti-resonance frequency changes. In order to examine effectiveness of the present method, damage identification of a clamped-clamped beam was conducted. As a result, it is shown that the damage identification method using resonance and anti-resonance frequency changes is effective in identifying damage locations and sizes.

REFERENCES

1.   A. S. Islam and K. C. Crag, "Damage Detection in Composite Structures using Piezoelectric Materials", Smart Materials and Structures, 3, (1994), 318-328.

2.   A. C. Okafor, K. Chandrashekhara and Y. P. Jiang, "Delamination Prediction in Composite Beams with Built-In Piezoelectric Devices using Modal Analysis and Neural Network", Smart Materials & Structures, 5, 3, (1996), 338-347.

3.   X. H. Jian, H. S. Tzou, C. J. Lissenden and L. S. Penn, "Damage Detection by Piezoelectric Patches in a Free Vibration Method", Journal of Composite Materials, 31, 4, (1997), 345-359.

4.   Y. Zou, L. Tong and G.P. Steven, "Vibration-Based Model-Dependent Damage (Delamination) Identification and Health Monitoring for Composite Structures - A Review", Journal of Sound & Vibration, 230, 2, (2000), 357-378.

Damage and toughening mechanisms of carbon fibers reinforced carbon matrix composites

Mohamed S. Aly-Hassan
Hatta Lab., Division of Space Propulsion
The Institute of Space and Astronautical Science(ISAS)
3-1-1 Yoshinodai, Sagamihara-shi,
Kanagawa-ken 229-8510, Japan
aly-has@pub.isas.ac.jp

Hiroshi Hatta
The Institute of Space and Astronautical Science (ISAS)
3-1-1 Yoshinodai, Sagamihara-shi,
Kanagawa-ken 229-8510, Japan
hatta@pub.isas.ac.jp

Shuichi Wakayama
Department of Mechanical Engineering ,
Tokyo Metropolitan University,
1-1 Minami-Ohsawa, Hachioji, Tokyo 192-0397, Japan
wakayama@ecomp.metro-u.ac.jp

Kiyoshi Miyagawa
IHI Aerospace Co. Ltd.,
900 Fujiki, Tomioka, Gunma 370-2398, Japan
ki-miyagawa@iac.ihi.co.jp

ABSTRACT

Static mechanical responses of two-dimensionally and three-dimensionally reinforced Carbon/Carbon composites, 2D- and 3D-C/Cs, were compared. Mechanical properties were examined in the present study include tensile, compressive, shear stress-strain, S-S, relations and fracture behavior on the basis of compact tension tests. Compared with 2D-C/Cs, the 3D-C/Cs were shown to possess similar tensile S-S relation, intensively nonlinear shear S-S curves and far notch insensitivity and higher fracture resistance. The shear and fracture resistance differences suggested to be derived from weak fiber/matrix interface, intra-bundle damage through the rich matrix regions and weak interfacial bonding between fiber bundles in 3D-C/Cs. These weak interface characteristics are brought by high value of plane residual stresses caused by three-dimensional fiber constraint of 3D-C/Cs due the mismatch of thermal expansion during the processing.

INTRODUCTION

Carbon/carbon composites, C/Cs, have received an increasing attention in recent years as potential high-temperature materials for practically advanced applications. This enlarged attention arises mainly from the unique and combined materials properties of C/Cs [1-3]. In addition, some of these superior properties not only can be maintained at very high temperature about 3000 K but also can be enhanced with increasing the temperature. High fracture toughness is one of important advantages of C/Cs [1]. Compared with the monolithic ceramics, C/Cs possesses nearly one order higher fracture toughness [4]. However, when C/Cs are actually used, there are various shortcomings need to overcome e.g., weakness against oxidation, poor matrix mechanical properties and low interlaminar shear strength (5 to 20 MPa) [1-3]. To overcome the latter deficit, three-dimensionally reinforced C/C composites, 3D-C/Cs, have often been introduced. However, mechanical properties of the 3D-C/Cs have not been sufficiently examined, probably due to extremely high material cost.

Several research works [5-19] have been carried out in last decade to clarify the mechanical behavior of C/Cs but most of these studies [5-12] were directed toward two-dimensionally reinforced C/Cs. According to recent literatures [5-7], it has been widely believed that the ability of C/Cs to redistribute the stress near notch tips is due to the shear band formation mechanism. This damage mechanism enhances crack extension resistance by partially shielding the crack tip from the applied load. The origin of this belief stems from the low shear/tensile strength ratio and weak fiber/matrix interface [5]. Although that, the shear bands have never been observed clearly in C/C composites neither in microscopic nor in macroscopic scales [8-11]. Only in the case of two-dimensionally reinforced (2D-) C/Cs with a filler addition (carbon blacks and colloidal graphite), that were introduced between the UD layers before the first infiltration of pitch, the shear bands were observed microscopically [12]. Whereas in the same type of C/Cs but without filler addition (conventional 2D-C/Cs), the shear bands were not detected [12]. The few studies of 3D-C/Cs have been centered on the kinking morphology [13], strength criterion under biaxial loading [14], damage and failure mechanisms under uniaxial tensile and shear loads [15], prediction of notched strength [16], modeling for the mechanical behavior and damage [17-18] and response under shock loading [19].

Hence, from the pervious review there is neither a comprehensive study for the mechanical properties of 3D-C/Cs nor a clear image about the effect of 3D fiber reinforcements on various C/C mechanical properties. In addition, the fracture toughness criterion has not been thoroughly investigated yet in 3D-C/Cs. Thus, 3D C/Cs at the present time are mostly used as low-stressed components not as load-bearing components in various advanced applications.

In the present work, a comparison between 2D and 3D-C/Cs for the tensile, compressive and shear stress-strain curves as well as fracture behavior was carried out. This attempt has been made not only to identify the damage growth mechanisms but also to rationalize the high fracture toughness behavior and notch insensitivity in terms of the static mechanical responses and damage observations.

EXPERIMENTAL PROCEDURES

Materials

The examined 2D-C/Cs, produced via preform yarn method by Across Co., had 0/90 lamination and fiber volume fraction of 50%. The reinforcing fiber of this C/C was Torayca M40 supplied by Toray Co. Two kinds of 3D-C/Cs with different void contents were used, produced by IHI Aerospace Co. These C/Cs were reinforced with pitch-based fibers and had a fiber volume fraction of 36 % (12% each direction). The source materials of these were unidirectionally reinforced CFRP rods, which were arranged into a preform reinforced in orthogonal directions. Then the preform was HIP-treated over 500 atm., where impregnation material was coal tar pitch, followed by heat treatment over 2000 °C. These HIP cycles were repeated several times to yield C/C with 1.9 and 2.0 gm/cm³. In this paper, the abbreviation LD and HD are used to indicate low density and high density of 3D-C/C respectively.

Characterization of Mechanical Properties

To characterize the materials, stress-strain relations under tensile, compressive and shear loading were obtained. The Iosipescu method was adopted to determine the shear responses. At least one test for each material was repeated loading and unloading to measure residual shear strain as a function of applied load. The compressive tests were performed according to JIS for carbon materials. Screw-driven mechanical testing machine, Autograph AG-5000A, was used for all the mechanical tests under cross head speed 0.05 mm/min.

Fracture Toughness Measurement

The compact tension (CT) tests were performed using the same above mechanical testing machine under a cross-head speed of 0.1mm/min to obtain the R-curve and to observe the damage growth of C/Cs. The configuration of the CT specimens used in the present study is given in Fig. 1, which is based on ASTM Standard E-399-72. In the CT specimens, crack length-to-size ratio, a/W, was set to 0.5. This pre-crack was introduced at first by diamond wheel cutter with a thickness of 0.4 mm, then the notch sharpness was increased by razor blade until 0.1 mm.

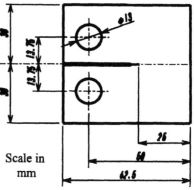

Scale in mm

Figure 1. CT specimen geometry

However C/C composite is not sensitive to the notch tip radius at least within that of 1-20 mm [9]. To observe damage extension process during CT test, the samples were polished using diamond-past of 15 µm diameter, degreased in ethanol, cleaned in an ultrasonic bath, dried at room temperature, and then thinly painted white. The damage process during CT tests was observed using a travelling optical microscope with a magnification of 25 times and CCD camera from both sides of the specimens. A clip gage

(UB-5, Tokyo-Sokki Crop, Japan) was used to measure the crack opening displacement (COD). Two acoustic emission sensors (AE-900M-WE, NF Electric Instruments Crop, Japan) were attached on the CT specimen to provide signals for the damage formation. Compact tension tests were conducted with repeating loading-loading/unloading cycles.The compliance method have been used to determine of the crack growth R-curve of CT test specimens by applying the following Steps:

a- Calculate edge/pin COD ratio by FEM to determine the experimental COD at the load axis of CT configuration.

b- Generation of the compliance curve (pin compliance, c, versus the measured crack length, a) from load-COD relation, where c was obtained during loading stages.

c- Applying the linear elastic fracture mechanics, LEFM, to estimate the energy release rate [8] by the following equation;

$$G_{IR} = \frac{P^2}{2B}\frac{dc}{da} \tag{1}$$

where P, B, c, and a are the load at each crack increment, CT specimen thickness, compliance measured at the loading point, and crack growth length, respectively.

Finite Element Modeling

The loading axis COD and compliance of the CT specimens during the tensile loading were determined using finite element method (FEM) under the assumption of the plane stress condition. This model simulates the half of CT configuration with a/W=0.5. The mesh of this model is composed of biquadratic elements with 8-nodes. The FEM calculations were carried out using a commercial code, ABAQUS version 5.8. For 3D-C/Cs nonlinear FEM using piecewise-linear constitutive approximation was adopted for the same above model.

RESULTS AND DISCUSSION

Micro-Structures

Although many transverse cracks and micro-voids can be seen in 2D-C/C, the interlaminar crack is hardly observed using an optical microscope as shown in Fig. 2. On the other hand in the 3D- LD, large scale debonding is observed along the interfaces between fiber bundles and pure matrices and large scale near-closed-voids are been in matrix pockets formed between fiber bundles running three dimensionally. In 3D-HD, the matrix pockets almost filled with matrix carbon. See Fig. 2.

Tensile and Compressive Properties

The comparison of tensile and compressive responses of 2D- and 3D-C/Cs are shown in Figs. 3 and 4 respectively. It can be seen in these figures that the tensile S-S curves of all the test materials exhibited linear relation. On the other hand compressive curves showed strong non-linearity. The tensile strength of 3D-LD was slightly higher than 3D-HD but the compressive strength showed opposite tendency. The tendency of the tensile strength is due to higher interfacial strength of 3D-LD than that of 3D-HD and compressive strength feature was probably derived from internal deficits.

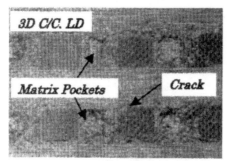

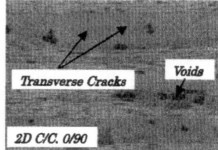

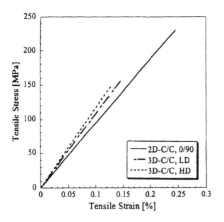

Figure 2. Microstructures of LD 3D-C/Cs, HD 3D-C/Cs and 0/90 2D-C/Cs respectively.

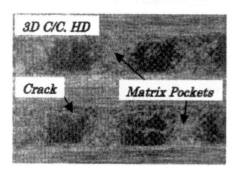

Figure 3. Tensile stress-strain curves.

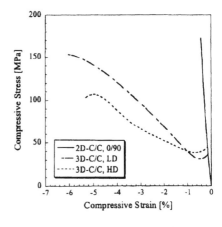

Figure 4. Compressive stress-strain curves.

Shear Properties

Shear S-S curves for all the test materials are compared in Fig.5. It follows from this figure that 2D-C/C has relatively high strength with low ultimate strain. On the other hand, 3D-C/Cs have low maximum stress and very large fracture strain about 7 %. It should be noted that the S-S curves for both 3D C/Cs almost coincided.

The repetition of shear loading and unloading yielded residual strain. The residual shear strains of 2D- and 3D-C/Cs are compared in Fig. 6 as a function of the maximum stress during each lording and unloading cycle. The residual strain was small in 2D-C/C,

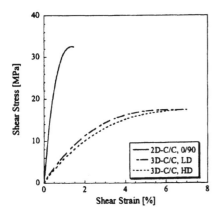

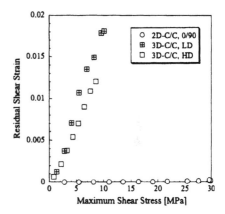

Figure 5. Shear stress-strain curves.

Figure 6. Residual shear strain vs. maximum shear stress.

but was large from the first cycle and increased rapidly in 3D-C/C. This result clearly shows that the shear deformation in 3D-C/C proceeded inducing defects and might due to the large-scale interfacial cracks induced in the processing stage.

Resistance Curve (R-curve)

Cycles of loading/unloading during the CT test were conducted to determine the energy release rate, G_R, as a function of crack growth length using a single CT test specimen. The load-COD curve of 2D is illustrated in Fig. 7. It is imperative to notice the 2D-C/Cs is exhibited slight non-linearity started before the maximum load, and load dropped abruptly at the maximum load of the each cycle. Figure 8 shows a typical fracture resistance versus the crack extension length for 2D-C/Cs. On the other hand, the load-COD curves of both types of 3D-C/Cs are wholly differed from that of 2D-C/Cs, i.e., 3D-C/Cs, has no abruptly load drop (no fiber failure). In other words, in 3D-C/Cs the R-curve approach based on the linear elastic fracture mechanics can not adopted because there are neither mode I fiber failure neither nor mode II fiber failure. This special damage mechanism, saturated load with increasing COD, absorbs more energy that provides those 3D-C/Cs with a significant improvement in toughness.

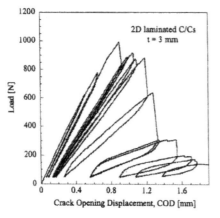

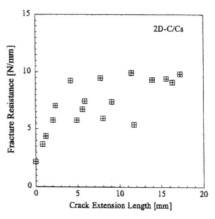

Figure 7. Typical Load-COD curve

Figure 8. Resistance curve (R-curve)

DISCUSSIONS

High Toughness Mechanisms of 3D-C/Cs

Main factor for high toughness behavior of 3D-C/Cs compared with 2D-C/C is easy deformation of 3D-C/Cs. This large deformation is induced mainly by shear deformation. Residual CODs during CT tests and residual shear strain for 2D-C/C and 3D-C/Cs were observed as a function of P. In these relations both residual deformations show similar trend and this tendency supports the statement that the large deformation during CT test was induced mainly by shear. As we discussed earlier, easy shear deformation of 3D-C/C is due to many large-scale interfacial cracks. On the other hand, it was already shown that thermal residual stresses in 3D composites are much higher than that in 2D composites [20-21]. The shear and fracture resistance characteristics suggested to be derived from weak fiber/matrix interface, intra-bundle damage through the rich matrix regions and weak interfacial bonding between fiber bundles in 3D-C/Cs. These weak interface characteristics are brought by high value of plane residual stresses caused by three-dimensional fiber constraint of 3D-C/Cs due the mismatch of thermal expansion during the processing. This high value of thermal residual stress induces the interfacial damage, which induce easy shear deformation. In addition three-dimensionally entangled reinforcement prevent early ultimate fracture to yield high deformation capability. Therefore we can conclude that 3D-C/Cs have easy shear deformation capability and exhibit high toughness behavior.

CONCLUSION

The following conclusions were obtained through comparison of fracture behavior of 2D and 3D-C/Cs based on the previous results.

1) Fracture resistance of 3D-C/Cs was higher than that of 2D-C/C.
2) 3D-C/Cs have extremely low shear resistance and high shear deformation capability.
3) High fracture resistance of 3D-C/Cs is attributed due to the shear deformation characteristic.

REFERENCES

1. G Savage, Carbon-Carbon Composites, Chapman & Hall, New York, NY.1993.
2. D.L. Schmidt, K.E. Davidson and S. Theibert "Unique Applications of Carbon-Carbon Composite Materials (Part One)", SAMPE J, 35(3), 1999, 27-39.
3. M. Lacoste, A. Lacombe, P. Joyez, R.A. Ellis, J.C.Lee and F.M. Payne "Carbon/Carbon Extendible Nozzles", Acta Astronautica, 50(6), 2000, 357-367.
4. A. Wanner, G Rizzo and K. Kromp, In Toughening Mechanisms in Quasi-Brittle Materials (Edited by S. P. Shah), Kluwer Academic Publishers, London. 1991, 405-423.
5. A.G Evans, F.W. Zok "Review: The Physics and Mechanics of Fibre-Reinforced Brittle Matrix Composites", J. Mater. Sci., Volume 29, 1994, 3857-3896.
6. T.J. Mackin, T.E. Purcell, M.Y. He and A.G Evans "Notch Sensitivity and Stress Redistribution in Three Ceramic-Matrix Composites", J. Am. Ceram. Soc., 78(7), 1995, 1719-1728.
7. M.Y. He, B. Wu, Z. Suo "Notch-Sensitivity and Shear Bands in Brittle Matrix Composites", Acta. Metall. Mater., 42(9), 1994, 3065-3070.
8. H. Hatta, Y. Kogo, H. Asano and H. Kawada "Applicability of Fracture Mechanics in Carbon-Carbon Composites", JSME Int. J., Series A, 42(2), 1999,265-271.
9. Y. Kogo, H. Hatta, H. Kawadaand T. Machida "Effect of Stress Concentration on Tensile Fracture Behavior of Carbon-Carbon Composites", J. Comp. Mater., 32(13), 1998, 273-1294.
10. K. Goto, H. Hatta, H. Takahashi and H. Kawada "Effect of Shear Damage on the Fracture Behavior of Carbon-Carbon Composites", J. Am. Ceram. Soc., 84(6), 2001, 1327-1333.
11. L. Denk, H. Hatta, A. Misawa and S. Somiya "Shear Fracture of C/C Composites with Variable Stacking Sequence2, Carbon, Volume 39, 2001,1505-1513.
12. G Chollon and et al. "Microstructure and Mechanical Properties of Coal Tar Pitch-Based 2D-C/C Composites with a Filler Addition", Carbon, Vol. 39, 2001, 2065-2075.
13. A.G Evans and W.F. Adler "Kinking as a Mode of Structural Degradation in Carbon Fiber Composites", Acta Metall. Mater., Volume 26, 1978, 725-738.
14. Z. Gu, O. Gao and W. Zhang "Nonlinear Bimodulus Model and Strength Criterion of 3D Carbon-Carbon Material", Journal of Composite Materials, 23(10), 1989,988-996.
15. O. Siron, and J. Lamon "Damage and Failure Mechanisms of a 3-Directional Carbon/Carbon Composite under Uniaxial Tensile and Shear Loads", Acta Materialia, 46(18), 1998, 6631-6643.
16. Z. Gu and L. Meng "Prediction of Notched Strength of 3D Carbon-Carbon Composites", Journal of Composite Materials, 23(10), 1989, 997-1008.
17. J. Pailhes, G Camus and J. Lamon "A Constitutive Model for the Mechanical Behavior of a 3D C/C Composite", Mechanics of Materials, 34(3), 2002,161-177.
18. P. Ladeveze, O. Allix and C. Cluzel "Damage Modling at the Macro- and Meso-Scales for 3D Composites", Damage in Composite Materials, Elsevier Science Publishers B.V. 1993
19. O. Allix, M. Dommanget, M. Gratton and P.L. Héreil "A Multi-Scale Approach for the Response of a 3D Carbon/Carbon Composite under Shock Loading", Composites Science and Technology, 61(3), 2001, 409-415.
20. H. Hatta, T. Takei and A. Morii "Thermo-Mechanical Properties of Resin Matrix 3D Fabric Composites", Materials System, Volume 10, 1991, 10:71-80.(In Japanese)
21. H. Hatta, T. Takei and M. Taya "Effect of Dispersed Microvoids on Thermal Expansion Behavior of Composite Materials", Mater. Sci. Eng., A285, 2000, 99-110.

Fragmentation behavior analysis in single carbon fiber composite

Shinji Ogihara and Akira Mochizuki
Department of Mechanical Engineering
Tokyo University of Science
2641 Yamasaki, Noda, Chiba 278-8510, Japan
ogihara@rs.noda.tus.ac.jp

Yasuo Kogo
Tokyo University of Science

ABSTRACT

Effects of fiber surface oxidization and sizing treatment on both fiber strength distribution and fragmentation behavior in single carbon fiber epoxy composites are investigated experimentally. Based on the above experimental data, the fiber/matrix interfacial properties are discussed.

INTRODUCTION

It is well known that effects of the properties of the fiber/matrix interface on the mechanical properties of fiber reinforced composites is very significant. In order to develop high performance composite materials, it is very important to understand the relation between the interface properties and composite macroscale properties. A fragmentation test is often used as a evaluation method of the fiber/matrix interface properties of fiber reinforced composite materials [1-3]. In the fragmentation test, single fiber composite materials are loaded in tension in fiber direction. During the fragmentation test, fiber multiple fractures occur and the fiber break saturates at high composite strain. The interfacial shear stress is evaluated from the average fiber fragment length when the fiber break saturates. However, the fragmentation test has two major problems. One of them is that the strain which the interfacial shear stress is evaluated is much higher than the fracture strain of the common unidirectional composite material in fiber direction. The other is that only the average fiber strength is used to evaluate the interfacial shear stress. The statistical fiber strength distribution should be taken into account.

Curtin presented an analysis which can predict the relation between the fiber break density and the composite strain during the fragmentation test considering the statistical fiber strength distribution [2]. The interfacial shear stress is assumed to be constant.

In the present study, both evaluations of statistical fiber strength distribution and fiber fragmentation behavior during the fragmentation test are conducted. By using Curtin's analysis, the interfacial shear stress is evaluated. The effect of fiber surface treatment on the fiber strength distribution and the interfacial shear stress is discussed.

EXPERIMENTAL PROCEDURE

Two kind of carbon fibers, UM46 and IM600, are used in the present study. Both fibers with and without surface oxidization and/or sizing   treatment are used for both. In the following, UM denotes UM46 carbon fiber without oxidization and sizing treatments, UMo UM46 with only oxidization treatment, UMs only sizing treatment and UMos with both oxidization and sizing treatment. The same is true for the IM, IMo, IMs, and IMos for the IM600 carbon fibers.

Tension tests are performed on the 8 kinds of carbon fibers to evaluate the fiber strength distribution. The gage length is 25mm.   An Instron type tensile testing machine is used.   The crosshead speed is 1mm/min.

For fragmentation tests, single carbon fiber composite specimens are fabricated. Bisphenol-A type epoxy resin (Epikote828; Japan Epoxy Resin Inc.) and triethylenetetramine (TETA) as hardener are used in 100:11 weight ratio as matrix resin. The resin is cured at 50°C for 60 min and post cured at 100°C for 80 min. Specimen size is 72 mm long, 16 mm wide and 1 mm thick.   Figure 1 shows the specimen configuration used in the fragmentation test. Fragmentation tests are performed using a loading device put on the stage of an optical microscope.   The number of fiber breaks is measured at every 0.1 % composite strain. The fiber break density is defined as the number of the fiber breaks per unit specimen length.

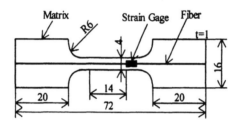

Figure 1 Specimen used in fragmentation test

RESULTS AND DISCUSSION

The fiber strength distribution obtained by the tensile test on the carbon fibers is assumed to be Weibull distribution.   Figures 2 and 3 show the Weibull plots obtained for UM46 and IM600 carbon fibers, respectively (Fig.2 (a) UM, (b) UMo, (c) UMs, (d) UMos, Fig.3. (a) IM, (b) IMo, (c) IMs, (d) IMos).   Weibull parameters (shape parameter ρ, scale parameter σ) obtained from the results are listed in Table 1.   It is seen that the effect of the fiber surface oxidization and the sizing treatments on the fiber strength distribution is small.

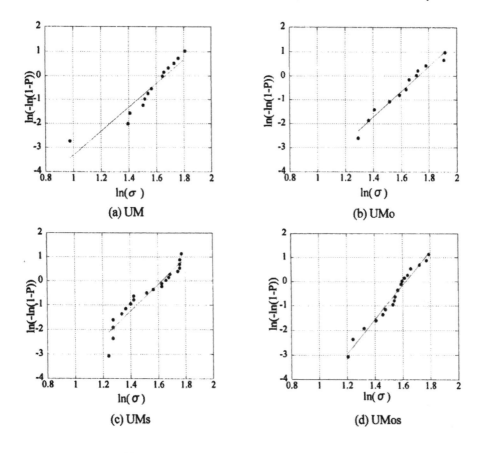

Figure 2   Weibull plot of fiber strength distribution (UM46)

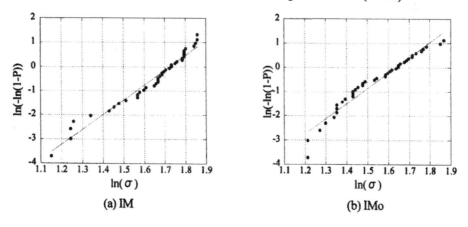

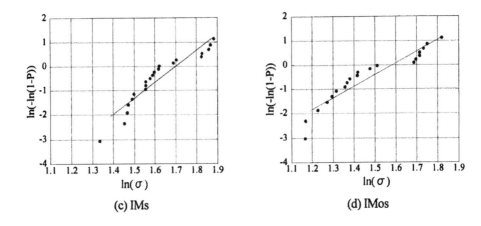

(c) IMs

(d) IMos

Figure 3   Weibull plot of fiber strength distribution (IM600)

Figures 4 and 5 show the fiber break density as a function of the applied composite strain obtained by the fragmentation test of UM46 and IM600, respectively (Fig.4 (a) UM, (b) UMo, (c) UMs, (d)UMos, Fig.5 (a) IM, (b) IMo, (c)IMs, (d) IMos).   It is seen that the fiber fragmentation behavior is different for the carbon fibers with different fiber surface treatments.

Predictions of the relation between the fiber break density and the composite strain using the Curtin's analysis are also shown in the figures. In the prediction, the experimentally-obtained fiber strength distribution is used. The interfacial shear stress is determined to fit the experimental results. The interfacial shear stresses obtained from the procedures are listed in Table 1. The interfacial shear stress is higher in the specimen with oxidation and sizing treatments.

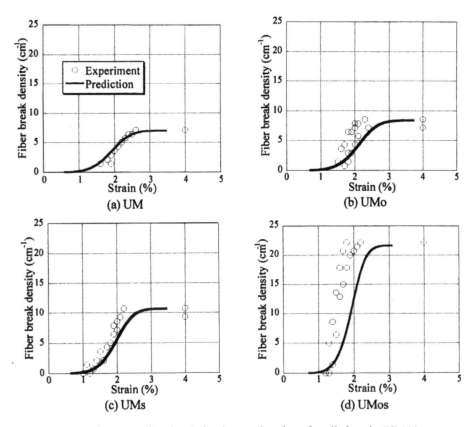

Figure 4   Fiber break density as a function of applied strain (UM46)

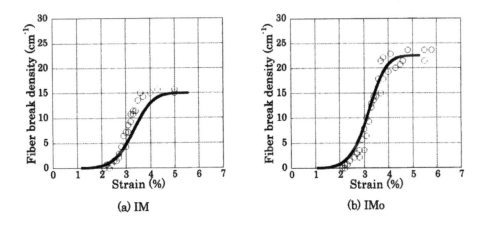

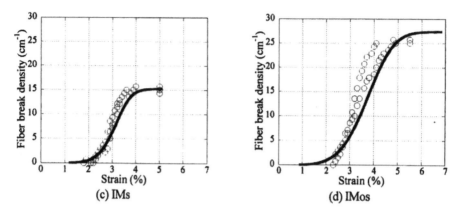

Figure 5   Fiber break density as a function of applied strain (IM600)

Table 1   Weibull parameters and interfacial shear strength

	UM46				IM600			
	UM	UMo	UMs	UMos	IM	IMo	IMs	IMos
ρ	4.97	5.29	5.46	7.00	6.13	6.31	6.62	4.83
σ(GPa)	5.13	5.61	5.10	5.06	5.61	5.14	5.47	4.87
τ(MPa)	14	18	20	40	30	46	32	70

CONCLUSION

Tensile tests on two kinds of carbon fibers with four different fiber treatments (with fiber surface oxidization and/or sizing treatment) are performed to evaluate the effect of the treatments on the statistical fiber strength distribution. It is found that the effects of the fiber treatments on the fiber strength is small.

Fragmentation tests for single fiber composites with the 8 kinds of carbon fibers and epoxy are also conducted. The experimentally-obtained fiber fragment behavior is compared with Curtin's analysis and the interfacial shear stress is evaluated. It is found that the interfacial shear stress is higher for the carbon fibers with surface oxidization and sizing treatment.

REFERENCES

1. J.A.Nairn and Y.C.Liu, On the use of energy methods for interpretation of results of single-fiber fragmentation experiments, Composite Interfaces, 4, 241-267, 1997

2. W.A.Curtin, Exact theory of fiber fragmentation in a single filament composite, Journal of Materials Science, 26, 5239-5252, 1991

3. P.W.J.Hauvel, B.Hogeweg and T.Peijs, An experimental and numerical investigation into the single fiber fragmentation test: stress transfer by locally yielding matrix, Composites Part A., 28, 237-249, 1997

SENSORS AND CONTROL

Modeling of Laminated Composite Structures with Piezoelectric Actuators

El Mostafa. Sekouri and Anh Dung Ngo
École de technologie supérieure (ETS)
Université du Québec
Montréal (Québec), Canada H3C 1K3

Yan-Ru. Hu
Directorate of Spacecraft Engineering
Canadian Space Agency
St-Hubert, Canada J3Y 8Y9

ABSTRACT

The objective of this paper is to develop the modeling of adaptive composite structures for active vibration control. The first order shear deformation beam theory is used for the analysis to ensure accurate bending solutions. Experimental, theoretical and numerical modeling approaches are conducted into active vibration control of laminate beam structures with bonded and embedded piezoelectric actuators. The assumed-modes method can predict the natural frequencies and mode shape of the laminate beam accurately. The modeling of the dynamic response of composites structures is presented. Modal analysis is done for determining the natural frequencies and mode shapes of the structures. The results obtained by the assumed-modes method and the theoretical solutions are validated through the experimental results obtained from (5). Finally, the effects of the number and locations of the actuators on the control system are also investigated.

463

INTRODUCTION

Smart structures have the ability to adapt to their environments through shape modifications. The applications of this technology are numerous and they include vibration and buckling control, shape control, damage assessment and active noise reduction. Because of their coupled mechanical and electrical properties, piezoelectric ceramics have recently attracted significant attention for their application as sensor for monitoring and actuator for controlling the response of the structures. The concept of using a network of actuators and sensors to form a self-controlling and self-monitoring "smart" system in advanced structural design has drawn considerable interest among the research community. This new technology could possibly be applied to the design of large-scale space structures, aircraft structures, satellites, and so forth. Additionally, the use of piezoelectric materials to control the vibration of structures has been extensively studied (1-5). However, their use to control the shape of structures has received less attention. With proper selection and placement of piezoelectric actuators, it is feasible to generate enough forces on a structure in order to control its shape. The changes in shapes of composite plate for specified applied voltage to the piezoelectric actuators have been study in reference (1). Recent advances design and manufacturing technologies have greatly enhanced the use of advanced fiber-reinforced composite structures for aircraft and aerospace structural applications. As a consequence, the integration of composite structural design with the "smart system" concept could potentially result in significant improvement in the performance of aircraft and space structures. The objective of this investigation was to develop an analytical method for modeling the mechanical–electrical response of fiber-reinforced laminated composite structures containing distributed piezoceramics under static as well as dynamic mechanical or electrical loadings. The assumed-modes method was also presented. The experimental results obtained from (5) using T300/976 composite and PZT G1195 piezoelectric ceramics were also conducted to verify the theory and the computer simulations. Finally, the effects of the number and locations of the actuators on the control system are also investigated.

STATEMENT OF THE PROBLEM

Consider a fiber-reinforced laminated composite beam containing distributed piezoelectric ceramics as actuators that can be bonded on the surfaces or embedded inside the structures. The ply orientation and thickness of the laminate can be arbitrary. It was desired to determine the mechanical response of the structure for given mechanical loading condition or electrical potential on the actuators.

TRANSVERSE VIBRATION OF LAMINATE BEAM

Following the beam theory of Timoshenko, the displacement field of the beam can be defined as

$$u(x,z,t) = u^0(x,t) + z\phi_x(x,t) \tag{1}$$

$$w(x,z,t) = w^0(x,t) \tag{2}$$

Where u, w are the longitudinal and the transverse displacements at any point z from the middle plane; u^0 and w^0 the longitudinal and transverse displacements of the beam middle plane, and ϕ_x the rotation of the beam section. The longitudinal strain ε_x and shear strain γ_{xz} are given by

$$\varepsilon_x = \frac{\partial u}{\partial x} = \frac{\partial u^0}{\partial x} + z\frac{\partial \phi_x}{\partial x} \tag{3}$$

$$\gamma_{xz} = \frac{\partial w}{\partial x} + \frac{\partial u}{\partial z} = \frac{\partial w}{\partial x} + \phi_x \tag{4}$$

For a one-dimensional composite beam with actuators, the width in the y-direction is assumed to be free of normal stresses, i.e., $\sigma_y = 0$, while $\varepsilon_y \neq 0$. Therefore the constitutive equation for the kth layer of the laminate beam can be written as

$$(\sigma_x)_k = (\bar{Q}_{11})_k (\varepsilon_x)_k \tag{5}$$

$$(\tau_{xz})_k = (\bar{Q}_{55} \gamma_{xz})_k \tag{6}$$

where: $\bar{Q}_{11} = Q_{11} \cos^4(\theta) + 2(Q_{12} + 2Q_{66}) \sin^2(\theta)\cos^2(\theta) + Q_{22} \sin^4(\theta)$

$\bar{Q}_{55} = Q_{44} \sin^2(\theta) + Q_{55} \cos^2(\theta)$

$Q_{11} = \frac{E_1}{1 - \nu_{12}\nu_{21}}, Q_{12} = \frac{\nu_{12}E_2}{1 - \nu_{12}\nu_{21}} = \frac{\nu_{21}E_1}{1 - \nu_{12}\nu_{21}}, Q_{22} = \frac{E_2}{1 - \nu_{12}\nu_{21}}, Q_{66} = G_{12}, Q_{44} = G_{23}, Q_{55} = G_{13}$

and σ_x is the axial stress in the x-direction; τ_{xz} the shear stress in x-z direction; E_i the modulus of elasticity; ν_{ij} the Poisson's ratios, G_{ij} (i, j=1, 2, 3) the shear modulus and the subscript k denotes quantities belonging to the kth layer, θ is the angle measured from the x-coordinate to the material coordinate 1.

The constitutive relations for the ith actuator can be written as

$$(\sigma_x(x,z))_i = (\frac{E_p}{1 - \nu_p^2}\varepsilon_x(x,z) - e_{31}E)_i \tag{7}$$

$$[\tau_{xz}(x)]_i = (G_p\gamma_{xz})_i \tag{8}$$

where E_p, G_p and ν_p are the modulus of elasticity, shear modulus and Poisson's ratio of the actuator, respectively; e_{31} is the piezoelectric z-x stress coefficient of the actuator; E the electric field intensity in z direction in the actuator, subscript p indicate piezoelectric.

The stress resultant-displacement relationships for the laminate beam can be obtained by integrating the stresses through the cross-sectional area of the beam.

$$\begin{bmatrix} N_x \\ M_x \end{bmatrix} = \begin{bmatrix} A_{11} & B_{11} \\ B_{11} & D_{11} \end{bmatrix} \begin{bmatrix} \frac{\partial u}{\partial x} \\ \frac{\partial \phi_x}{\partial x} \end{bmatrix} - \chi \begin{bmatrix} N_p \\ M_p \end{bmatrix} \tag{9}$$

$$[Q_{xz}] = [A_{55}\gamma_{xz}] \tag{10}$$

where $\{A_{11}, B_{11}, D_{11}\} = \sum_{k=1}^{n+n_p} \int_{z_{k-1}}^{z_k} (\bar{Q}_{11})_k (1, z, z^2)dz$; $A_{55} = k_s \sum_{k=1}^{n+n_p} \int_{z_{k-1}}^{z_k} \bar{Q}_{55}dz$

$$\{\{N^P, M^P\} = \sum_{j=1}^{n_L} \int_{z_{k-1}}^{z_k} \left(\frac{E_p}{1-v_p^2} d_{31} E\right)_j (1, z) dz \text{, where } d_{31} = e_{31} \frac{(1-v_p^2)}{E_p} \text{ is the piezoelectric z-x}$$

strain coefficient; where N_x, M_x are the longitudinal force and the moment per unit width of the beam; $\chi = 0$ for beam sections without actuators and $\chi = 1$ for sections with actuators, n denotes number of layers, n_p number of actuators and $k_s = 5/6$ is the shear correction factor. Here we consider the bending of symmetrically laminated beams according to CLPT. For symmetric laminates, the equations for bending deflection are uncoupled from those of the stretching displacements. If the in-plane forces are zero, the in-plane displacements (u_0, v_0) are zero, and the problem is reduced to one of solving for bending deflection and stress. We assume $M_{yy} = M_{xy} = 0$ everywhere in the beam.

From the equation of symmetric laminate $B_{ij} = 0$ we have

$$\frac{\partial \phi_x}{\partial x} = D_{11}^* M_{xx}, \quad \frac{\partial w_0}{\partial x} + \phi_x = A_{55}^* Q_{xz} \tag{11}$$

or $M(x) = b M_{xx}$ and $Q(x) = b Q_{xz}$ \hfill (12)

the equation of motion

$$\frac{\partial Q_{xz}}{\partial x} + q = I_0 \frac{\partial^2 w_0}{\partial^2 t} \tag{13}$$

$$\frac{\partial M_{xx}}{\partial x} - Q_{xz} = I_2 \frac{\partial^2 \phi_x}{\partial t^2} \tag{14}$$

where $(I_0, I_1, I_2) = \int_{\frac{h}{2}-h_p}^{\frac{h}{2}+h_p} \rho(1, z, z^2) dz$, h and hp are the thickness of beam and actuators

respectively, A^*, D^* denote the inverse matrix of A and D respectively and b is the width of beam. From Eq.(13) and (14), the equations of motion can be write in terms of the displacement function

$$A_{55}^* b\left(\frac{\partial^2 w_0}{\partial x^2} + \frac{\partial \phi_x}{\partial x}\right) + qb = b I_0 \frac{\partial^2 w_0}{\partial t^2} \tag{15}$$

$$\frac{b}{D_{11}^*} \frac{\partial^2 \phi_x}{\partial x^2} - A_{55}^* b\left(\frac{\partial w_0}{\partial x} + \phi_x\right) = b I_2 \frac{\partial^2 \phi_x}{\partial t^2} \tag{16}$$

Bending and Natural Frequencies

When the laminated beam problem is such that the bending moment M(x) and Q(x) can be written readily in terms of known applied loads (in statically determinate beam problems), Equation (11) can be utilized to determine ϕ_x, and than w_0. When M(x) and Q(x) cannot be expressed in terms of known loads, Equations (15) and (16) can be used to determine $w_0(x)$ and $\phi_x(x)$.

For modal analysis, we assume that the applied axial force and transverse load are zero and the motion is periodic. For a periodic motion, we assume solution in the form

$$w_0(x,t) = W(x)e^{i\omega t}, \quad \phi_x(x,t) = \chi(x)e^{i\omega t}, \quad i = \sqrt{-1} \tag{17}$$

where ω is the natural frequency of vibration, and Equations (15) and (16) become

$$\frac{b}{D_{11}^{\bullet}}\frac{d^4W}{dx^4}+(\frac{b}{A_{55}^{\bullet}}\frac{I_0}{D_{11}^{\bullet}}+bI_2)\omega^2\frac{d^2W}{dx^2}-(1-\frac{b}{A_{55}^{\bullet}}\omega^2I_2)bI_0\omega^2W=0 \tag{18}$$

or $p\dfrac{d^4W}{dx^4}+q\dfrac{d^2W}{dx^2}-rW=0$ (19)

The general solution of equation (19) is:

$W(x)=c_1\sin\lambda x+c_2\cos\lambda x+c_3\sinh\mu x+c_4\cosh\mu x$ (20)

where $\lambda=\sqrt{\dfrac{1}{2p}(q+\sqrt{q^2+4pr}})$, $\mu=\sqrt{\dfrac{1}{2p}(-q+\sqrt{q^2+4pr}})$ and c_i i=1, ...,4 are constants,

which are to be determined using the boundary conditions. Note that we have

$(2\lambda^2 p-q)^2=q^2+4pr$ or $p\lambda^4-q\lambda^2-r=0$

Alternatively, Equation (18) can be written, with W given by Equation (19) in terms of ω as: $P\omega^4-Q\omega^2+R=0$ (21)

Where $P=A_{55}^{\bullet}I_2$, $Q=1+(\dfrac{A_{55}^{\bullet}}{D_{11}^{\bullet}}+\dfrac{I_2}{I_0})\lambda^2$, $R=\dfrac{\lambda^4}{D_{11}^{\bullet}I_0}$ (22)

Hence, there are two (sets of) roots of this equation (when $I_2\neq 0$)

$(\omega^2)_1=\dfrac{1}{2P}(Q-\sqrt{Q^2-4PR})$, $(\omega^2)_2=\dfrac{1}{2P}(Q+\sqrt{Q^2-4PR})$ (23)

It can be shown that $Q^2-4PQ>0$ (and $PQ>0$), and therefore the frequency given by the first equation is the smaller of the two values.

NUMERICAL APPROACH: ASSUMED-MODES METHOD

The assumed-modes method consists of assuming a solution of free vibration problem in the form of a series composed of linear combination of admissible functions ϕ_i, which are function of spatial coordinates, multiplied by time-dependent generalized coordinates $q_i(t)$. These admissible functions satisfied the natural boundary condition. For the laminate beam, the transverse displacement is approximately expressed as:

$$w(x,t)=\sum_{i=1}^{n}\phi_i(x)q_i(t) \tag{24}$$

where $\phi_i(x)=(\dfrac{x}{L})^{i+1}$,Using Lagrange's equation and Eq.(24), the equation of motion as

$[M]\{q\}+[K]\{q\}=[B_v]\{v\}$ (25)

where [M] and [K] are the masse, and stiffness matrices ,$[B_v]$ is the input matrix which used to apply forces to the structure by piezoelectric actuators. Vector q represents the beam response modal amplitudes and v is the vector of applied voltage on piezoelectric.

RESULTS AND COMPARISONS

Numerical and analytical results are presented to show the static and dynamic behaviour of laminated beams with piezoelectric actuators. In order to verify the proposed and the assumed-modes approach, numerical calculations were generated from a cantilevered laminated composite beam. A cantilever composite beam with bonded the upper and lower surfaces symmetrically bonded by piezoelectric ceramics, as shown in Fig. 1, is considered. The dimensions are also shown in the same figure. The beam is made of AS/3506 graphite-epoxy composites and the piezoceramic is PZT BM532. The adhesive layers are neglected. The material data is given in Table 1. The staking sequence of composite beam is $[0/\pm 45]_s$. A constant voltage (equal amplitude) with an opposite sign was applied to the piezoelectric on each side of the beam. Due to the converse piezoelectric effect, the distributed piezoelectric actuators contract or expand depending on negative or positive active voltage. In general, for an upward displacement, the upper actuators need a negative voltage and the lower actuators need a positive one. The control of static deformation and modal analysis for the beam under the distribution piezoelectric are analyzed. Figure 2 shows the shape of the cantilever beams for various specified voltages. It is observed that the deformation of the laminate beams increases with increases in applied voltage. A uniformly distributed load of 2 (N / m^2) is applied to beam. It is clear from Fig.3 that the structure comes back to the undeformed position as the specified voltages are increased. To investigate the effect of the number and the placement of the actuator pairs on the deformation control, three sets of the actuator pairs are considered: three pairs, (the left, the middle and the right); two pairs (the left and the middle ones, the left and the right) and one pairs located at the end of the beam. The comparison of the figure 3-6 reflects the fact that a lower voltage is needed to eliminate the deflection caused by the external load when more actuators are used (Fig.4-6). When one pair of actuators is used, as shown in Figure3, a very high active voltage is needed to delete the deformation and the beam is also not smoothly flattened. Figure 4 shows that, under a certain active voltage, the beam can be flattened quite smoothly by two pairs of actuators placed in the left and the middle position of the beam. Than is not necessary to cover structures entirely with piezoelectric. Figure 7 shows the calculated centreline deflection of the composite beam with one pair of the actuators at different positions. It seen that the location of the actuators has a significant effect on the control of the deformation.

Table 1 Material Properties

	PZT-G_1195N	PZT- (BM532)	T300/976	AS/3501
E1 (Gpa)	63	71.4	150	144.8
E2 (Gpa)	63	71.4	9	9.65
G12 (Gpa)	24.8	26.8	7.10	7.10
v_{12}	0.3	0.33	0.3	0.3
G13 (Gpa)	24.2	--	7.10	7.10
G23 (Gpa)	24.2	--	2.50	5.92
Density (Kg/m^3)	7600	7350	1600	
d31 (pm/V)	254	200		

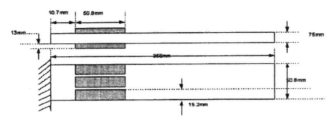

Figure1 Cantilever beam with surface bonded piezoceramics

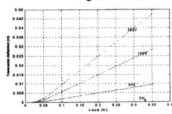

Figure2 Effect of actuator voltage
on transverse deflection.

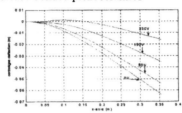

Figure3 Beam subjected to uniformly distributed
load with 6 actuators located at the left end.

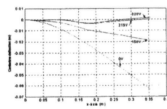

Figure4 Two pairs of actuators located at
Left end-middle of the beam.

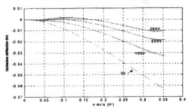

Figure5 Two pairs of actuators:
located at: left end- right end

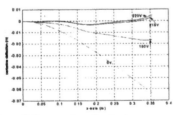

Figure6 three pairs of actuators.
left end, $L/4 -b/2 < x < L/4 +b/2$ and
$3L/4 -b/2 < x < 3L/4 +b/2$

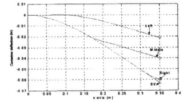

Figure7 One pair of actuators
at different position(220V)

Natural frequencies for composite beam without any piezo-elements and with bonded piezo-elements are presented in tables 2 and 3, respectively. The beam is made of T300/976 graphite-epoxy composites and the piezoceramic is PZT G1195N. This example is taken from Ref.(5). The discrepancy is 12.6% for composite beam with bonded piezoactuators, while it is only about 1.6% for beam without any piezoelectric.

One can notice the better agreement between the numerical results and analytical ones than that of the experimental results and analytical ones. The numerical results listed in tables 2 and 3 show that the assumed-modes method could accurately predict the natural frequencies of an integrated piezo-actuator system.

Table 2 Natural frequencies (Hz)of the beam without piezoelectric

Modes	F.E 1	Theory-2	Assumed-modes -3	%Error (1-2)	%Error(1-3)
1	7.99	8.02	8.02	0.2	0
2	49.82	50.66	50.57	1.6	0.1

Table 3 Natural frequencies of the beam with 6 piezoelectric

Modes	Exper.-1	Theo.-2	Ass.-modes 3	%Err.1-2	%Err.1-3	%Err.2-3
1	9.25	9.42	9.52	1.8	2.8	1.0
2	51.9	59.4	56.44	12.6	8.0	5.2

CONCLUSIONS

An analytical method for modeling the mechanical–electrical response of fiber-reinforced laminated composite structures containing distributed piezoceramics under static as well as dynamic mechanical or electrical loadings is presented. The first order shear deformation beam theory is used for the analysis to ensure accurate bending solutions. The assumed-modes method was also presented. Experimental results obtained from Ref.(5) using T300/976 composite and PZT G1195 piezoelectric ceramics were also conducted to verify the theory and the computer simulations. Finally, the effects of the number and locations of the actuators on the control system are also investigated. The investigation shows that in designing smart structures with distributed piezoelectric actuators, the number and the location of the actuators must be considered carefully.

REFERENCES

1. D.B. Konis, L.P. Kollar and G.S. Springer "Shape Control of Composite Plates and Shells with Embedded Actuators, I. Voltage Specified", J. Composite Mater., 28 ,1994, 415-458.
2. J.N. Reddy, Mechanics of Laminated Composite Plates, Theory and Analysis, CRC Press, Inc.1997.
3. P. Donthireddy and K. Chandrashekhara, "Modeling and Shape Control of Composite Beams with Embedded Piezoelectric Actuators", Composite Structures 35, 1996, 237-244.
4. E.F. Crawley, "Induced Strain Actuation of Isotropic and Anisotropic Plates", AIAA Journal, vol. 29, no. 6, 1989, 944-951.
5. Sung Kyu Ha, Charles Keilers, and Fu-Kuo Chang, "Finite Element Analysis of Composite Structures Containing Distributed Piezoceramic Sensors and Actuators", AIAA Journal, Vol.30, no. 3, march 1992.

Further Development of Structural Health Monitoring of Smart Composite Systems in Japan

Nobuo Takeda, Yoji Okabe
Graduate School of Frontier Sciences
The University of Tokyo
c/o Komaba Open Laboratory, The University of Tokyo
4-6-1 Komaba, Meguro-ku, Tokyo 153-8904, Japan
ntakeda@k.u-tokyo.ac.jp

Naoyuki Tajima, Tateo Sakurai
R&D Institute of Metals and Composites for Future Industries
Bridgestone Toranomon Bldg. 3F
3-25-2 Toranomon, Minato-ku, Tokyo 105-0001, Japan
tajima@rimcof.or.jp, sakurai@rimcof.or.jp

ABSTRACT

Recent activities of the structural health-monitoring of smart composite systems in Japan is summarized, as a university-industry collaboration program at the University of Tokyo along with 10 research organizations. The research themes include: (a) development of high-performance sensor system technology with newly-developed sensors, (b) development of a damage detection and self-diagnosis system for structural integrity based on micro-mechanical damage identification, and (c) development of application technology for model structures. The sensor technology includes: (a) development of small diameter optical fiber sensor, (b) damage suppression in composite laminate systems with embedded shape-memory alloy films, and (c) development of maximum strain memory sensors with electrically conductive composite systems. The sensor output is correlated with the underlying damage evolution in the structure and its theoretical micro-mechanical modeling. Based on the development of the structural health-monitoring technologies at the structural element level, a concept demonstrator is being constructed to demonstrate our developed health monitoring technologies in a composite fuselage structure. The present status of the concept demonstrator is also reviewed.

INTRODUCTION

In Japan, a "R&D for Smart Material/Structure System (SMSS)" project started in October 1998 as one of the Academic Institutions Centered Program supported by NEDO (New Energy and Industrial Technology Development Organization), Japan. This 5-year SMSS project includes four sub-themes: (a) structural health monitoring, (b) smart manufacturing, (c) active/adaptive structures, and (d) actuator materials development. The author acts as a group leader in the structural health monitoring group, which consists of 10 research organizations with the University of Tokyo. Our structural health monitoring group (SHMG) is currently developing a health monitoring system, which conducts a real-time damage detection and self-diagnosis as well as damage control in light-weight composite structural systems. The research themes include: (a) development of high-performance sensor system technology with newly-developed sensors, (b) development of a damage detection and self-diagnosis system for structural integrity based on micro-mechanical damage identification, and (c) development of application technology for model structures or substructures. The sensor output is correlated with the underlying damage evolution in structures such as aircrafts, satellites, high-speed trains and large-scale civil infrastructures, as described by Takeda (1). Some of the latest results in SHMG research efforts are reported for each research theme, and a concept demonstrator under development is reviewed to demonstrate our developed health monitoring technologies in a composite fuselage structure.

DEVELOPMENT OF HIGH-PERFORMANCE SENSOR SYSTEM TECHNOLOGY IN COMPOSITE STRUCTURES

The following is the list of individual research themes under progress: (a) Detection of Transverse Cracks and Delamination in CFRP Laminates with FBG Sensors – Univ. Tokyo (Okabe et al., 2-4, S. Takeda et al., 5), (b) Development of Small-Diameter Optical Fiber Sensors – Hitachi Cable (Satori et al., 6), (c) Identification of Impact Damage Parameters in Composites Using Embedded Optical Fiber Sensors – Kawasaki Heavy Industries (Tsutsui et al., 7), (d) Strain and Damage Monitoring of CFRP Space Satellite Structures Using Optical Fiber Sensors – Mitsubishi Electric (Kabashima et al., 8), (e) Damage Suppression and Control in CFRP Laminates with Embedded SMA Films – Fuji Heavy Industries (Ogisu et al., 9), (f) Quantitative Evaluation of Electric Properties of CFRP and CFGFRP Hybrid Composites as a Maximum Strain Memory Sensor – Toray (Park et al., 10), (g) Self-Diagnosis of Electrically-Conductive FRP Containing Carbon Particles – JFCC (Okuhara et al., 11), (h) Impact Damage Monitoring of Composite Structures Using Integrated Acoustic Emission (AE) Sensor Network – EADS-CCR (Saniger et al., 12), (i) Damage Detection of Transparent Composites Using Light Transmission and Reflection Measurement for High-Speed (Magrev) Train – Hitachi (Aoyama et al., 13), (j) Integrated Global and Local Strain Measurement Using Distributed BOTDR (Brillouin Optical Time Domain Reflectometry) and FBG Sensors – Mitsubishi Heavy Industries (Yari et al., 14), and (k) Development of FBG Sensor Elements and Real-Time Monitoring Systems for Large-Scale Infrastructures – Shimizu (Iwaki et al., 15). Some notable research accomplishments are presented in the following.

Detection of Transverse Crack and Delamination in Composite Laminates with Embedded Small-Diameter FBG Sensors

Fiber Bragg Grating (FBG) optical fiber sensors have been developed to measure strain, temperature and so on, through the shift of the wavelength peak in the reflected light. A uniform strain within the gage section (typically 10 mm) is normally assumed. FBG sensors are very sensitive to non-uniform strain distribution along the entire length of the gratings. The strain distribution deforms the reflection spectrum from the FBG sensors. Taking advantage of the sensitivity, the authors applied FBG sensors for detecting transverse cracks that caused non-uniform strain distribution in CFRP laminates (2). However, the cladding of common optical fibers is 125 μm in diameter, which is almost the same as the normal thickness of one ply in CFRP laminates. Thus, when the normal FBG sensors are embedded into CFRP composites, there is a possibility that the optical fibers might deteriorate the mechanical properties of the laminates. In order to prevent the deterioration, small-diameter FBG sensors have recently been developed by the authors and Hitachi Cable Ltd. (6). The outside diameter of the polyimide coating is 52 μm, and the cladding is 40 μm in diameter. As shown in Fig. 1, resin-rich regions cannot be found around the small-diameter optical fiber embedded within one ply, but clearly exist around the conventional optical fiber.

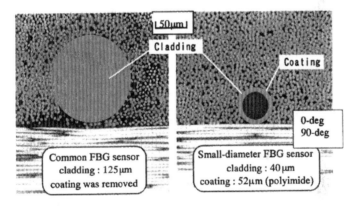

Fig.1 Conventional and small-diameter FBG sensors embedded in CFRP cross-ply laminates

The small-diameter FBG sensors can also detect transverse cracks in CFRP cross-ply laminates (3) as well as in quasi-isotropic laminates. A small-diameter FBG sensor was embedded in a CFRP T800H/3631 [45/0/-45/90]s laminate, where the sensor was located in the −45° ply on the border of the 90° ply, parallel to carbon fibers in the −45° ply. Figure 2 shows the reflection spectra measured at various levels of tensile strain ε and crack density ρ. The crack density was defined as the number of transverse cracks per unit length along the loading direction in the 90° ply. While there was no transverse crack, the spectrum kept its shape and the center wavelength shifted corresponding to the strain. After transverse cracks appeared, the reflection spectrum deformed and became broad with an increase in the crack density ρ. For confirmation that the change in the form of the spectrum was caused by transverse cracks, the spectrum was calculated theoretically. In the calculation, it was assumed that the FBG sensor was affected only by the axial strain. At first, the non-uniform strain distribution in the FBG sensor was calculated using 3-D FEM analysis with ABAQUS code. Next, the distributions of the grating period and the average refractive index of the FBG were calculated from the axial strain distribution. Then the reflection spectrum was simulated from the distributions using the software "IFO_Gratings". The calculated results of reflection spectra are shown in Fig. 3. These spectra correspond to those in Fig. 2. The change in the form of the calculated spectrum is similar to that of the measured one. These results show that the change in the spectrum is caused by the non-uniform strain distribution due to the occurrence of the transverse cracks. Thus, the transverse cracks in quasi-isotropic laminates can also be detected from the deformation of the reflection spectrum.

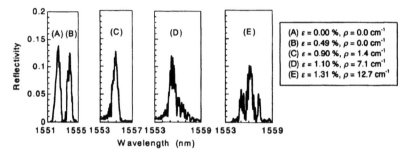

Fig. 2 Reflection spectra measured at various levels of tensile strain ε and crack density ρ.

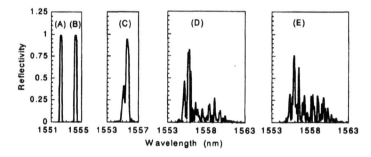

Fig. 3 Calculated reflection spectra, which correspond to the measured spectra in Fig. 1.

The above technique was also applied to the detection of delamination (S. Takeda et al., 5). The specimen was CFRP T800H/3631 [$90_{10}/0_4/90_{10}$] cross-ply laminate. As shown in Fig. 4, the FBG sensor was embedded in $0°$ ply to be parallel to carbon fibers and in contact with $90°$ ply. A strip type delamination was grown along a $0°/90°$ interface by four-point bending test. For the delamination onset from the tip of a transverse crack, a vertical notch was introduced at the mid-span of the specimen. The transverse crack occurred from the root of the notch and reached the $0°/90°$ interface. An end of the FBG sensor was set on the tip of the transverse crack so that the induced delamination propagates in one direction within the region of the grating. Figure 5 shows the reflection spectra measured at various steps of the delamination progress. These spectra were normalized by the intensity of the highest component. When there was only a transverse crack before the occurrence of the delamination, the reflection spectrum had only one sharp narrow peak as shown in Fig. 5(a). After the delamination was initiated from the crack tip, another peak appeared at longer wavelength. The intensity of the longer wavelength peak increased relatively with an increase of the

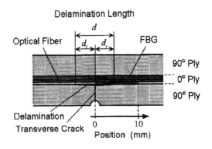

Fig. 4 Embedded FBG sensor for detecting delamination growth.

delamination length. This phenomena can be well reproduced and predicted by the theoretical calculation (4).

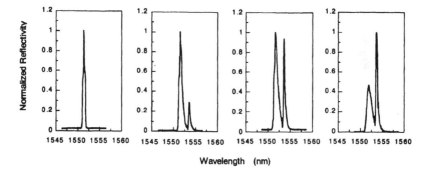

Fig. 5 Reflection spectra measured at various steps of delamination progress: (a) $d = 0.0$ mm, $d_t = 0.0$ mm; (b) $d = 5.4$ mm, $d_t = 2.8$ mm; (c) $d = 8.4$ mm, $d_t = 4.2$ mm; (d) $d = 13.0$ mm, $d_t = 6.4$ mm.

Identification of Impact Damage Parameters in Composite Laminates Using Embedded Small-Diameter Optical Fiber Sensors

The objective is the development of the technology for the detection of barely visible impact damage (BVID) in composite structures by embedded small-diameter optical fiber sensors. Compared with the results on plate specimens (Tsutusi et al., 7), the stiffened composite panels, which are the representative structural elements of airplanes, are characterized by different impact damages from that of the plate specimens. In this study, single-mode and multi-mode optical fibers were used as a sensor for detecting impact load and impact damage in flat and curved stiffened composite panels. The impact tests of the flat stiffened panels (Fig. 6) revealed different distributions of damage in the mid-bay skin, the stiffener flange and web subjected to impact load, and it was found possible to detect impact damage by the optical fiber sensors embedded in appropriate positions for each damage type.

Fig. 6 A stiffened composite panel with
embedded small-diameter optical fiber
sensors.

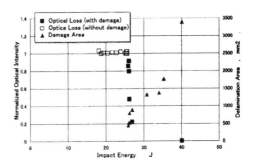

Fig. 7 Relationship among impact energy, normalized optical
intensity and delamination area.

Figure 7 indicates that normalized optical intensity decreases as impact energy is over about 25J and the intensity is almost zero as impact energy is 40J. At this impact energy, it was found that impact damage occurred in the composites. In other words, it was supposed that optical loss was induced by the bending deformation of optical fibers and damage of optical fibers. In fact, the existence of the delamination and damaged fibers was confirmed by the observation of the cross section.

The relationship between the delay of the arrival time of strain responses measured by each FBG sensor and strain gauge, and the distance from impact location to each sensor is plotted in Fig. 8. The delay time can be expressed as a linear function of the distance, and the velocity of stress propagation is approximately 1200 m/s. Using the difference of the delay time, the position of three sensors and the experimental velocity of stress propagation, the impact location was predicted. An example of the result predicted by using three sensors selected from four sensors is shown in Fig. 9. It is found that the predicted point is well correlated with the location of impact damage.

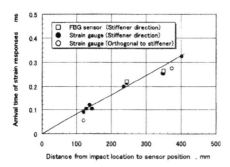

Fig. 8 Relationship between delay of arrival time of strain
and distance from impact location to sensor position.
(Stiffener flange impact)

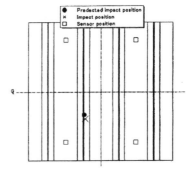

Fig. 9 An example of detection of impact damage
point in a stiffened panel.
(Stiffener flange impact)

Damage Suppression and Control in CFRP Laminates with Embedded SMA Films

Shape memory alloy (SMA) films were embedded and used to suppress and control microscopic damages in CFRP laminates such as transverse cracks and delamination (Ogisu et al., 9). Improvement of interlaminar shear strength (ILSS) between SMA films and CFRP laminas was made using spattering, sol-gel, ion-plating and anodic-oxidation. A high ILSS was obtained similar to that of CFRP laminates alone. SMA films were stretched into the phase transformation region. Then, they were embedded in CFRP laminates with the deformation kept by the fixture jig during the fabrication in order to introduce the shrinking stress in 90-degree plies. Such shrinking stresses were found to suppress the evolution of transverse cracks in quasi-isotropic laminates (Fig. 10). A theoretical method was developed to formulate the stress distribution in each layer and predict the transverse crack evolution in CFRP laminates with SMA films (Ogisu et al., 9).

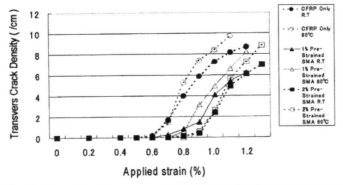

Fig. 10 Suppression of transverse cracks in CFRP T300/F593 [45/0/-45/90]$_s$ laminate with embedded pre-strained SMA films.

Integrated Global/Local Strain Measurement Using Distributed BOTDR and FBG Sensors

Improvement of spatial resolution and dynamic strain measurement in the BOTDR technique were made. Integrated BOTDR and FBG strain measurement systems including the temperature compensation were made for aerospace structures (Yari et al., 14). Figure 11 illustrates an example of BOTDR measurement for the strain and temperature distribution in a CFRP liquid hydrogen cryogenic tank.

DEVELOPMENT OF COMPOSITE FUSELAGE DEMONSTRATORS

In order to integrate the developed sensor and actuator elements into a smart structure system and show the validity of the system, two demonstrator programs have been established. Both demonstrators are CFRP stiffened cylindrical structures with 1.5 m in

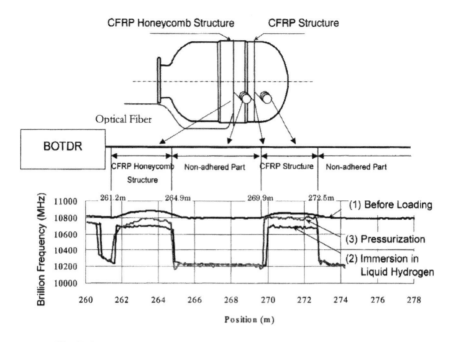

Fig. 11 An example of BOTDR measurement for strain and temperature distribution
in a CFRP liquid hydrogen cryogenic tank

diameter and 3 m in length. The first demonstrator integrates several technologies for damage
detection and suppression in composite structures. The second one is for demonstrating the
suppression of vibration and acoustic noise generated in the composite cylindrical structure.
The detailed design of the demonstrator was made and the testing program has been planned
to minimize the time and the cost for the demonstration.

Preliminary design of test article for the Damage Detection and Damage Suppression
Demonstrator is summarized below, and the outline of the test article is shown in Fig. 12
(Tajima et al., 16).

- Test article: consists of CFRP quasi-isotropic laminates, simulating an aircraft fuselage
  with a length of 3m and diameter of 1.5m.
- Structure: a build-up structure with composite skin-stringer panels and aluminum alloy
  frames. The panels are divided into four along the circle, and also the support and the load
  structure at both ends are also divided into four corresponding to the panels. The
  bulkhead panel can be freely removed/mounted, allowing a fastener joint to be connected
  to the load structure section.
- Arrangement: The frame and stringer have a pitch of about 500mm and 150-200mm,
  respectively. Internally, the test article has a working floor inside of the fuselage for test
  preparations.

Material: The external panel and stringer use CFRP material. The aluminum alloy used for the frames mainly consists of 2024 and 7075 alloys. The load and support structures at both ends are made of steel.

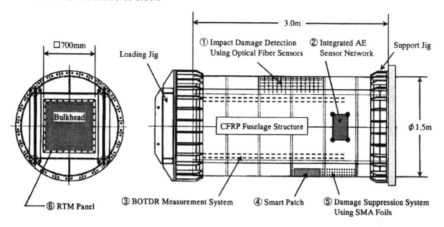

Fig. 12 Damage detection and damage suppression demonstrator.

The demonstrator test will be performed late in 2002. The obtained results will be presented later.

ACKNOWLEDGEMENTS

This research was conducted as a part of the "R&D for Smart Materials Structure System" project within the Academic Institutions Centered Program supported by NEDO (New Energy and Industrial Technology Development Organization), Japan.

REFERENCES

1. N. Takeda, "Summary Report of Structural Health Monitoring Project for Smart Composite Structure Systems", *Adv. Comp. Mater.*, Vol. 10, No. 2-3, 2001, 107-118.
2. Y. Okabe, S. Yashiro, T. Kosaka, and N. Takeda, "Detection of Transverse Cracks in CFRP Composites Using Embedded Fiber Bragg Grating Sensors," *Smart Mater. Struct.*, Vol. 9, No. 6, 2000, 832-838.
3. Y. Okabe, T. Mizutani, S. Yashiro and N. Takeda, "Detection of Microscopic Damages in Composite Laminates with Embedded Small-Diameter Fiber Bragg Grating Sensors", *Compos. Sci. Tech.*, 2002, in press.
4. Y. Okabe, S. Yashiro, R. Tsuji, T. Mizutani and N. Takeda, "Effects of Thermal Residual Stress on the Reflection Spectrum from Fiber Bragg Grating Sensors Embedded in CFRP Laminates", *Composites: Part A*, 2002, in press.

5. S. Takeda, Y. Okabe and N. Takeda, "Delamination Detection in CFRP Laminates with Embedded Small-Diameter Fiber Bragg Grating Sensors", *Composites: Part A*, 2002, in press.
6. K. Satori, K. Fukuchi, K. Kurosawa, A. Hongo, and N. Takeda, "Polyimide-Coated Small-Diameter Optical Fiber Sensors for Embedding in Composite Laminate Structures," Proc. SPIE 8th Int. Symp. Smart Struc. Mater., Vol. 4328, 2001, 285-294.
7. H. Tsutsui, A. Kawamata, T. Sanda and N. Takeda, "Impact Damage Monitoring of Composite Panels by Embedded Small-Diameter Optical Fibers", *Mater. Sci. Res. Int.*, Special Tech. Pub., Part 2, 2001, 121-125.
8. S. Kabashima, T. Ozaki and N. Takeda, "Structural Health Monitoring Using FBG Sensor in Space Environment", Proc. SPIE 8th Int. Symp. Smart Struc. Mater., Vol. 4332, 2001, 78-87.
9. T. Ogisu, N. Ando, J. Takaki, T. Okabe and N. Takeda, "Improved Surface Treatment of SMA Foils and Damage Suppression of SMA-Foil Embedded CFRP Laminates", J. Intelligent Mater. Struct., Vol. 12, 2001, 265-270.
10. J.B. Park, T. Okabe, N. Takeda and W.A. Curtin, "Electromechanical Modeling of Unidirectional CFRP Composites under Tensile Loading Condition", Composites: Part A, Vol. 33, 2002, 265-275.
11. Y. Okuhara, S.-G. Shin, H. Matsubara, H. Yanagida and N. Takeda, "Development of Conductive FRP Containing Carbon Phase for Self-Diagnosis Structures", Proc. SPIE 8th Int. Symp. Smart Struc. Mater., Vol. 4328, 2001, 314-322.
12. J. Saniger, L. Reithler, D. Guedra-Degeorges, J.-P. Dupuis and N. Takeda, "Development of an adapted Structural Health Monitoring System based on Acoustic Emission", Proc. SPIE 8th Int. Symp. Smart Struc. Mater., Vol. 4332, 2001, 88-97.
13. H. Aoyama, K. Tanaka, and N. Takeda, "Health Monitoring Technology for Alumina FRP Using Transmitted Light", Mater. Sci. Res. Int., Special Tech. Pub., Part 2, 2001, 161-165.
14. T. Yari, T. Shimizu, K. Nagai, and N. Takeda, "Structural Health Monitoring System for Aerospace Structures Using Optical Fiber Distributed Sensor," Proc. 3rd Int. Workshop on Structural Health Monitoring, 2001, 355-362.
15. H. Iwaki, H. Yamakawa, K. Shiba, and A. Mita, "Structural Health Monitoring System Using FBG-Based Sensors for a Damage Tolerant Building", Proc. 3rd Int. Workshop on Structural Health Monitoring, 2001, 584-593.
16. N. Tajima, T. Sakurai, N. Takeda and T. Kishi, "Progress in Demonstrator Program of Japanese smart Material and Structure System Project", Proc. SPIE 9th Int. Symp. Smart Struct. Mater., Vol. 4701 2002, in press.

Cure and health monitoring of FW GFRP pipes by using FBG fiber optic sensors

Tatsuro Kosaka, Katsuhiko Osaka , Masaya Sando and Takehito Fukuda
Department of Intelligent Materials Engineering, Osaka City University, Sugimoto 3, Sumiyoshi, Osaka, 558-8585, Japan

ABSTRACT

FBG sensors were applied to cure and health monitoring of filament winding (FW) glass/epoxy pipes by using embedded FBG fiber optic sensors. Three kinds of pipes, which are wound by 45° helical and 45° helical + 15° parallel winding methods, were used for experiments. FBG sensors were embedded along axis direction of pipes or fiber reinforcing direction. From the results of cure monitoring, it was found that FBG sensors could be used for measurement of curing shrink as well as thermal residual strain. It also appeared that curing strain of FW pipes during heating showed complex behavior due to the winding force. The results of tensile tests showed that strain from FBG sensors had very good agreement with that from attached strain gauges. The FBG sensors which were embedded in helical + parallel winding pipes could monitor transverse crack by observing FBG spectral shape. From the results of a fatigue test, instability of strain output from FBG sensors were shown after crack initiation. From the results obtained in this study, it can be concluded that embedded FBG sensor in FW FRP pipe have high functions of real-time cure and health monitoring.

INTRODUCTION

Cure and health monitoring by using built-in fiber optic sensors is a new approach for enhancement of reliability of fiber reinforced plastic (FRP) structures. In this technique, internal state of materials or structures is monitored in real time and then manufacturing and operational costs can be reduced due to simplified inspections. Therefore, many applications of this technique have been studied in this decade (1). Several fiber optic sensors can be applied only to cure monitoring. But fiber optic strain sensors can be applied both cure and health monitoring. Fiber optic strain sensors such as extrinsic Fabry-Perot interferometer (EFPI) sensors, fiber Bragg grating (FBG) sensors and Brillouin optical time domain reflectometric (B-OTDR) sensors can be used for cure and health monitoring since they are composed of single optical fiber. EFPI sensors are especially suited for curing monitoring since they have low temperature sensitivity (2,3). Since FBG is an optical grating written in a core of a fiber, they need no mechanical part. In addition, FBG sensors are independent upon optical loss and can be multiplexed easily Therefore, FBG sensors are suited for embedding in large composites. Several papers addressed applications of FBG sensors to cure and health monitoring of FRP (4-7).

We have applied EFPI sensors to cure monitoring of filament winding (FW) FRP (3). From the study, it was found that EFPI sensors have a good performance of strain monitoring during cure. However, when considering their applications to large structures, EFPI sensors are not so suitable. Then, in the present study, FBG sensors were applied for cure and health monitoring of FW FRP pipes. Internal strain measurements of specimens were conducted during cure, in tensile tests and in a fatigue test. From the experimental results, self-monitoring function of developed FW FRP pipes was discussed.

FIBEROPIC SENSORS

In this paper, mainly FBG sensors were used for strain monitoring but EFPI sensors were also employed for comparative study. Figure 1 illustrates constructions of FBG sensors. FBG sensors have optical sensing parts (Bragg grating), in which refractive index of a fiber core periodically changes. FBG sensors act as a narrow-band optical filter and a center wavelength of the reflected light indicates a strain-temperature coupled value. The strain output ε_3 of FBG sensor can be obtained from Bragg wavelength shift $\Delta\lambda$ and a temperature variation ΔT by the following equation (8).

$$\frac{\Delta\lambda}{\lambda_0} = \left[1 - \frac{n_0^2}{2}\left\{p_{12} - v_s\left(p_{11} + p_{12}\right)\right\}\right]\left(\varepsilon_3 - \alpha_s\Delta T\right) + \left(\alpha_s + \frac{1}{n_0}\frac{dn_0}{T}\right)\Delta T, \qquad (1)$$

where, λ_0 is initial Bragg wavelength, p_{11} and p_{12} are Pockels constants, v_s is Poisson ratio, n_0 is refractive index at $\Delta T = 0$. All values belong to FBG sensor. In this paper, a sensitivity of wavelength shift to ε_3 and ΔT was already obtained before the strain measurement. Then, the following relation was used for calculation of strain in this paper.

$$\varepsilon_3 = \left(\Delta\lambda/\lambda_0 - 6.9032\times10^{-6}\times\Delta T\right)/0.7368 + \alpha_s\Delta T \qquad (2)$$

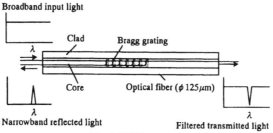

Fig.1 FBG sensors

Two types of sensing systems were used for strain measurement. One is the OSA system which is composed of a super luminescence diode (SLD), an optical spectrum analyzer (OSA; Anritsu MS9710) and an optical directional coupler. This system was used for acquisition of reflected spectrum from FBG and EFPI sensors. The measured spectra were sent to PC and then strain was computed by using equation (2). This system is especially useful for checking state of sensors by observation of the reflected spectral shape. Another system is the FBG-IS system (FBG-IS, Micron optics, Inc. and NTT-AT, Inc.). It reports only Bragg wavelengths reflected from FBG sensors.

SPECIMENS

Three types of FW GFRP pipes as illustrated in Fig.2 were manufactured. The Pipe A and B had 45°helical winding configurations and the Pipe C had 15° parallel + 45° helical winding configuration. Fiber optic sensors were embedded axially in the Pipe A and C but in 45° direction in the Pipe B. In this paper, notation of winding such as $[\pm 15P_2/90S/\pm 45H_4]$, where P and H denote respectively parallel and helical winding methods, S means fiber optic sensors, angles are values from radial direction and subscription is number of layers. Length of all specimens was 300 mm and the inner radius was 15 mm. Thicknesses of the Pipe A and B was 2 mm and that of the Pipe C was 1.5 mm. Fiber optic sensors were embedded at a position of about 150 mm from end of pipes and they were drawn from their ends.

E-glass fiber strands (Nippon Sheet Glass, Co. Ltd.) and epoxy resin (EPIKOTE 807, Yuka Shell Epoxy, Co. Ltd.) were used to fabricate FW GFRP pipes. Glass fibers were impregnated with resin, which was defoamed in a vacuum chamber for 10 minutes, were wound on an aluminum pipe. After the winding, thermocouple was attached to specimens for temperature measurement. Finally, Then, the specimens were cured in a furnace and then sensors were connected their respective data acquisition systems. The curing condition was composed of heating up to 100 °C for 40 minutes, holding for 120 minutes, heating up to 175 °C for 30 minutes, holding for 3 hours and natural cooling. After the manufacturing, their both ends were cut because reinforcing direction was random there. Then, loading jigs adhered on inner surface of pipes. In mechanical loading test, foil strain gages adhered on the surface of the specimens.

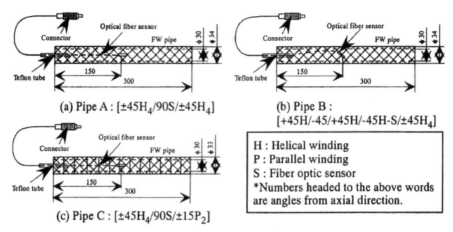

(a) Pipe A : [±45H₄/90S/±45H₄]

(b) Pipe B :
[+45H/-45/+45H/-45H-S/±45H₄]

H : Helical winding
P : Parallel winding
S : Fiber optic sensor
*Numbers headed to the above words
are angles from axial direction.

(c) Pipe C : [±45H₄/90S/±15P₂]

Fig.2 Constructions of specimens.

EXPERIMENTAL RESULTS AND DISCUSSIONS

Cure Monitoring

Internal strain of FW pipes was measured during curing process. Figure 3 shows histories of strain outputs obtained from FBG sensors embedded in FW pipes during cure. Strain outputs measured by an EFPI sensor for the Pipe A, the outputs of epoxy resin and temperature were also plotted in the same figure. In the figure, it can be seen that the FBG sensors detected both curing strain about from 3 to 7 hours and thermal strain during natural cooling. By comparing results of the Pipe A with the Pipe C, it appears that strain of the Pipe A was smaller than that of the Pipe C since difference of their CTEs (coefficient of thermal expansion) and stiffness. As for strain of the Pipe C, it is found that the strain was smaller than that of the Pipe A. Especially the value was much smaller with regard to curing strain since the sensor was embedded in reinforcing fiber direction.

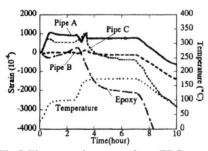

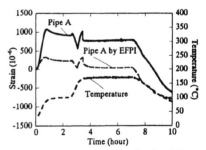

Fig.3 Time - strain curves from FBG sensors embedded in Pipe A, B, C, and epoxy resin during curing process. Temperature data are also plotted.

Fig.4 Strains of Pipe A obtained from FBG and EFPI sensors. Temperature data are also plotted.

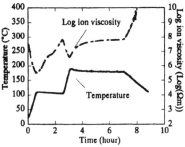

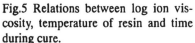

Fig.5 Relations between log ion viscosity, temperature of resin and time during cure.

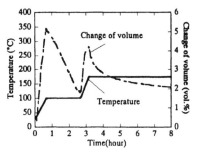

Fig.6 Histories of Change of volume of epoxy and temperature during cure.

From these results, it was found that embedding directions of sensors strongly affect their sensitivity to curing strain of base FW FRP. During cooling, thermal shrink corresponding to macro CTEs of respective FRP was measured.

Figure 4 shows results of an EFPI sensor embedded in the Pipe A with that of the FBG sensor. It is found that behavior of strain measured by FBG sensors was reliable since strain behavior of the EFPI sensor qualitatively agreed with that of the FBG sensor. However curing strain obtained from the EFPI sensor was a little smaller than that from the FBG sensor due to difference of effective stiffness of these sensors.

To discuss curing strain in detail, two kinds of curing properties of epoxy were measured. One is log ion viscosity and another is change of volume. Log ion viscosity was measured by dielectric measurement system (Eumetric 100A: Micromet Instruments, Inc.) and change of volume was measured by using a dilatometer. The results of log ion viscosity of resin during cure are plotted against time with temperature in Fig.5. The Figure shows that viscosity of resin decreased in heating and increased when temperature was static. On the other hand, relations between change of volume of resin, temperature and time are shown in Fig.6. In the figure, it can be seen that change of volume increased in heating and decreased when temperature was static.

It is apparent that increase of change of volume contains thermal expansion and curing shrink but the decreasing value indicates only curing shrink due to condition of static temperature. However from the figure 3, it is found that strain of fiber optic sensors decreased and then increased unless the temperature was increasing in the second heating process. By considering that strain behavior of epoxy was different from FBG sensor outputs and viscosity decreased in the second heating process, it can be concluded that the strain measured by embedded fiber optic sensors in FW FRP affected by winding force applied by winding of reinforcements. In the second temperature-constant process, strain behavior quantitatively agreed well with results of change of volume. Therefore, with regard to this resin, it is concluded that outputs of FBG sensors can be used for detection of curing shrink in the temperature-constant process.

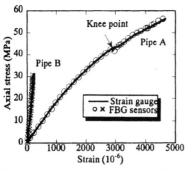

Fig.7 Stress-strain curves of the Pipe A and B by using both sensors in tensile tests.

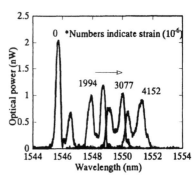

Fig.8 spectra obtained from FBG sensors of the Pipe A in tensile tests.

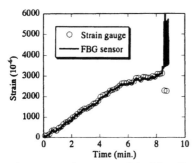

Fig.9 Time-strain curves of the Pipe C by using FBG sensors in tensile tests.

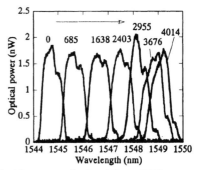

Fig.10 spectra obtained from FBG sensors of the Pipe C in tensile tests.

Health Monitoring in Tensile Tests

Internal strain measurement of FW pipes by using FBG sensors were conducted in tensile tests. Loads were applied under displacement control of 0.1mm/min. An universal loading test machine (Instron FastTrack8801, Instron, Inc.) was used for tensile tests. Internal and surface strains were measured by using FBG sensors and strain gauges, respectively. Figure 7 shows stress-strain curves of the Pipe A and B by using both sensors. From the figure, it appears that the results of FBG sensors agree well with those of strain gauges. Therefore, it can be concluded that embedded FBG sensors in FW FRP pipes have a very good function of internal strain measurement in spite of their embedding directions. In Fig.7, a clear knee point can be seen in the curve of the pipe A. At this knee point, initiation of cracks in reinforcing directions was observed. Figure 8 shows spectra obtained from FBG sensors of the Pipe A. It can be seen that cracks affected spectral shapes of the reflected light from the FBG sensor. However, since the transition is very minor, it can be concluded that it is difficult to detect cracks in ±45° helical winding FW pipes from FBG spectral shape. Only initiation of cracks can be detected from strain behavior due to high precision of FBG sensors.

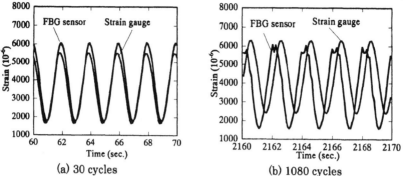

(a) 30 cycles   (b) 1080 cycles

Fig.11 Time-strain curves obtained from an FBG sensor and a strain gauge of the Pipe A in a fatigue test.

Figure 9 shows results of strain measurements of the Pipe C by using the FBG sensor and a strain gauge in tensile tests. From the figure, it appears that the FBG sensor and a strain gauge have a good agreement with each other up to 3000με. After the strain, both the sensors output errors due to initiation of transverse cracks in 15° layers. Figure 10 shows FBG spectra of the Pipe C. In this figure, it can be seen that the spectral shape shifted until the strain reached about 3000με and then the shape suddenly changed. Therefore, it can be concluded that transverse cracks occurs in parallel layers of FW FRP can be detected by monitoring FBG spectral shape.

Health Monitoring in a Fatigue Test

A fatigue test was conducted by using the same testing machine used in the tensile tests. In this tests, sinusoidal load, which had 0.5 Hz frequency and load from 2kN to 12 kN, was applied under load control. Figures 11 (a) and (b) show time-strain curves obtained from embedded FBG sensor and a strain gage at 30 and 1080 cycles, respectively. From the figure, it is found that outputs of FBG sensor became unstable at 1080 cycles. In this test, matrix cracks were generated about 1000 cycles. Then, it can be considered that this behavior resulted from deterioration of FBG spectrum caused by strain concentration around the FBG sensor. FBG sensor was broken at about 1300 cycles due to development of cracks. More detail study corresponding with quantitative damages is necessary to apply this technique for quantitative damage monitoring of FW pipes.

CONCLUSIONS

FBG sensors were applied to cure and health monitoring of FW GFRP pipes. Three configurations of winding and sensor-embedding were used in this study. From the results of cure monitoring, it was found that embedded FBG sensors can monitor curing shrink as well as thermal residual strain in spite of winding and sensor-embedding configurations of the specimens. It appeared that curing strain behavior of FW Pipes was different from that of resin in the second heating process due to winding force of reinforcing fibers. In tensile tests, FBG sensors show good performances of monitoring internal strain. Especially for parallel + helical winding specimens, it was shown that transverse cracks could be monitored by observing reflected FBG spectral shape. In a fatigue test, fatigue damages of a helical winding specimen could be monitored qualitatively. From the results obtained in this study, it can be concluded that embedded FBG sensor in FW FRP pipe have high functions of real-time cure and health monitoring. More quantitative observation of damages is necessary to enhance function of damage monitoring of FW FRP pipes with built-in FBG sensors.

REFERENCES

1.   T. Fukuda and T. Kosaka, "Cure and Health Monitoring, Encyclopedia of Smart Materials", Edited by Mel Schwartz, John Wiley & Sons, Inc., New York, in press, ISBN: 0-471-17780-6, 2002

2.   K. Osaka, T. Kosaka, Y. Asano and T. Fukuda, "Measurement of Internal Strains in FRP Laminate with EFPI Optical Fiber Sensor during Autoclave Molding; Measurement in Off-axis Directions", Proc. the 2nd Asian-Australasian Conf. on Composite Materials (ACCM-2000), Kyongju, Korea, 1117-1122, 2000

3.   T. Kosaka, K. Osaka, M. Sando and T. Fukuda, "Curing Strain Monitoring of FRP FW Pipe with EFPI Fiberoptic Sensors", Materials Science Research International- Special Technical Publication Volume 2, 100-104, 2001

4.   V.M. Murukeshan, P.Y. Chan, L.S. Ong, and L.K. Seah, "Cure Monitoring of Smart Composites Using Fiber Bragg Grating Based Embedded Sensors", Sensors and Actuators: A Phys. Volume 79, 153-161, 2000

5.   T. Kosaka, K. Osaka, Y. Asano and T. Fukuda, "Cure and Health Monitoring of RTM Molded FRP by Using Optical Fiber Strain Sensors", a book of extended abstracts of 5th International Conference on Durability Analysis of Composite Systems (DURACOSYS 2001), Tokyo, Japan, 138-142, 2001

6.   Y. Okabe, S. Yashiro, T. Kosaka and N. Takeda, "Detection of Transverse Cracks in CFRP Composites Using Embedded Fiber Bragg Grating Sensors", Smart Mater. Struct., Volume 9, 832-838, 2000

7.   X. Tao, L. Tang, W.C. Du and C.L. Choy, "Internal Strain Measurement by Fiber Bragg Grating Sensors in Textile Composites", Composites Sci. and Tech., Volume 60, 657-669, 2000

8.   Y.J. Rao, "Recent Progress in Applications of In-Fibre Bragg Grating Sensors", Opt. Lasers Eng., Volume 31, 297-324, 1999

NANOCOMPOSITES, NATURAL FIBER COMPOSITES

Effect of temperature on the preparation of epoxy nanocomposites with nanoclays

Minh-Tan Ton-That and Kenneth C. Cole
National Research Council Canada, Industrial Materials Institute
75 De Mortagne Blvd., Boucherville, Quebec, Canada J4B 6Y4
kenneth.cole@nrc.ca; minh-tan.ton-that@nrc.ca

Suong Van Hoa and Dafu Shen
Concordia Center for Composites
Department of Mechanical Engineering
Concordia University
Montreal, Quebec, Canada H3G 1M8
hoasuon@vax2.concordia.ca

ABSTRACT

Epoxy nanocomposites based on layered silicates have received much attention recently. Unlike thermoplastic-based nanocomposites, thermoset-based nanocomposites are subject to certain constraints in their preparation because of the high reactivity of the monomers or oligomers used in their formation. This paper concentrates on the effects of mixing temperature and time on the formation of epoxy-based nanocomposites. Different commercial organo-nanoclays were used at different concentrations, and both solvent-based and direct mixing processes were used to prepare the nanocomposites. The sample prepared with the use of solvent was taken as a reference for this study. Differential scanning calorimetry (DSC) and Fourier transform infrared (FT-IR) spectroscopy were used to characterize the mixtures obtained.

It was found that organo-clay treated with primary amine can play a role as a catalyst for the homopolymerization of the epoxy resin, whereas organo-clay treated with quaternary amine does not do so. It is important to take this into account in preparing the resin-clay mixtures. Thus the quaternary-amine-treated clay can be mixed with epoxy resin at temperatures up to 140°C without risk of polymerization. In the case of the primary-amine-treated clay, however, mixing at 140°C should be limited to less than one hour to avoid significant polymerization, although mixing at 100°C implies little risk.

Further on-going experiments will answer additional questions on the effect of mixing conditions on the exfoliation of the nanoclays and the performance of the resulting nanocomposites.

Effect of temperature on the preparation of epoxy nanocomposites with nanoclays

Minh-Tan Ton-That and Kenneth C. Cole
National Research Council Canada, Industrial Materials Institute
75 De Mortagne Blvd., Boucherville, Quebec, Canada J4B 6Y4
kenneth.cole@nrc.ca; minh-tan.ton-that@nrc.ca

Suong Van Hoa and Dafu Shen
Concordia Center for Composites
Department of Mechanical Engineering
Concordia University
Montreal, Quebec, Canada H3G 1M8
hoasuon@vax2.concordia.ca

ABSTRACT

Epoxy nanocomposites based on layered silicates have received much attention recently. Unlike thermoplastic-based nanocomposites, thermoset-based nanocomposites are subject to certain constraints in their preparation because of the high reactivity of the monomers or oligomers used in their formation. This paper concentrates on the effects of mixing temperature and time on the formation of epoxy-based nanocomposites. Different commercial organo-nanoclays were used at different concentrations, and both solvent-based and direct mixing processes were used to prepare the nanocomposites. The sample prepared with the use of solvent was taken as a reference for this study. Differential scanning calorimetry (DSC) and Fourier transform infrared (FT-IR) spectroscopy were used to characterize the mixtures obtained.

It was found that organo-clay treated with primary amine can play a role as a catalyst for the homopolymerization of the epoxy resin, whereas organo-clay treated with quaternary amine does not do so. It is important to take this into account in preparing the resin-clay mixtures. Thus the quaternary-amine-treated clay can be mixed with epoxy resin at temperatures up to 140°C without risk of polymerization. In the case of the primary-amine-treated clay, however, mixing at 140°C should be limited to less than one hour to avoid significant polymerization, although mixing at 100°C implies little risk.

Further on-going experiments will answer additional questions on the effect of mixing conditions on the exfoliation of the nanoclays and the performance of the resulting nanocomposites.

EXPERIMENTAL

Materials

The two most widely recommended organo-nanoclays for epoxy resins were chosen for study; they differ in the chemistry of the onium ion. The first one is Cloisite 30B, which was kindly supplied by Southern Clay Products, Inc., and the second one is Nanomer I30E as obtained from Nanocor, Inc. Cloisite 30B consists of montmorillonite treated with a methyl bis-2-hydroxyethyl tallow quaternary amine, while Nanomer I30E consists of montmorillonite treated with a primary amine based on the octadecyl group.

The resin and hardener selected for this study were Shell Epon 828 and Shell Epicure 3046, respectively. This is one of the most popular systems for a low curing temperature process, recommended for production of a large range of semi-rigid products including coatings, adhesives, and composites.

Preparation of the Nanocomposites

The nanoclay content in the epoxy was varied between 1 and 4 wt%. Two different processes were used to prepare the nanocomposites. In the solvent process, the nanoclay was dispersed in methyl ethyl ketone (at a concentration of 2 wt%) at 50°C for 2 h, then the mixture was subjected to ultrasonification at the same temperature for 30 min, then it was cooled to room temperature, and finally the epoxy resin was introduced over a period of 30 min. The product was dried at room temperature in a vacuum oven to remove solvent and stored at −10°C. Low temperatures were used for the later steps in order to avoid chemical reaction of the epoxy group with the organo-nanoclays. In the direct mixing process, the nanoclay was dispersed directly into the epoxy matrix by means of mixing at various temperatures between 80 and 140°C for different periods of time ranging between 0.5 and 4 h. Table 1 describes the samples prepared.

Sample Characterization

The samples were analyzed by Fourier transform infrared (FT-IR) spectroscopy and differential scanning calorimetry (DSC). The FT-IR spectra were acquired on a Nicolet Magna 860 instrument with a resolution of 4 cm^{-1} by means of the attenuated total reflection (ATR) technique. The DSC curves were measured on a Perkin-Elmer DSC7 instrument under nitrogen atmosphere. For non-isothermal scans, the sample was heated from 0°C to 250°C at different heating rates between 2 and 20°C·min^{-1} ("first scan") to follow the chemical reaction happening in this temperature range. Then it was cooled to 20°C at 20°C·min^{-1} to minimize the relaxation in the second heating scan. Finally it was reheated to 220°C at 20°C·min^{-1} ("second scan") in order to determine the glass transition temperature (T_g) and the post-curing reaction if any. The onset temperature of the curing peak in the first scan was used to choose the curing temperature for the study of curing under isothermal conditions. In the isothermal scans, at first the sample was heated to the designated temperature at a heating rate of 20°C·min^{-1} and then it was kept at this temperature for 30 min. Finally the sample was cooled and reheated as in the non-isothermal scan.

Table 1 – Description of Samples

Sample	Process	Mixing temp. (°C)	Mixing time (min)	Clay type	%Clay (wt%)
Sol-30B-3	Solvent			30B	3
Sol-I30E-1	Solvent			I30E	1
Sol-I30E-2	Solvent			I30E	2
Sol-I30E-3	Solvent			I30E	3
Sol-I30E-4	Solvent			I30E	4
30B-4-100-30	Direct	100	30	30B	4
30B-4-100-60	Direct	100	60	30B	4
30B-4-100-240	Direct	100	240	30B	4
30B-4-140-30	Direct	140	30	30B	4
30B-4-140-60	Direct	140	60	30B	4
30B-4-140-240	Direct	140	240	30B	4
I30E-4-100-30	Direct	100	30	I30E	4
I30E-4-100-60	Direct	100	60	I30E	4
I30E-4-100-240	Direct	100	240	I30E	4
I30E-4-140-30	Direct	140	30	I30E	4
I30E-4-140-60	Direct	140	60	I30E	4
I30E-4-140-150	Direct	140	150	I30E	4

RESULTS AND DISCUSSION

The first DSC scans for samples Sol-30B-3 and Sol-I30E-3, done at a heating rate of $10°C \cdot min^{-1}$, are given in Figure 1a. For Sol-30B-3 no exothermic or endothermic peak was observed up to 250°C, but for Sol-I30E-3 a strong exothermic peak (543 $J \cdot g^{-1}$) appeared in the range between 170 and 230°C. Thus, no chemical reaction occurred in the Sol-30B-3, while the exothermic peak for Sol-I30E-3 suggests that a curing reaction took place during the first scan. From this graph, T_{onset}, the temperature at which the chemical/physical phenomenon became significant, was found to be about 216°C for Sol-I30E-3. In order to verify the effect of the heating process on the physical properties of the sample, the second scan was done. The results showed no change for Sol-30B-3, but for Sol-I30E-3 a new glass transition step ($T_g = 102°C$) appeared as shown in Figure 1b. The appearance of the transition peak proves that the epoxy liquid has been polymerized to transform to the solid state; the final product was in the form of a brown powder. This can be explained by the homopolymerization of epoxy groups initiated by the onium ion of the organo-nanoclay I30E. Based on the chemistry of the onium ions in the two clays, it can be concluded that the primary amine used in I30E can play a catalyst role for the homopolymerization of epoxy, while the quaternary amine used in 30B is inactive. Homopolymerization of epoxy by organo-nanoclay has also been reported by Lan et al. [2], who observed that the extent decreased on going from primary to quaternary amine.

A difference in this study, however, is that polymerization was completely absent for the quaternary amine in 30B, even at the low heating rate of 2°C·min⁻¹.

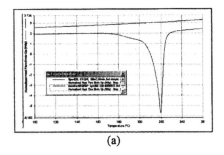
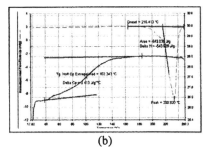

(a)	(b)

Figure 1 – DSC curves: (a) the first scans of the samples at 10°C·min⁻¹ (upper = Sol-30B-3, lower = Sol-I30E-3); (b) the first and second scans of Sol-I30E-3, at 20°C·min⁻¹.

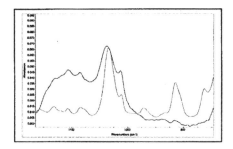

Figure 2 – FT-IR spectra of the Sol-I30E-3 sample before and after the first DSC scan.

In order to verify the nature of the reaction, the sample Sol-I30E-3 was analyzed by FT-IR. Figure 2 compares the spectra of the uncured material and the product from the first DSC scan. The most significant change is the disappearance of the epoxide ring band at 915 cm⁻¹ upon curing. This is accompanied by the growth of a broad band around 1170–960 cm⁻¹ that can be assigned to stretching of the C–O–C ether linkages formed by reaction of the epoxide rings. The spectrum thus indicates that the exothermic peak in the DSC curve of the first scan can be assigned to the homopolymerization of the epoxy groups initiated by the onium ion of the organo-nanoclay. It was found that the commercial clay I30E contains some unbound onium ions. This raises the question as to whether the initiation of the reaction is due to the onium ion bound on the nanoclay surface or to the unbound onium ion. In order to check this, the clay I30E was purified by washing it several times with hot deionized water. The purified product behaved in a similar manner to the unpurified one, indicating that the bound onium ion also plays a catalyst role for the homopolymerization of the epoxy.

Since the Cloisite 30B does not have any effect on the polymerization of epoxy, further experiments focused only on the Nanomer I30E. The effect of the nanoclay concentration on the homopolymerization is shown in Figure 3. Multiple overlapping exothermic peaks were observed for the first-scan DSC curves for all the concentrations studied. Lan et al. also observed two separate peaks in the DSC curves [2]. Because the first peak significantly increased as a function of clay concentration, while the second greatly decreased, they assigned the first peak to polymerization inside the clay galleries and the second peak to reaction outside the galleries. However, this is in contradiction with the observations in this work, namely that there are more than two peaks in the DSC curve and the first peak (maximum around 180°C) becomes less obvious with increasing nanoclay concentration, even at low (2°C·min⁻¹) and high (20°C·min⁻¹) heating rates. Curve fitting techniques were used to separate the peaks and it was found that the onset of the first peak remains unchanged for all three clay concentrations at a given heating rate. The explanation for the presence of the multiple peaks may be the contributions of different chemical reactions in the system, including the reaction of the onium ion to form intermediate products followed by reaction of these intermediate products with epoxy and then with hydroxyl groups, etc. In addition, the reaction can also be affected by diffusion control, both in the early stages when the epoxy is diffusing into the clay galleries and in the later stage after gelation. However, because of the complexity of the reaction process, further investigations must be carried out to obtain a better understanding.

Figure 3a also shows that the increase of nanoclay concentration shifted the main peak of the curve to the left and made it narrower, indicating that the reaction occurs more easily and at a faster rate. This shows a positive effect of the clay concentration on the polymerization of epoxy, which is in contradiction with any expectations based on a hindering effect of the nanoclays. The slopes of the curves in Figure 3b show that the polymerization reaction rate increases significantly with the clay concentration.

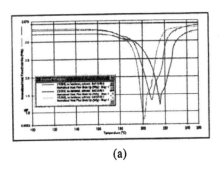   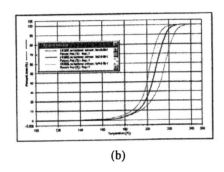

(a)   (b)

Figure 3 – (a) DSC curves of the first scan (5°C·min⁻¹) and (b) transformation curves of the Sol-I30E-4 (left), Sol-I30E-2 (middle) and Sol-I30E-1 (right).

In all cases, no exothermic peak was observed in the second DSC scan, indicating that the polymerization was completed in the first scan. In addition, the glass transition step was clearly apparent for each curve. For all samples, similar results were obtained at different heating rates between 2 and 10°C·min⁻¹, except for the first peak of the first scan. The T_{onset}, the total reaction enthalpy (ΔH), and the T_g values of the cured samples are shown for two different heating rates in Table 2. The T_{onset} for both peak 1 and peak 2 are independent of clay concentration but vary with heating rate, while the ΔH and T_g are independent of both the clay concentration and the heating rate.

Table 2 – DSC data for different samples

Sample	Heating rate of 2°C·min⁻¹				Heating rate of 5°C·min⁻¹			
	1ˢᵗ scan			2ⁿᵈ scan	1ˢᵗ scan			2ⁿᵈ scan
	T_{onset} (1ˢᵗ peak)	T_{onset} (2ⁿᵈ peak)	ΔH ($J \cdot g^{-1}$)	T_g (°C)	T_{onset} (1ˢᵗ peak)	T_{onset} (2ⁿᵈ peak)	ΔH ($J \cdot g^{-1}$)	T_g (°C)
Sol-I30E-1	159	194	543	104	168	206	570	103
Sol-I30E-2	158	189	556	105	167	196	561	104
Sol-I30E-4	160	187	536	103	168	193	558	102

Since the T_{onset} of peak 1 is above 160°C, for the isothermal scan the temperature was fixed slightly higher than this value, at 170°C. Figure 4a shows that polymerization starts off quite fast at this temperature but slows down dramatically after 15 min. The enthalpy of reaction during the isothermal cure was only 30 $J \cdot g^{-1}$, which is less than 7% of the total expected enthalpy. When the cooled sample was rescanned dynamically at 20°C·min⁻¹, the polymerization reaction resumed above 180°C (Figure 4b). Even though the reaction was not complete at the upper limit of 250°C, the enthalpy of the peak in the second scan is greater than 350 $J \cdot g^{-1}$. From this observation, it seems that the polymerization involves at least two separate steps. Further experiments are required to obtain a better understanding of this reaction.

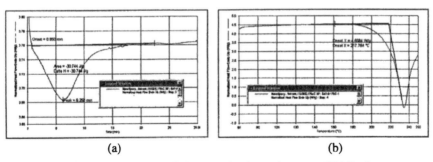

(a)   (b)

Figure 4 – DSC curves for Sol-I30E-3: (a) isothermal scan at 170°C; (b) subsequent dynamic scan to 250°C.

Samples were also prepared by a direct mixing procedure, without solvent, at different temperatures between 100 and 140°C. With the Cloisite 30B, all the mixtures with epoxy showed no significant increase in viscosity even at high mixing temperatures and long mixing times (up to 4 h). In the case of the Nanomer I30E, however, the situation was different. Although there was no noticeable increase in the viscosity for samples mixed at 100°C, substantial changes were observed at 140°C (Figure 5). The increase in viscosity is due to polymerization of the epoxy groups. Figure 6 shows that the DSC curves of I30E samples prepared at 100°C remained unchanged up to 240 min mixing time, while the exothermic peak of the mixtures prepared at 140°C showed a dramatic decrease after only 150 min. By comparison of the area of the exothermic curing peak with that of a sample mixed very briefly, the degree of polymerization achieved during mixing can be easily evaluated, and the results are given in Figure 5. An excellent correlation between the increase in viscosity and the level of polymerization is observed. In the case of the 30B samples, no significant changes were seen in any of the DSC curves, indicating that no polymerization took place.

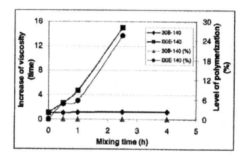

Figure 5 – Relative viscosity and level of polymerization for mixtures of clays 30B and I30E with epoxy as a function of mixing time at 140°C.

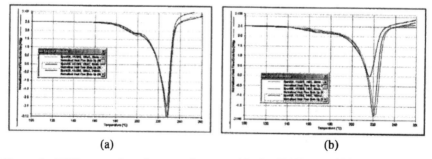

(a)   (b)

Figure 6 – DSC curves for mixtures of epoxy with clay I30E at (a) 100°C, mixing times of 30, 60, 240 min, and (b) 140°C, mixing times of 30, 60, 150 min.

CONCLUSIONS

It can be concluded that:

(1) Organo-clay treated with primary amine (Nanomer I30E) can play a role as a catalyst for the homopolymerization of the epoxy resin, whereas organo-clay treated with quaternary amine (Cloisite 30B) does not do so.

(2) With Cloisite 30B, mixing can be done up to 140°C with negligible risk of polymerization. With Nanomer I30E, mixing at higher temperatures must be carefully controlled; at 140°C the limit is one hour.

(3) Since organo-nanoclay treated with primary amine can promote the polymerization of epoxy groups, it is important to take this reaction into account in developing products and processes designed for high-temperature curing and involving mixtures of epoxy resin, nanoclay, and various types of hardeners. Further on-going experiments will answer additional questions concerning the effect of mixing conditions on the exfoliation of the nanoclays and the performance of the resulting nanocomposites.

REFERENCES

1. P. B. Messersmith and E. P. Giannelis, Chem. Mater., 1994, 6, 1719-1725.
2. T. Lan, P. D. Kaviratana, and T. J. Pinnavaia, J. Phys. Chem. Solids, 1996, 57, 1005-1010.
3. J. M. Brown, D. Curliss, and R. A. Vaia, Chem. Mater., 2000, 12, 3376-3384.
4. Lü Jiankun, Ke Yucai, Qi Zongneng, and Yi Xiao-Su, J. Polym. Sci., Polym. Phys., 2001, 39, 115-120.
5. A. S. Zerda and A. J. Lesser, J. Polym. Sci., Polym. Phys., 2001, 39, 1137-1146.
6. P. Butzloff, N. A. D'Souza, T. D. Golden, and D. Garrett, Polym. Eng. Sci., 2001, 41, 1794.
7. X. Kornmann, H. Lindberg, and L. A. Berglund, Polymer, 2001, 42, 4493-4499.
8. X. Kornmann, H. Lindberg, and L. A. Berglund, Polymer, 2001, 42, 1303-1310.
9. I.-J. Chin, T. Thurn-Albrecht, H.-C. Kim, T. P. Russell, and J.Wang, Polymer, 2001, 42, 5947-5952.

Microstructure and Performance of Injection Molded Polyamide 6 /Clay Nanocomposite and its Glass Fiber Composite

Johanne Denault, Martin N. Bureau, Kenneth C. Cole and Minh-Tan Ton-That
National Research Council Canada, Industrial Materials Institute
75 De Mortagne Blvd., Boucherville, Quebec, Canada J4B 6Y4
Johanne.denault@nrc.ca; martin.bureau@nrc.ca; kenneth.cole@nrc.ca; minh-tan.ton-that@nrc.ca

ABSTRACT

Polymer nanocomposites (PNC) are emerging as a new class of industrially important materials that offer improvements over conventional composite systems. The high aspect ratio of the nano particles leads to higher reinforcement efficiency and therefore higher specific modulus, strength and HDT. The low organoclay content (< 5% by weight) also guarantees good processabilty as well as higher recyclability potential. In this work, the crystalline structure, mechanical behavior and the injection-molding of polyamide 6 (PA6) reinforced with 2 wt.% polymer-intercalated nano-layered silicate (montmorillonite) is studied. X-ray diffraction and differential scanning calorimetry show that the presence of layered silicate in PA6 nanocomposites lead to the formation of a different crystalline structure when compared with unmodified PA6. Considerably improved tensile strength and modulus obtained from the nano-layered silicates while maintaining the ductility of the PA6 matrix indicate that strong matrix-filler ionic interactions and very high specific area and aspect ratio of the polymer intercalated-layered silicates characterized this nanocomposite. In the conditioned state, while an elasto-plastic fracture with extensive tearing is observed in unmodified PA6, a linear-elastic fracture is observed in the PA6 nanocomposite.

INTRODUCTION

Thermoplastic polymer nanocomposites have recently been the subject of several research studies and reviews (1-5). These emerging materials consist of filled polymers in which at least one dimension of the dispersed phase is in the nanometer range. Among these, polymer-layered crystal nanocomposites, generally based on nano-layered silicate fillers, in which the polymer matrix is intercalated to achieve proper exfoliation, have attracted much interest due to the availability of the layered silicates (montmorillonite, hectorite, saponite, etc.) and to their well-known intercalation chemistry (1,5).

More specifically, PA6/layered silicate nanocomposites received most attention due to their high expected level of matrix-filler interaction (ionic bonds with the silicate layers, ref. 1). Owing to their very high specific area and aspect ratio, additions in PA6 of low levels of these nano-layered silicates lead to significantly improved mechanical, thermal and optical properties when compared to the unmodified base polymer. However, much work is needed to fully understand the benefits and limitations of this new class of materials.

The objective of this contribution is to report on the present state of comprehension of the crystalline microstructure and mechanical behavior of injection-molded PA6/layered silicate nanocomposites. Potential of use of nanocomposite matrix to the development of improved fiber reinforced polyamide is also discussed.

EXPERIMENTAL

Two materials supplied by UBE Industries were studied: PA6 nanocomposite (UBE's 1015C2), a PA6 matrix modified with 2 wt.% of layered silicates (montmorillonite), and for comparison purposes the unmodified PA6 (UBE's 1015B) used as the matrix in the nanocomposite. The as-received materials were injection-molded into dog-bone and plate specimens. The melt temperature was fixed at 255°C and three mold temperatures of 50, 65 and 80°C were used to investigate the effect of the cooling rate on the development of the crystalline structure.

X-ray diffraction (XRD) patterns were obtained from the skin (0-200 μm in depth) of the injection-molded specimens using a Scintag X2 Powder X-ray Diffractometer with CuKα radiation. The thermal properties were analyzed using a Perkin-Elmer differential scanning calorimeter (DSC-7). Thermograms between 30 to 250°C at a heating rate of 20°C/min were obtained. Sampling from the skin (0-200 μm in depth), sub-skin (700-900 μm in depth) and core (1400-1600 μm in depth) of the injection-molded specimens was done using a Leitz-Wetzlar microtome.

Tensile testing was done according to ASTM D638 at 23°C using a computer-controlled Instron mechanical tester with a longitudinal extensometer. A crosshead speed of 25 mm/min was used. Injection-molded Type I dog-bone specimens with a thickness of 3.2 mm were tested. A minimum of five tests was performed for each reported value and the standard deviation was less or equal to 5%, with the exception of the strain at break showing a standard deviation of 10-15%.

Fracture toughness measurements in mode I were obtained from compact tension (CT) specimens according to ASTM E1820 using a computer-controlled mechanical tester. A crosshead speed of 5 mm/min was used. The CT specimens with a nominal width of 50 mm and a notch depth of 22.5 mm were machined from 5-mm thick injection-molded plates. A razor blade cut was introduced manually at the notch tip just prior to fracture testing. The CT specimens were loaded parallel to the melt-flow direction during injection molding. When the material tested exhibits a linear-elastic behavior characterized by a sudden fracture (plane strain conditions), the fracture toughness reported is expressed in terms of critical mode I stress intensity factor, K_{Ic}, given in equation 1:

$$K_{Ic} = \frac{P_c}{B\sqrt{W}} f(a/W) \qquad (1)$$

where P_c is the critical load at crack propagation, B is the specimen thickness, W is the specimen width, a is the crack length, $f(a/W)$ is a empirical function of the normalized crack length based on the specimen geometry (cf. ASTM E1820).

Scanning electron microscopic (SEM) observations of fracture surfaces were obtained from gold-palladium coated specimens using a JEOL JSM-6100 microscope.

RESULTS AND DISCUSSION

The XRD patterns obtained from the skin of the injection-molded specimens show that the PA6 nanocomposite differs significantly from the unmodified PA6 in terms of crystallinity (Fig. 1). The unmodified PA6 shows a broad diffraction peak associated to the α- and γ-phases of PA6 (6). On the contrary, a sharp and narrow diffraction peak associated to the γ-phase (6) is observed for the PA6 nanocomposite, with traces of the α phase also visible. The XRD patterns also show that when the mold temperature is lowered from 80°C to 65°C and to 50°C, the sharp diffraction peaks observed in the nanocomposite remains virtually unchanged, whereas it becomes progressively narrower in the unmodified PA6. These results indicate that a different crystalline structure is obtained in the presence of polymer-intercalated layered silicate in PA6.

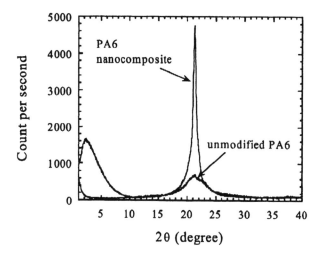

Figure 1 – XRD patterns of the unmodified PA6 and of the PA6 nanocomposite (mold temperature of 80°C).

The DSC thermograms of samples obtained from the skin, sub-skin and core of the injection-molded PA6 nanocomposites are presented in Figure 2. These thermograms confirm that the microstructure of these specimens varied across their thickness. Considering that silent crystallization of the PA6 matrix occurs during the heating scan, leading to the formation of the α-phase (7), the thermograms in Figure 2 indicate that the crystallinity and the amount of γ-phase present in the PA6 nanocomposite are higher in the samples taken from the core of the injection-molded specimens. In the unmodified PA6, this effect is considerably more important and the presence of the γ-phase becomes significant only in the core region of the injection-molded specimens.

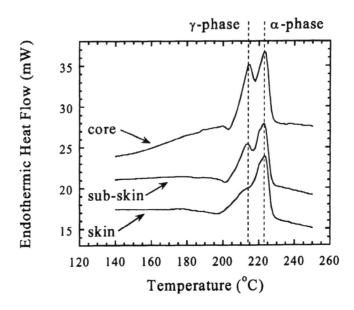

Figure 2 – Thermograms obtained from the skin, sub-skin and core of the injection-molded PA6 nanocomposites (mold temperature of 80°C).

The DSC and XRD results thus indicate that, in comparison to the unmodified PA6, the crystallinity of the injection-molded PA6 nanocomposite specimens is less sensitive to the cooling rate employed. They also indicate that the presence of the polymer-intercalated layered silicate favors the formation of the γ-phase in PA6. A change in the crystallization mechanisms in the PA6 nanocomposite is thus expected.

Tensile results obtained from the unmodified PA6 and the PA6 nanocomposite in the dried and conditioned state for different mold temperatures are shown in Table 1. An example of the tensile curves is shown in Figure 3. These results indicate that the humidity levels in both the unmodified PA6 and the PA6 nanocomposite significantly affected their mechanical behavior, changing from a ductile behavior in the conditioned state to a semi-ductile behavior in the dried state. The tensile results also indicate that the use of 2 wt.% of layered silicate as a filler in PA6 leads to important improvements in strength and modulus, as a result of the high level of matrix-filler interaction (ionic interactions) and very high specific area and aspect ratio of the polymer intercalated-layered silicates. In the dried state, the Young's modulus and the ultimate tensile strength are respectively 40% and 25% higher in the PA6 nanocomposite than in the unmodified PA6. In the conditioned state, the Young's modulus and the ultimate tensile strength are respectively 125% and 40% higher in the nanocomposite, while the elongation at break remains almost unaffected.

Table 1. Tensile Properties of the unmodified PA6 and PA6 nanocomposite (dried and conditioned; cd).

Materials	T_{mold} (°C)	Young's Modulus (GPa)		Ultimate Tensile Strength (MPa)		Stain at Ultimate Tensile Strength (%)		Strain at Break (%)	
		dried	cd	dried	cd	dried	cd	dried	cd
PA6	50	3.0	0.8	85	37	5	25	15	700
	65	2.9	0.8	80	38	5	35	60	700
	80	2.7	0.8	74	39	5	40	175	700
PA6 nano	50	3.9	1.7	99	50	4	20	< 10	600
	65	3.9	1.8	99	53	4	20	< 10	600
	80	4.1	1.8	100	52	4	20	< 10	600

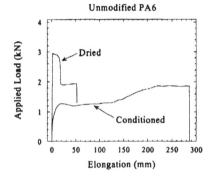
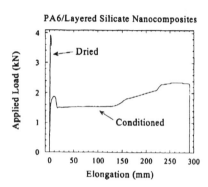

Figure-3 Typical tensile curves of the unmodified PA6 and PA6 nanocomposite in the dried and conditioned state (mold temperature of 80°C).

Finally, the tensile results of the unmodified PA6 are significantly affected by the mold temperature, whereas no significant effects are noted in the PA6 nanocomposites. This latter observation is in agreement with the XRD and DSC results showing that, in comparison to the unmodified PA6, the crystallinity of the injection-molded PA6 nanocomposite specimens is less sensitive to the cooling rate employed and that it is in part associated to the presence of the polymer-intercalated layered silicates.

The fracture toughness results obtained from the unmodified PA6 and PA6 nanocomposite for a mold temperature of 80°C are shown in Table 2. An example of the fracture curves (load vs. displacement) is shown in Figure 4 for both materials tested in the dried and conditioned state. These results reveal the well-known effect of humidity on the fracture behavior of unmodified PA6, changing from linear-elastic with shear yielding in the dried state, to elasto-plastic with generalized tearing in the conditioned state. In the latter, the fracture toughness, in terms of critical energy J_c, increases by 150% (from 8,100 J/m^2 in the dried state to 20,400 J/m^2 in the conditioned state). This large increase is attributed to the extensive tearing corresponding to the area under the curve, A_{pl}, shown in Figure 4.

Table 2. Fracture toughness values of the unmodified PA6 and PA6
nanocomposite in the dried and conditioned state.

Materials	K_{Ic} (MPa√m)		J_c (J/m²)	
	Dried	conditioned	dried	conditioned
Unmodified PA6	4.9 ± 0.1	---*	8,100 ± 400	20,400 ± 250
PA6 nanocomposite	1.8 ± 0.2	3.0 ± 0.2	750 ± 180	4,500 ± 570

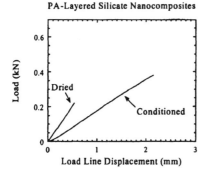

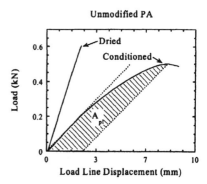

Figure 4 – Fracture toughness testing curves of the unmodified PA6 and PA6
nanocomposite in the dried and conditioned state (mold temperature of 80°C).

However, the fracture results in Table 2 and Figure 4 indicate a different effect of
the humidity content in the PA6 nanocomposite (at the same equilibrium humidity level
than in the unmodified PA6). In both the dried and conditioned nanocomposites, the
fracture behavior is linear-elastic, despite the large increase in fracture toughness J_c of
500% (from 750 J/m² in the dried state to 4,500 J/m² in the conditioned state). Thus, in
the conditioned state, a completely different fracture behavior is observed in the PA6
nanocomposite, when compared to conditioned unmodified PA6, changing from elasto-
plastic in the latter to linear-elastic in the nanocomposite.

The fracture surface of the PA6 nanocomposite in the dried and conditioned state
is shown in Figure 5. In the dried state (Fig. 5a), a large subcritical crack growth zone,
followed by a transition zone and a final fracture zone is observed. Crazing, with
characteristic voids separated by remnants of broken ligaments visible at higher
magnification (Fig. 6a), is observed in the subcritical zone. The sudden final fracture
occurs at craze instability. However, half-dimples in the final fracture zone (Fig. 6b), less
than 1 µm in width, indicate that fracture occurred by shear yielding at very high speed.

In the latter, a very large number of ridge lines, indicating multiple simultaneously propagating cracks, are observed (Figs. 5a and 6b). In the conditioned state (Fig. 5b), a near 45°-oriented crack plane is at first observed, progressively changing to normal after 300-500 μm of propagation. At higher magnification, similar features than those observed in the died state (Fig. 6b) indicate that fracture also occurred by shear yielding in the conditioned state. Numerous ridge lines, indicating multiple simultaneously propagating cracks, are also observed in the latter. Crack propagation thus occurs in PA6 nanocomposite by shear yielding.

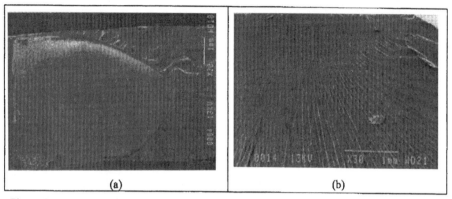

(a)   (b)

Figure 5 – Fracture surface observations at low magnification of the PA6 nanocomposite: a) in the dried state with crack propagation from left to right; b) in the conditioned state with crack propagation from top to bottom. Numerous ridge lines are observed in final fracture zone in a) and b).

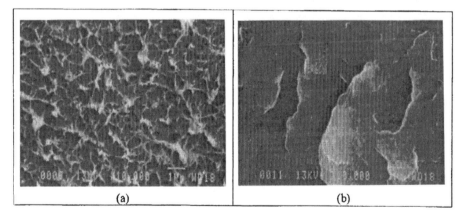

(a)   (b)

Figure 6 – Fracture surface observations at high magnification of the PA6 nanocomposite in the dried state: a) in the craze region and b) in the shear yielding region. Macroscopic crack propagation is from top to bottom..

Short glass fiber composites with of 7, 14, 20 & 27 % by weight glass fiber content were also prepared using both pure PA-6 and 2% organoclay/PA-6 PNC. The mechanical study of these materials shows that the modulus of injection molded glass fiber/PA-6 based composites can be improved by the addition of nanoclays to the matrix. The results also reveal that the latter modification of the PA-6 matrix lead to a complete change in the matrix crystalline form, changing from the α form for the unmodified PA-6 to the γ form for PNC PA-6 (Fig.7). The results, however, showed that an increase in tensile strength could not be obtained by the addition of nanoclays to PA-6 in the composites.

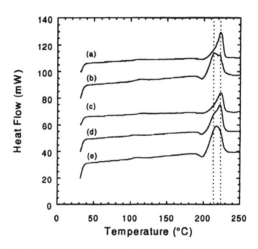

Figure 7 – DSC melting thermograms of injection molded samples of (a) PA-6, (b) PNC PA-6, (c) glass fiber PA-6, (d) glass fiber PA-6-PNC PA-6, (e) glass fiber PNC PA-6.

CONCLUSIONS

From the results obtained, it can be concluded that the presence of 2 wt.% of matrix-intercalated layered silicate (montmorillonite) used as a filler in PA6 leads to a modification of the PA6 crystalline structure generally obtained, favoring the formation of the γ crystalline phase. The higher values of Young's modulus and strength obtained in the nanocomposite are attributed to its different crystallinity, high level of matrix-filler interaction (ionic interactions) and to the very high specific area and aspect ratio of the polymer intercalated-layered silicates. While the unmodified PA6 exhibits a change from a linear-elastic fracture in the dried state to a elasto-plastic fracture in the conditioned state, the nanocomposite shows a linear-elastic fracture in both dried and conditioned state. In the PA6 nanocomposite, subcritical crack growth associated to crazing followed by shear yielding is observed in the dried state, whereas shear yielding only is observed in the conditioned state. Crack propagation thus occurs by shear yielding in the PA6 nanocomposite.

Results of this study also showed that the modulus of injection molded glass fiber/PA-6 based composites can be improved by the addition of nanoclays to the matrix. However, an increase in tensile strength could not be obtained by the addition of nanoclays to PA-6 in the composites. Further studies are required to obtain a better understanding of the behavior of this new class of materials.

REFERENCES

1. Kojima, Y., Usuki, A., Kawasumi, M., Okada, A., Fukushima, Y., Karauchi, T. and O. Kamigaito, *J. Mater. Res.*, 6 1185-1189 (1993)

2. Akkapeddi, M. K., *Polymer Composites '99*, SPE Topical Conference, October 6-8 1999, 169-182 (1999)

3. A Gianellis, E. P., *Adv. Mater.*, 8 29-35 (1996)

4. Alexandre, M. and P. Dubois, *Mater. Sci. Eng. Reports*, 28 1-63 (2000).

5. Ogawa, M. and K. Kuroda, *Bull. Chem. Soc. Jpn.*, 70 2593-2618 (1997).

6. Sibila, J. P, Murthy, N. S., Gabriel, M. K., McDonnell, M. E., Bray, R. G. and S. A. Curran, *Nylon Plastic Handbook*, Chap. 4, M. I. Kohan (editor), Hanser/Gardner Publ., 69-106 (1995).

7. Khanna, Y. P., and W. P. Kuhn, *J. Polym. Sci.: Part B: Polym. Phys.*, 35 2219-2231 (1997).

Development of Lumber Substitutive Materials from Fiber Wastes

T.KIMURA and Y.NAITOU
Advanced Fibro-Science
Kyoto Institute of Technology
Matugasaki,Sakyo-ku,Kyoto
606-8585,Japan
tkimura@ipc.kit.ac.jp

S.HATTA
Kyoto Municipal Textil Research Institute
Kamidacyuri-agaru,karasuma-dori
Kamigyo-ku,Kyoto
601-0898,Japan
hatta@city.kyoto.jp

K.KADOKURA
Kadokura Trading Company CO,LTD
Kaigan-dori
Chuo-ku,Kobe
650-0024,Japan
kenzo@kadoco.co.jp

ABSTRACT

The compression moldig of lumber substitutive materials was carried out by using fiber wastes(Rag). The felt with polypropylene(PP) binder was pre-molded by using the shoddy of various Rag. The substitute lumbers with various density and also with various volume fraction of PP were molded by compression method. The mechanical properties such as bending strength, holding strength of screw and pulling resistance of nail increased with increasing density of molded lumbers. The cutting ability of saw was also examined for the molded lumber,and the optimum density and PP contents were cleared. The results suggest that the compression molding method described here shows promise in contributing toward the material recycle of fiber wastes(Rag) as substitute lumber.

INTRODUCTION

To construct a recycling-type society, various attempts have been carried out in our textile industry. We are required to cope with the used fibers that accounts for two percent of urban and industrial wastes. It is estimated that the used fibers(rag and fiber wastes) are released two million tons annually in Japan. Although two hundred thousand tons thereof have been recycled or reused as the used clothes,the industrial wiping cloths,the shoddy and the felt,almost forty percent of recovered waste fibers are destroyed by fire and buried underground, due to the hollowing-out of Japanese industry by rising yen or competition from developing countries. These conditions have raised the concerns about the necessity of finding innovative usage of used fiber(rag and fiber wastes) and new recycling technologies.

Therefore,the compression molding of porous material such as substitute lumber was performed by using waste fibers (rag) in this paper. In order to mold the lumber,the thermoplastic fiber was used for the binder material. Especially the attention was focused upon the molding of packinglumber here.

EXPERIMRNTAL METHOD

Compression Molding method of Substitute Lumber

The compression molding of substitute lumber was carried out by using the felt with PP fibers as a binder material. The felt was made from rag. As a raw material of felt, the rag was separated in two groups of cellulose and wool. The contents of PP fiber as a binder material were 30% and 50 %. Table 1 shows the raw materials and their contents for 4 types of felt used in the present work.

Table1 The materials and there components

The wool group(1.3kg/m^3)			The cellulose group(1.3kg/m^3)		
	Content(%)			Content(%)	
PP	28~35	42~55	PP	27~33	48~53
wool	48~53	30~38	cellulose fibers		
acrylic	8~12	7~12	(cotton, flax, ramie	45~55	23~28
nylon	5~10	4~8	and rayon)		
another fibers	2~3	2~3	polyesters	15~18	10~15
			another fibers	4~10	2~5

Firstly, the pre-molded felt was heated in a furnace with forced convection air flow at 190℃ during 30 minutes in order to melt the binder material(PP). After the heating process, the felt was compressed and cooled in the die. Three types of dies with different dimensions were used for the molding such as 900x150x15mm, 900x100x15mm and 900x85x15mm. The density of substitute lumbers was adjusted from 0.3g/cm^3 to 0.8g/cm^3 by changing the sheet number of felt being staffed in the die.

Measurement of property

The measurements of bending strength, holding strength of screw, and pulling resistance of nail were carried out for the molded substitute lumber. The cutting ability of saw was also examined for the substitute lumber. These examinations were carried out in accordance with Japan Industry Standard (JIS) for wood materials. Moreover, the mechanical properties under the wet condition for the molded lumber were examined against the condition of 2 hours-boiling and 1 hour-cooling before testing.

The properties of molded lumber were compared with those of various commercial lumbers such as pine tree produced in New Zealand, medium density fiber board and paper lumber.

RESULTS AND DISCUSSION

Figures 1 and 2 depict the micrographs of typical fracture surfaces of the felt before and after compression molding, respectively. As can be seen in Fig.2, PP fibers melt perfectly and the block of melted PP adheres the neighboring fibers. Fig.3 show the aspects of micrographs of surface of middle density fiberboards(MDF). Fairy good matrix cellulose fibers adhesion was obtained for the composites. These results show that the MDF lumber have not become the porous board.

Fig.1 Micrographs of fracture surface of felt (cellulose group) before the compression.

Fig.2 Micrograph of the felt after molding.

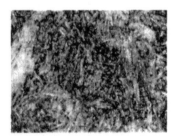

Fig.3 Micrograph of surface of MDF.

Figure 4 shows the bending strength of the molded lumbers as a function of density. Notice that the bending strength appears to increase with increasing density. Especially, the bending strength increases largely with increasing the density(ρ) in the rang (ρ)>0.5g/cm^3. It should be noted here that the strength under the wet condition is slightly smaller than that under the dry condition. These results agree well with those of wool group as shown in Fig.5.

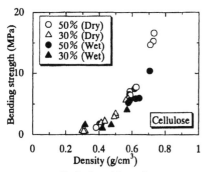

Fig.4   Relationship   between bending strength and density (cellulose group)

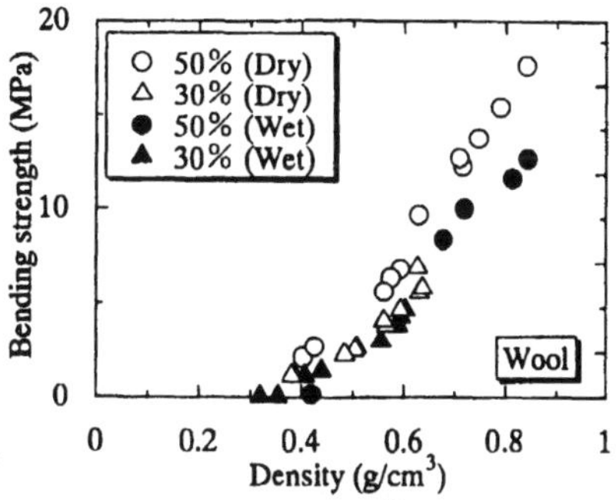

Fig.5   Relationship   between bending strength and density (wool group)

Figures 6 and 7 show the holding strength of screw held in the molded lumbers for both cellulose and wool groups, respectively. As expected, the holding strength of screw increases with increasing density similar to the tendency of said bending strength. It should be noted, however, that the pulling resistance of nail exhibits the different tendency from that of said holding strength. Namely, as shown in Fig.8, the pulling resistance of nail tested under the wet condition results the higher value in comparison with the result obtained under the dry condition.

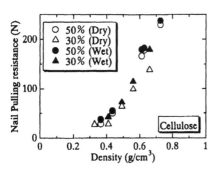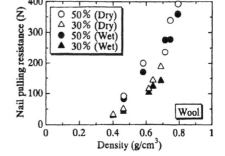

Fig.6 Relationship between nail pulling resistance and density (cellulose group)

Fig.7 Relationship between nail pulling resistance and density (wool group)

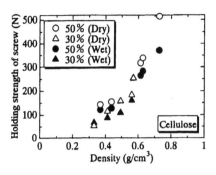

Fig.8 Relationship between holding strength of screw and density (cellulose group)

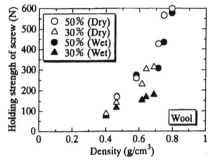

Fig.9 Relationship between holding strength of screw and density (wool group)

Especially, this tendency is clearly seen in the cellulose group. These results can not be seen in the wool group. This result may be caused by the facts that the strength of cellulose fibers such as cotton, flax, ramie and rayon increases in the wet condition as compared with the dry condition.

The mechanical properties of molded lumber and the cutting ability of saw are concluded in Table 2 together with those of commercial lumbers. It is cleared from the table, the bending strength, holding strength of screw and pulling resistance of nail of present molded lumber are slightly smaller than those of pine tree under the wet condition. It should be noted however that the holding strength for molded lumber of cellulose group is larger than that of pine tree under the wet condition. In the case of MDF lumber, the properties decrease largely under the wet condition. In the case of paper lumber, the skin layer of lumber peels under the wet condition, and then the lumber can not be used outdoors where are caught in a rain. From the results described above, it is pointed out that the mechanical behavior of substitute lumber is stable under the wet condition.

The results of cutting tests suggest that the substitute lumber molded in this paper can be cut easily by using the usual saw machine.

Table2 The mechanical property and the cutting ability of saw

	Density	Bending strength (dry) (MPa)	Bending strength (wet) (MPa)	Holding strength of screw (dry)(N)	Holding strength of screw (wet)(N)	Nail pulling resistance (dry)(N)	Nail pulling resistance (wet)(N)	The cutting ability of saw
Pain tree	0.55	15	14	640	440	240	210	◯
Substitute lumber (cellulose)	0.4	4	3	300	250	40	50	◔
Substitute lumber (cellulose)	0.6	6	5.5	480	400	120	150	◯
Substitute lumber (cellulose)	0.7	10	7.5	550	500	140	180	△
Medium density fiber lumber	0.63	12	6	340	170	240	140	◯
Paper lumber	0.73	10	×	310	×	130	×	◯

CONCLUSION

    In the present work, a new substitute lumber composite material has been developed by using Rag as a molding raw material and the mechanical properties were evaluated. It is concluded that the bending strength, the holding strength of screw and the pulling resistance of nail of molded materials are good for the substitute lumber. The results suggest that the compression molding method described here shows promise in contributing toward the material recycle of fiber waste (Rag) as a substitute lumber. More examination will be expected to obtain more light weight lumber with more stronger properties in taking account of the combination and length of fibers, content of binder and fiber orientation in the lumber   materials.

APPLICATIONS

Sheet molded Natural Fiber Composites for Automotive Applications

Muhammad Pervaiz and Mohini M. Sain
Associate Professor, Earth Science center &Forestry
University of Toronto 33 Willcocks Street
Toronto, Ontario, Canada M5S 3B3

ABSTRACT

The use of natural fibers as reinforcing filler in thermoplastics is a relatively new innovation and has great potential to replace glass fiber products in automotive industry. However, most of the research in this area has been focused primarily on flax fiber alone. In this research work, hemp fiber non-woven mats derived from local annual crop of Ontario has been used exclusively in combination with polypropylene matrix to study the mechanical properties of NMT (natural fiber mat thermoplastic) without using any binder aid. Film stacking method was used and the results showed that hemp based NMT have compatible or even better strength properties as compared to conventional flax based thermoplastics. A value of 63MPa for flexural strength is achieved at 64% fiber content by weight. The influence of compression ratio on mechanical properties and density of NMT is also reported in this study. A definite trend in increase in strength is observed with corresponding increase in compression along with a much more uniform density profile. Impact energy values (84-154 J/m) are also very encouraging and overall results of this study indicate that hemp based NMT can have very effective utility in auto applications where high specific stiffness is required. In future work, interfacial modification with binder is planned to further enhance the mechanical performance of NMTs.

INTRODUCTION

Over the last few years, ecological concern has initiated a considerable interest in using natural materials to produce green products. Consumer awareness and tough environmental legislation has resulted in implementing LCA (life cycle assessment) approach to most of the products and manufacturing processes.

Traditionally, glass fibers have been extensively used as reinforcement in thermoplastics and played a pivotal role so far in interior and exterior applications in automobiles due to their better impact strength properties. However, GMTs (Glass fiber Mat Thermoplastics) have several disadvantages: Glass fibers are obtained from non-renewable resources and a lot of energy is consumed in their production. These fibers are abrasive to process equipment and have some potential health risks to production workers due to skin irritation and attack on respiratory tract/lungs by tiny particles. During collision or accident these composites may transform into sharp splints causing injury. Moreover, GMTs are non recyclable and their incineration generates clinker like mass that is hard to dispose off.

On the other hand, natural fibers are being explored more extensively by research institutions and automobile companies as environmental friendly alternative for the use of glass fibers. Most of the bast fibers being studied are obtained from naturally growing plants of flax, hemp, and kenaf. These fibers are renewable, non-abrasive to process equipment and can be incinerated at the end of their life cycle for energy recovery as they posses a good deal of calorific value. They are also very much safe during handling and less suspected to affect lungs during processing and use.

Automotive applications represent the best opportunity for natural fibers thermoplastics due to some of distinctive advantages over glass fibers, like, low weight (35-40% less as compared to glass fiber), low price, better crash absorbance and sound insulation properties. Some of the potential applications in this field are: Door and instrument panels, package trays, glove boxes, arm rests, and seat backs.

In spite of the several benefits of natural fibers, the mass scale production of NMTs has not yet been optimized and comprehensive research is still required to improve some critical characteristics of these products like, overall mechanical behavior, dimensional stability and fire resistance.

Although some papers have discussed the morphology of natural fibers in detail but unfortunately, not much data is available on the physical properties of fibers itself. B.Van Voorn [1] has reported interesting comparison about tensile strength of various natural fibers and regenerated cellulose (rayon) as a reference. The tensile strength of Flax and Hemp falls between 600 to 800 MPa, which is much higher than other natural fibers. In terms of density [2], natural fibers weigh only 40% that of glass fibers, thus having competitive advantage in automobile applications.

Unfortunately, most of the work in this area has been done on flax fibers only and several researchers have explored its use as reinforcement in thermoplastics. B.van Voorn [1] has compared the mechanical properties of Glass and Flax fiber sheet molded composites based on wet-laid process and it is shown that stiffness of Flax fiber composites is comparable or even better whereas, flexural and tensile strength properties are slightly lower. However, impact strength of flax fiber composite is only 3-7kJ/m^2 as compared to 40 kJ/m^2 for glass fiber composite. Similar comparisons have been reported by other researchers [2,6] while using film-stacking technique.

This study has been focused primarily on native hemp fiber of Ontario and main objective was to investigate mechanical properties of finished NMT. The effect of compression ratio on strength and consolidation has also been reported in this work.

2. EXPERIMENTAL

2.1 Materials

The polypropylene (PP) matrix used in this work was obtained from a local manufacturer and has the ultimate tensile along TD 37 MPa whereas secant modulus 0.8GPa.

The non-woven needle punched mats of hemp fiber, were supplied by Hempline and Flaxcraft Inc. New Jersey. The hemp fiber content by weight was 80% in the mat.

2.2 Manufacturing of composite sheets

Representative samples of composites were manufactured by *film-stacking method*. First the blanks of hemp non-woven mats were cut and dried. Immediately after this pre-dried non-woven fiber mats and PP films were stacked alternately for impregnation in a hydraulic press, Wabash, having 50-ton capacity with air/water cooling arrangement. The molding time and temperature were optimized. To study the effect of compression different pressures were used and few samples were made by using spacers between the press plates. The blanks were cooled at the end of heating/impregnation cycle in the same press before completing the compression molding process.

2.2 Testing

2.2.1   Flexural and Tensile Testing

Flexural test (3-point bending) was conducted on Zwick-z100 tester according to standard test procedure D-790. Specimens were 12.5mm wide and 150 mm long. The thickness varied from 1.1 to 2.9mm and machine speed was set at 10mm/min. The tensile testing was done on same machine and specimens were prepared according to test method D-638 while the crosshead speed was maintained at 2mm/min.

2.2.2   Impact Energy

Izod "notched" impact testing was carried out on Tinius Olsen (92T Impact T.M) test machine according to ASTM D 256. The specimens had depth of 12.7mm and length of about 64 mm. The testing machine had inbuilt rocessor to calculate absorbed energy in J/m. However, impact energy in J/m^2 was also calculated by dividing the net value of absorbed energy with product of measured width and depth under the notch.

2.2.3 Density Profiling

An X-ray density profiler, QMS (QDP-01X) was used to study the effect of different compression ratios on molded composites and have accurate density measurements. This machine has capability of scanning 1/1000 of an inch and specimens had dimensions of 50X50mm with varying thickness from 1.1 to 2.9 mm. The scanning was done through five zones across the thickness of specimen.

2.3 Results

2.3.1 Mechanical Properties

Significant results have been observed by achieving maximum flexural and tensile strength of 63 and 47MPa respectively, figure 1, without using any binder aid. Unfortunately no reference data is available for hemp based NMTs to compare this result, however, similar values are reported [1, 2, 6] for flax based composites while using 3-5% maleic-anhydride binder in some cases.

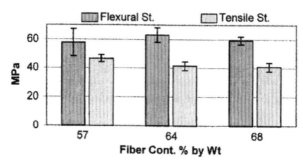

Figure 1. Flexural and tensile properties of hemp based NMTs

More interestingly, these results correspond to density values from 0.90 to 0.96 g/cm^3, which is about 56 % less as compared to glass fiber composites (11). It means that specific strength and stiffness properties of these NMTs can be very compatible to traditionally used GMTs in auto industry with an added advantage of low weight.

2.3.2 Effect of Compression

A definite trend of increase in tensile/flexural values has been shown in figure 2 with corresponding increase in molding compression for two different fiber contents while keeping the other conditions almost same. A tremendous increase of 23 times was observed in flexural strength for 67% fiber content NMT while increasing the compression ratio from 42 to 77%. For 65% fiber content NMT, an increase of 6 times in flexural value was obtained against an increase of compression ratio from 55 to 79%.

Compression ratio = $(t_i - t_f)/t_i$

Where t_i is initial combined thickness of hemp mats and PP films before consolidation and t_f is final thickness of NMT sample.

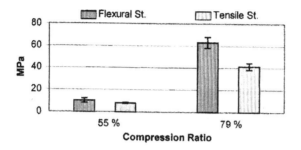

Figure 2. Compression effect on flexural and tensile properties for 65% hemp fiber NMT.

Similarly, higher compression produced a more uniformed density product as shown in figure 3. All this can be attributed to void spaces present in original fiber mat and poor fiber-to-fiber bonding under low compression. However, further testing is in progress to identify optimum values of *pressure* and *fiber content* for maximum strength properties.

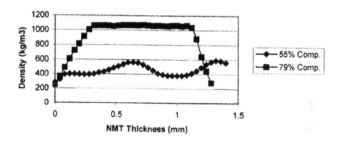

Figure 3. Compression effect on density for 65% hemp fiber NMT.

Impact Energy: We have been able to get impact energy of 84,111 and 154 J/m for NMT samples containing 68, 57 and 70% hemp fiber by weight respectively under different impregnation pressures as shown in figure 4. Similar values have been reported (2) in literature for flax fiber composites while using 5% binder. Therefore, it is anticipated that with further optimization and use of proper binder, we will be able to enhance interfacial bonding between fiber and matrix yielding much better results in overall mechanical behavior.

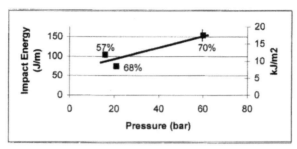

Figure 4. Effect of Compression and Fiber Content on Impact Energy

A comparison of tensile strength has been shown in figure 5 for hemp, flax and glass reinforced mat thermoplastics. It may be noted that except for hemp based NMT, all the other data is taken from literature.

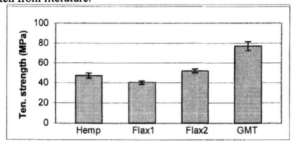

Figure 5. Tensile strength of different NMTs and GMT

The flax1 (Danflax) and GMT data is taken from work of Kristina Oksman [6]. In both these cases the fiber content is 50% by weight. The flax2 data belongs to S.K.Garkhail [2] in which fiber volume fraction is about 0.45. This comparison shows that hemp based NMT has compatible strength to that of flax based thermoplastics. However higher strength of GMT can be attributed to better strength of glass fibers, 3450 MPa [4], compared to hemp and flax fibers. Further, NMTs tend to be more anisotropic due to inherited physiological characteristics of natural fibers.

CONCLUSIONS

Film stacking technique was used in this work to produce representative samples of hemp based NMT. The main parameters studied were fiber content and compression effect on different mechanical properties without using any binder aid. Pre-prepared non-woven needle punched hemp mats and PP films were used as main materials. The strength properties achieved at 57% fiber content are compatible to most of literature data available for flax based NMTs. As expected, compression has significant effect on interfacial bonding between fibers and matrix and there is a dire need to further optimize this relationship. Density profiling by sophisticated equipment has shown that uniform consolidation is very much dependent on impregnation pressure. It is anticipated that after fiber treatment and interfacial modification the mechanical performance of hemp based NMTs can be further improved and there exists a good potential for these thermoplastics to replace GMTs in various applications especially auto sector.

The consumption of plastic and composites has experienced tremendous growth of 1300% [5] during the last four decade with main area of application being land transportation [6]. The projected demand of thermoplastics in auto industry alone by the end of 2005 is expected to reach 100 million lbs [7] in North America. The future use of natural fibers in composites looks even brighter after signing of Koyoto Protocol by Canadian government recently. The main advantages of NMTs is tremendous saving in energy consumption, less cost, low weight and environmental friendly life cycle as compared to glass fiber products.

Acknowledgement

Authors gratefully acknowledges the financial support offered by Network of Center of Excellence Auto 21st century (C3-Polymer Composites) ; Authors also thanks to Atofina Canada, Copol Int., Hempline and Flaxcraft Inc. for supplying polymer and fiber materials.

References

1.   B.Van Voorn, H.H.G.Smit, R.J.Sinke and B.de Klerk, "Natural fiber reinforced sheet molding compound", Composites: Part A 32, 2001, 1271-1279.

2   S.K.Garkhail, R.W.H.Heijenrath and T.Peijs, "Mechanical Properties of Natural-Fiber-Mat-Reinforced Thermoplastics based on Flax Fibers and Polypropylene", Applied Composite Materials, Volume 7, 2000, 351-372.

3.   Kristina Oksman, "Mechanical Properties of Natural Fiber Mat Reinforced Thermoplastic", Applied Composite Materials, Volume 7, 2000, 403-414.

4.   William D. Callister, Jr. "Materials Science and Engineering-An Introduction". Fifth Edition.

5.   Gerard Van Erp, "Engineered Fiber Composites-Opportunities for a New Smart Industry in South East Queensland", Proceedings of 17th Queensland Regional Development Conference, Australia.

6.   John Summerscales, Ian Salusbury, Tim Gutowski and Ignazio Crivelli Visconti, "Editorial", Composites Manufacturing, Volume 6, Number 3-4, 1995, 115-116.

7.   K.Line and Company, "Opportunities for natural fibers in plastic composites", Proceedings of 6th International Conference on Woodfiber-Plastic Composites, May 14, 2001.

Materials and Structural Design for Lightweight Primary Mirrors

C.J. Hamelin, and J. Beddoes,
Department of Mechanical & Aerospace Engineering,
Carleton University,
Ottawa, Canada, K1S 5B6
jbeddoes@mae.carleton.ca

J. Lo
CANMET,
Natural Resources Canada,
568 Booth St.,
Ottawa, Canada, K1A 0G1

ABSTRACT

A deflection analysis for seven potential primary mirror materials is presented for loading due to gravitational or external forces. Based on this analysis a finite element model of a plano-plano mirror is developed for three mirror materials including aluminum, Zerodur and an Mg particulate-reinforced composite. Comparison of mirror deflection according to the finite element model to interferometer-measured deflections of a Zerodur mirror indicate good accuracy is obtained from the model. Using the calibrated finite model various geometrical light weighting options are evaluated. These include the fabrication of between three and six circular pockets on the back face of the mirror. It can be concluded that while both pocket depth and diameter contribute to mass reduction of the mirror, mass reduction through increases in pocket depth are more advantageous than increases in pocket diameter, as this results in greater mirror stiffness for the same mass reduction. Also, an increase in the number of light weighting pockets for the same mass reduction has less impact on the mirror optical performance.

INTRODUCTION

There are two methods used to reduce primary mirror weight. The first involves geometrical optimization – the shape of the mirror is chosen to reduce weight while enhancing overall stiffness. Designs examples include contoured-back, radially ribbed, open-back, sandwich, and foam core structures [1]. The second method of light weighting is via materials selection. During the past 40 years a continuous advance in the design of new lightweight optical materials has occurred. While many mirrors have been constructed of aluminum, other materials including low-expansion glass, beryllium, and silicon carbide are available. Recently, several state-of-the-art optical mirrors have been fabricated from beryllium, Zerodur, Ultra-Low Expansion (ULE) glass, silicon carbide, and metal-matrix composites (MMCs).

The mechanical, physical, optical, metallurgical, and fabrication properties of candidate materials can influence their probability of selection [2], along with a significant impact of cost and availability. Recently, in Canada, aluminum and Zerodur have been selected for reflective telescope mirror fabrication [3], with beryllium mirrors selected military applications. By careful selection of the individual components of a composite material system, the physical properties of composites can to some extent be tailored for particular application requirements. As such, there is value to examining composites with good stiffness-to-weight ratios.

By optimizing the geometry and materials selection the primary mirror best suited for a particular application can be specified. However, if this optimization and selection is performed through prototyping, the time and cost of producing multiple prototype mirrors of various geometry and material is excessive. Therefore, a simple, rapid, and accurate method is needed for the optimization of primary mirror design and materials selection. Given the foregoing, the purpose of the work described in this paper is to develop a finite element model to successfully predict the behaviour of plano-plano mirrors with no light weighting for various candidate mirror materials. The accuracy of the model is established through calibration with interferometric analysis, and subsequently used to examine the benefits of structural light weighting.

MATERIAL SELECTION

The relevant physical properties of six materials frequently used for primary mirrors are listed in Table I, with a relative comparison of cost, manufacturing ease and optical performance given in Table II. Included in these tables is preliminary data for a proprietary MMC consisting of a particulate-reinforced magnesium matrix. From these tables it is immediately apparent why aluminum dominates primary mirror fabrication. Nevertheless, the excellent performance of beryllium has been utilized for several critical applications, such as the primary mirror of the Mars Orbiter Laser Altimeter [4].

By applying thin plate bending theory [5] and using aluminum as a baseline, the primary mirror surface deflection is compared in Figure 1, when subjected to either external or gravitational loading, with larger deflections having a greater impact on the

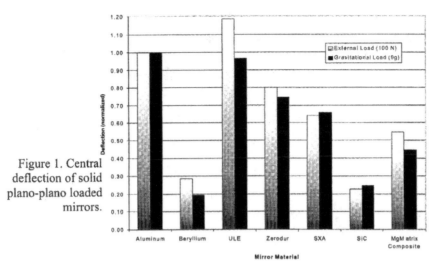

Figure 1. Central deflection of solid plano-plano loaded mirrors.

optical performance. Typically, external loads are most important during manufacturing, with gravitational loads encountered in service (for space-based applications, particularly upon launch). Beryllium and silicon carbide offer the best optical performance with aluminum and ULE glass performing less well. For the purposes of the current investigation a detailed finite element model was applied to mirrors fabricated from aluminum, Zerodur and the magnesium MMC.

Table I. Mechanical and thermal properties of lightweight mirror materials [3].

Material	ρ kg/m³ 10³	CTE ppm/K	Specific Heat J/kg-K	Thermal Conductivity W/m-K	E GPa	Thermal Distortion Coefficient µm/W	Specific Stiffness MH/kg 10⁶
Al	2.7	25	899	237	76	0.105	28.1
Be	1.85	11.4	1880	216	303	0.053	164
ULE	2.20	0.03	708	1.3	67	0.023	30.4
Zerodur	2.55	0.15	820	6.0	90	0.025	35.3
Al-SiC MMC	2.96	12.2	-	120	130	0.102	40
SiC	3.21	2.4	700	250	466	0.010	145
Mg matrix composite	2.2				125		56.8

ANALYSES & RESULTS

To validate the analysis, the results of a finite element analysis (FEA) applied to mechanically loaded plano-plano mirrors are compared to experimental interferometric data obtained for the same plano-plano mirror. This FEA is described in the next section and compared to the interferometric data in the subsequent section.

Table II. Comparison of mirror materials for candidate selection.

Material	Cost/Availability	Ease of Manufacture	Optical Performance
Aluminum	Very good	Very good	Poor
Beryllium	Poor	Very poor	Very good
ULE glass	Good	Poor	Very poor
Zerodur	Good	Good	Good
SXA	Very poor	Poor	Good
Silicon carbide	Poor	Very poor	Very good
Mg matrix composite	Good	Poor	Good

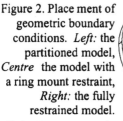 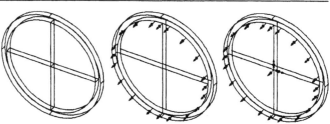

Figure 2. Place ment of geometric boundary conditions. *Left:* the partitioned model, *Centre* the model with a ring mount restraint, *Right:* the fully restrained model.

Finite Element Analysis

The simplified I-DEAS™ finite element model of the plano-plano mirror is shown in Figure 2. The model mirror diameter is 108.2 mm with a thickness of 6 mm, giving a diameter-to-thickness ratio of approximately 18. Typically, smaller mirrors such as these have a ratio of 6:1, but for the current work the maximum load that could be applied during interferometric analysis limited the mirror thickness. A concentric ring of 97.28-mm diameter supports the mirror, such that the model is constrained in the thickness direction at this location (shown in the centre. Figure 3). Consistent with the loading during interferometric analysis, a 36.95-N load is applied to the bottom centre of the mirror (shown at right, Figure 3). A finite element mesh consisting of 10-node tetra-hedrons is used throughout the modelled mirror. To improve the capability of the model to simulate various more complex lightweight configurations, solid elements are used instead of plate elements and axisymmetric simplifications were not applied. Varying the thickness normalized element length between 0.4 and 1 resulted in virtually no change to the predicted deflection of the top surface centre. Furthermore, this deflection compared well with that calculated based on thin plate bending theory [5]. Therefore, an element length of 0.5 is applied for all analyses, corresponding to the mesh of 11,305 elements shown in Figure 3. Three models were constructed for aluminum, Zerodur, and the Mg matrix composite. The resulting top surface central deflection is listed in Table III.

Table III. FEA predicted top surface central-point deflection for three different mirror materials.

Material	Deflection (μm)
Aluminum	3.125
Zerodur	2.492
Mg matrix composite	1.763

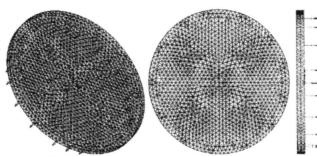

Figure 4. Finite element mesh of just over 11,000 tetrahedral elements used for all analyses.

Interferometric Analysis

Interferometric analyses utilized a Zygo GPI XPS interferometer and a Zerodur mirror. The central load was applied to the centre of the back surface. Figure 5 illustrates the interferometric deflection data for the Zerodur mirror subjected to a 36.95-N load, with a corresponding peak to valley (P-V) deflection of 3.345 μm. The interferometer software filters the results to minimize errors that occur as a direct result of off-axis mounting or initial surface sag of the mirror. Application of the software filters results in a central peak to valley deflection of 1.948 μm. Table IV compares the FEA-predicted deflection with this interferometric analysis for the Zerodur mirror. The FEA-predicted deflection is between the interferometric values obtained with and without application of the software filters. It is believed that filtering eliminates the effect of concavity that appears as an initial condition or state of the mirror: however, the load application at the centre of the mirror will produce deflection effects similar to natural concavity without loading. Thus, that the FEA-predicted deflection is between the two interferometric values is reasonable, as this would be the approximate deflection of a completely defect - free Zerodur mirror. Regardless, the error between either interferometer result and the FEA values is satisfactory given the very small displacements modelled.

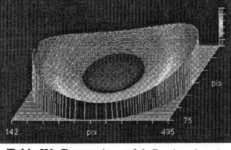

Figure 5. Interferometer measured profile of Zerodur mirror reflecting surface at 36.95 N load. Peak to valley defection of 3.037 + 2.248 = 5.285 × wavelength of laser or 3.345 μm.

Table IV. Comparison of deflection data between FEA and Interferometer results

Software Filter Statu s	Interferometer P-V Deflection (μm)	FEA Deflection (μm)	Deviation (μm)	% Error
Filtering Not Applied	3.345	2.492	+ 0.853	+ 34
Filtering Applied	1.948	2.492	- 0.544	- 22

STRUCTURAL LIGHT WEIGHTING

The first structural light weighting step was to create circular pockets in the back of the mirror – an easy to remove material during manufacture. To evaluate this option a detailed FEA parametric study to examine the effect of multiple circular pockets with variations in pocket depth and diameter is presented. For this analysis, more than 100 lightweight configurations were completed with three to six circular light weighting pockets in a Zerodur mirror.

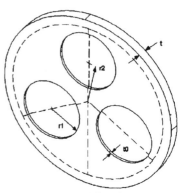

Three Circular Lightweight Pockets

Figure 5 shows an example of the structure imposed on the back of the mirror for a three pocket design. Variations to the dimensions of this model take two forms: the diameter of

Figure 5. Lightweight mirror, consisting of three light weighting pockets on the back of the mirror. Dashed lines represent partitions used for FEA operations.

each pocket is varied, and the depth of the pocket ($t0$) is varied for each pocket radius ($r1$). In each case – given a specified pocket depth and diameter – the mass of the mirror and the mirror deflection are calculated for a load of 36.95 N applied to the back surface. This load allows comparison of the deflections between plano-plano and lightweight designs. Figure 6 shows mass reduction and deflection data for the three pocket design, as both pocket depth ($t0$) and diameter ($r1$) is varied. The term $r2$ refers to the radius from the mirror centre to the centre of the light weighting pocket: for all models, $r2$ is equal to one quarter the mirror diameter (27.05 mm) and the mirror thickness (t) is 6 mm. For the three pocket lightweight design, combining a large pocket diameter with sufficient depth (i.e. for large percentages of mass removal) results in deflection values greater along the radial direction of the pocket centres. This illustrates the need to better distribute the mass reduction, leading to the analyses of lightweight designs with a larger number of pockets.

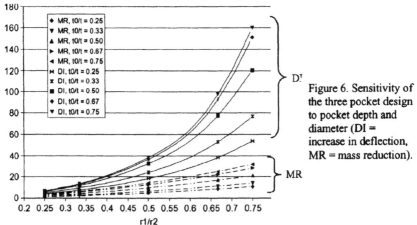

Figure 6. Sensitivity of the three pocket design to pocket depth and diameter (DI = increase in deflection, MR = mass reduction).

Four To Six Pocket Designs

To allow comparison of the various designs, sets of 25 analyses were carried out for each number of pockets, with variation of both the pocket depth and diameter. For all analyses, the finite element models contained approximately 10,000 to 12,000 elements. From Table V it is evident that as the number of pockets is increased, the deflection at a given mass is noticeably reduced. Also included in Table V is a comparison of two four pocket designs for which the mass reduction is nearly identical, but the pocket depth and diameter is varied, and as a result the deflection data is significantly different.

Comparison of light weighting through an increase of pocket depth and diameter reveals that for the same degree of mass reduction, the increase in deflection that arises due to pocket diameter increase is greater than that when pocket depth is increased. However, there is a limit to the depth of a light weighting pocket, and hence a suitable pocket diameter must be chosen. Table V illustrates this result – comparing two lightweight, four pocket designs with nearly identical values in mass reduction, illustrates that the design with a smaller pocket diameter and a larger pocket depth resulted in less mirror deflection under the same loading conditions. This characteristic is invaluable when attempting to reduce the mass of a mirror by a given amount – to maximize the extent of light weighting requires use of the largest pocket depth and diameter permissible. However, maintaining optical performance of the mirror dictates how aggressively geometrical light weighting can be applied. Clearly, a trade-off exists between the degree of light weighting possible while meeting optical performance criteria.

The advantages of using a smaller pocket diameter are further exemplified by comparison of three, five, and six pocket designs, illustrated in Table V. As the number of pockets is increased, the pocket depths and diameters are adjusted such that the mirrors of Table V experience approximately the same degree of mass reduction, by reducing pocket diameter and increasing pocket depth as the pocket number increases. It can be seen however that while the mass reduction values are similar, the six pocket design has the smallest deflection at the same loading conditions – the deflection is almost half that of the three pocket design. Clearly therefore, it is beneficial to increase the number of light weighting pockets in the mirror design.

SUMMARY & CONCLUSIONS

Finite element analysis and interferometric data indicate that the finite element method can be used to successfully model mirror deflections. The deflection of the mirror centre predicted by FEA deviated <1 μm from the deflection measured by interferometry. Using the validated FEA model, parametric mirror light weighting studies have been undertaken. It can be concluded that while both pocket depth and diameter contribute to mass reduction of the mirror, mass reduction through increases in pocket depth are more advantageous than increases in pocket diameter, as this results in greater mirror stiffness for the same mass reduction. Also, an increase in the number of light weighting pockets for the same mass reduction has less impact on the optical mirror performance.

Table V. Comparison of selected deflection data for multiple pocket designs.

Number of pockets		t0/t	r1/r2	% Mass Reduction Compared to Plano-Plano Mirror	% Deflection Increase Compared to Plano-Plano Mirror
3		0.50	0.75	21.1	120.7
5		0.67	0.55	25.2	101.5
6		0.75	0.45	22.8	67.4
4		0.50	0.70	24.5	155.4
4		0.67	0.60	24.0	95.4

ACKNOWLEDGEMENTS

The authors thank the late Martin High for his cooperation and guidance of this research, without which it would not have been possible. One of the authors (CJH) is grateful for financial support through the Ontario Graduate Scholarship for Science and Technology.

REFERENCES

1. Vukobratovich, D., "Lightweight laser communications mirrors made with metal foam cores," *Proc. SPIE Vol. 1044*, 1989, pp. 216-226.
2. Paquin, R.A., "Selection of Materials and Processes for Metal Optics," *Proc. SPIE Vol. 65, Metal Optics*, 1975, pp. 12-19.
3. CANMET/APS, "Advanced Manufacturing Technology for Space Telescopes", 1999.
4. Generie, P. & Hayden, B., "Estimation of the on-orbit distortion of the Mars Orbiter Laser Altimeter (MOLA II) Primary Mirror," *Proc. SPIE Vol. 2857*, 1996, pp. 45-56.
5. Timoshenko, S., Theory of Plates and Shells, McGraw-Hill, 1940.

Author Index

Printed and bound by CPI Group (UK) Ltd, Croydon, CR0 4YY

21/10/2024

01777086-0016